Pathogenic Fungi
Structural Biology and Taxonomy

Edited by

Gioconda San-Blas
Instituto Venezolano de Investigaciones Científicas
Caracas, Venezuela

Richard A. Calderone
Georgetown University, Washington DC, USA

Caister Academic Press
32 Hewitts Lane
Wymondham
Norfolk NR18 0JA
England

www.caister.com

British Library Cataloguing-in-Publication Data

A catalogue record for this book is available from the British Library

ISBN: 0-9542464-7-0

Description or mention of instrumentation, software, or other products in this book does not imply endorsement by the author or publisher. The author and publisher do not assume responsibility for the validity of any products or procedures mentioned or described in this book or for the consequences of their use.

Printed and bound in Great Britain

Contents

Section 2: New Taxonomic Tools

Books of Related Interest

- **Pathogenic Fungi: Host Interactions and Emerging Strategies for Control** (2004)
  Edited by: Gioconda San-Blas and Richard A. Calderone
- **Malaria Parasites: Genomes and Molecular Biology** (2004)
  Edited by: Andrew P. Waters and Chris J. Janse
- **Peptide Nucleic Acids: Protocols and Applications,** 2nd Edition (2004)
  Edited by: Peter E. Nielsen
- **Ebola and Marburg Viruses**: Molecular and Cellular Biology (2004)
  Edited by: Hans-Dieter Klenk and Heinz Feldmann
- **Metabolic Engineering in the Post-genomic Era** (2004)
  Edited by: Boris N. Kholodenko and Hans V. Westerhoff
- **MRSA:** Current Perspectives (2003)
  Edited by: Ad C. Fluit and Franz-Josef Schmitz
- **Tuberculosis: The Microbe Host Interface** (2003)
  Edited by: Larry S. Schlesinger and Lucy DesJardin
- **Genome Mapping and Sequencing** (2003)
  Edited by: Ian Dunham
- **Regulatory Networks in Prokaryotes** (2003)
  Edited by: Peter Dürre and Bärbel Friedrich
- **Bioremediation:** A Critical Review (2003)
  Edited by: I.M. Head, I. Singleton and M. Milner
- **Bioinformatics and Genomes**: Current Perspectives (2003)
  Edited by: Miguel A. Andrade
- **Frontiers in Computional Genomics** (2003)
  Edited by Michael Y. Galperin and Eugene V. Koonin :
- **Multiple Drug Resistant Bacteria:** Emerging Strategies (2003)
  Edited by: Carlos F. Amábile-Cuevas
- **Vaccine Delivery Strategies** (2003)
  Edited by: Guido Dietrich and W. Goebel
- **Probiotics and Prebiotics:** Where Are We Going? (2002)
  Edited by: Gerald W. Tannock
- **Genomic Technologies:** Present and Future (2002)
  Edited by: David J. Galas and Stephen J. McCormack
- **Genomics of GC-Rich Gram Positive Bacteria** (2002)
  Edited by: Antoine Danchin

Full details of these and all our books at:

www.horizonpress.com

Contributors

O. Ahrazem,
Escuela Superior de Ingenieros Agrónomos,
Universidad Politécnica de Madrid,
Avda. Complutense, s/n,
Ciudad Universitaria,
28040 Madrid,
Spain

Peggy E. Alvarez,
Unidad de Biotecnología e Ingeniería Genética,
Centro de Investigación y
de Estudios Avanzados del IPN,
Apartado Postal 629,
36500 Irapuato,
Gto. Mexico

Judith Berman,
Department of Genetics,
Cell Biology and Development,
University of Minnesota,
Minneapolis
MN 55455,
USA
Email: judith@cbs.umn.edu

M. Bernabé,
Instituto de Química Orgánica,
CSIC,
Juan de la Cierva, 3,
28006 Madrid,
Spain

***Richard A. Calderone**,
Georgetown University Medical Center,
Department of Microbiology & Immunology,
SE 312 Med Dent Building,
3900 Reservoir Rd., NW,
Washington,
DC 20057,
USA
Email: calderor@georgetown.edu

***Tamaki Cho,**
Section of Infection Biology,
Department of Functional Bioscience,
Fukuoka Dental College,
2-15-1 Tamura Sawara-ku,
Fukuoka 814-0193,
Japan
Email: chot1@college.fdcnet.ac.jp

M. Victoria Elorza,
Departament de Microbiología,
Facultat de Farmacia,
Universitat de València,
Vicent Andrés Estellés s/n,
46100 Burjassot,
València,
Spain

***Neil A.R. Gow,**
School of Medical Sciences,
Institute of Medical Sciences,
University of Aberdeen,
Aberdeen AB25 2ZD,
UK
Email: n.gow@abdn.ac.uk

Idit Hazan,
Department of Biological Chemistry,
University of California-Irvine,
Irvine,
CA 92697-1700,
USA

***J. A. Leal,**
Centro de Investigaciones Biológicas,
CSIC,
Ramiro de Maeztu, 9,
28040 Madrid,
Spain,
Email: aleal@cib.csic.es

***Haoping Liu**,
Department of Biological Chemistry,
University of California-Irvine,
Irvine,
CA 92697-1700,
USA
Email: H4LIU@UCI.edu

* indicates corresponding author

***Leonel Mendoza,**
Medical Technology Program,
Microbiology and Molecular Genetics,
Michigan State University,
East Lansing
MI 48824-1031,
USA
Email mendoza9@msu.edu

***Juan Murgich**,
Centro de Química,
Instituto Venezolano
de Investigaciones Científicas (IVIC),
Apartado 21827,
Caracas 1020A,
Venezuela
Email: jmurgich@ivic.ve

Gustavo Niño-Vega,
Instituto Venezolano
de Investigaciones Científicas (IVIC),
Centro de Microbiología y Biología Celular,
Apartado 21827,
Caracas 1020A,
Venezuela
E-mail: gnino@ivic.ve

***Rajendra Prasad,**
School of Life Sciences,
Jawaharlal Nehru University,
New Delhi-110067,
India
Email: rp47@hotmail.com

A. Prieto,
Centro de Investigaciones Biológicas,
CSIC,
Ramiro de Maeztu, 9,
28040 Madrid,
Spain

***José Ruiz-Herrera**,
Unidad de Biotecnología e Ingeniería
Genética,
Centro de Investigación
y de Estudios Avanzados del IPN,
Apartado Postal 629,
36500 Irapuato,
Gto. Mexico
Email: jruiz@ira.cinvestav.mx

***Gioconda San-Blas**,
Instituto Venezolano de Investigaciones
Científicas (IVIC),
Centro de Microbiología
y Biología Celular,
Apartado 21827,
Caracas 1020A,
Venezuela
E-mail: gsanblas@ivic.ve

***Rafael Sentandreu,**
Departament de Microbiología,
Facultat de Farmacia,
Universitat de València,
Vicent Andrés Estellés s/n.
46100 Burjassot,
València,
Spain
Email: rafael.sentandreu@uv.es

Victor Silva,
Microbiology and Mycology Program,
Biomedical Sciences Institute,
School of Medicine,
University of Chile,
Independencia 1027,
Santiago,
Chile
Email: vsilva@machi.med.uchile.cl

Eulogio Valentín,
Departament de Microbiología,
Facultat de Farmacia,
Universitat de València,
Vicent Andrés Estellés s/n,
46100 Burjassot,
València,
Spain

***Jianping Xu,**
Department of Biology,
McMaster University,
1280 Main Street West,
Hamilton,
Ontario,
L8S 4K1,
Canada
Email: jpxu@mcmaster.ca

Preface

During the past decade we have witnessed a mushrooming of papers in the area of medical mycology; detailing major advances in areas such as genomics, molecular and cellular biology, molecular epidemiology, immune response and vaccine development, and strategies to combat infections in humans. This sheer volume of information makes it extremely difficult for the busy research scientist and/or teacher of medical mycology to keep abreast of all the latest advances. This book, together with its companion volume "Pathogenic Fungi: Host Interactions and Emerging Strategies for Control", brings together expert international authors who critically review current topics, and through the provision of extensive reference sections positively encourage readers to pursue the subject in greater detail.

The book is divided into two sections: "Fungal Dimorphism and Pathogenicity" and "New Taxonomic Tools". The first section focuses on morphogenesis, the cell cycle, and the cell wall of human pathogens. These play a major role in elucidating fungal relationships, both with the environment and with the host. Experts in fungal structural biology contribute in-depth reviews on a variety of topics with a focus on molecular and biochemical analysis. The final chapter in this section presents a fascinating review of how mathematical modelling can be used to understand the building of three-dimensional cell structures in the morphogenetic process. The second section, entitled "New Taxonomic Tools", presents novel approaches to aid the understanding of strain variability, the significance of environmental and patient strains, and the relatedness of "uncultured" fungi. In addition the use of molecular tools for the taxonomic classification of previously unclassifiable fungi is featured.

The co-editors wish to thank each author for their valuable contributions and Horizon Scientific Press for allowing us to develop the text freely while offering their total support for our cause.

Gioconda San-Blas
Instituto Venezolano de Investigaciones Científicas
Caracas, Venezuela

Richard Calderone
Georgetown University, Washington, DC, USA

December 2003

Fungal
Dimorphism
and Pathogenicity

From: Pathogenic Fungi: Structural Biology and Taxonomy
Edited by: G. San-Blas and R.A. Calderone

Chapter 1

The Structure and Composition of the Fungal Cell Wall

Rafael Sentandreu,
M. Victoria Elorza, Eulogio Valentín,
and José Ruiz-Herrera

Abstract

The cell wall is probably one of the most characteristic structures of fungi and can be sued to define this group of organisms. The cell wall has many functions including the protection of the fungal cell from the harsh physical, chemical and biological conditions. It is also responsible for the shape of the organism. But, the cell wall also is the structure responsible for all of the interactions of fungi with their environment. Cell surface biomolecules are responsible for attachment of fungi to several types of surfaces, including host cells in the case of pathogenic fungi. The cell wall also acts as an immunogen, cause immunosuppression, or act as a virulence factor. Enzymes present in the cell wall degrade large molecules, such as proteins, polysaccharides, and lipids, to provide nutrients for cell growth. Chemically the fungal cell wall is mostly composed of polysaccharides. Proteins are less abundant, followed by lipids, and other minor components such as pigments and inorganic salts. These components are held together as a coherent structure by non-covalent bonds, mainly hydrophobic, hydrogen bonds, and by covalent bonds. Thus, the wall is a three-dimensional structure that serves its multiple functions. Structurally, the fungal cell wall appears as a biphasic system consisting of two kinds of components, structural polysaccharides, such as chitin and glucans, that are organized as microfibril bundles that provide rigidity for the cell, and an amorphous component, primarily glycoproteins and other polysaccharides that cover the inner, microfibrillar layer. The amorphous layer contains the adherence glycoproteins and other functional biomolecules associated with protection against the environment.

1. Introduction

External to the plasma membrane, fungi are surrounded by a rigid structure, the cell wall. The wall is probably the structure of the cell that is the most characteristic of fungi, defining the kingdom itself. The cell shape of fungi, at each developmental stage, with the exception of the naked zoospores of some aquatic fungi, is due to the cell wall. Besides this morphogenetic role, the wall is also the structure that protects the integrity of the organism by providing a rigid covering for the inner cell or protoplast. The cell wall is necessary to withstand the large differences in osmotic pressure of the surrounding media; it also provides protection against foreign enzymes and membrane-perturbing compounds, including toxic metals. But, we must not only think of the cell wall as a static and rigid structure. Its composition and arrangement change during cell growth. Thus, the wall has also important roles that permit the cell to interact with its environment. At its outmost surface, the wall contains specific molecules, such as adhesins, lectins and hydrophobins that specifically function in the attachment of the fungus to different surfaces of host origin, in the case of the human pathogenic fungi. Adherence to host cells is the first step in the invasion by pathogenic fungi. A large number of enzymatic activities are found in the wall, many of them involved in the degradation of complex, impermeable nutrients that are thus degraded into substrates that are taken up by cells and used as nutrients. Some of these enzymes are required for the assembly of the cell wall, and others are important in the cleavage of the protective superficial structures of host cells and tissues, permitting their invasion. Specific wall components act as antigens and immunomodulators, and their effects can alter the balance between the fungi and their hosts or predators.

In *Saccharomyces cerevisiae* about one fifth of its genetic information is used in the biosynthesis and construction of the three-dimensional structure of the cell wall (De Groot *et al.,* 2001). The cell wall makes up close to 30 % of the cell dry weight (Valentín *et al.,* 1987; Fleet, 1991), 80-90 % of which is polysaccharide, while the rest is represented by proteins, lipids, and other minor components (Bartnicki-García, 1968). The cell wall consists of structural compounds that are responsible for the resistance to unfavorable growth conditions such as high osmolarity, pressure and temperature. The most important structural components of the cell wall are chitin, made of N-acetylglucosamine (GlcNAc) units in β-1,4-linkages and β-glucans composed of glucose units of β-1,3- and β-1,6-linkages. These components exist in the form of microfibrils immersed in cementing material, mostly glycoproteins (Bartnicki-García, 1968). This type of organization provides rigidity for the cell wall and is critical to growth of these organisms. This is not surprising taking into consideration that chitin has probably the highest tensile strength of natural substances (see Ruiz-Herrera, 1992). Young's moduli of *S. cerevisiae* cells have been calculated in 112 and 107 Mpa in the exponential and stationary phases, respectively (Smith *et al.,* 2000), probably due to the high levels of glucans, and rather low proportion of chitin. The mechanical properties of the nascent and mature cell walls are different. The former is elastic, whereas the latter is viscoelastic (see review by Ruiz-Herrera, 1992). These differential mechanical properties are important to the growth of the cell wall growth at specific sites.

2. Chemical Composition of the Cell Wall

As described above, the fungal cell wall is a composite of many different substances that interact among themselves to give rise to a coherent structure. This composite provides physical and chemical protection to the cell, and is responsible for its morphology. The fungal wall accounts for about 20 % of the dry weight of the cell. Some of the wall components are linked together by covalent bonds, whereas other components are linked to the wall by hydrogen bonds, hydrophobic interactions, or salt associations; the latter components can be removed from the wall by treatment with detergents or salt solutions. These compounds are sometimes not considered as true wall components. Nevertheless, this philosophy represents a strictly structural concept, since many are proteins with enzymatic activity and/or host recognition or adhesion (see Ruiz-Herrera, 1992 for conceptual considerations). A similar operational concept has been proposed for the definition of a functional ribosome.

The fungal cell wall is primarily composed of polysaccharides, that make up about 80 % of the wall dry weight. The protein content is lower, around 3 to 20 %, whereas lipids, pigments, and inorganic salts are present in smaller amounts. Glucose-containing polysaccharides are the most abundant wall components in all fungal taxa, followed by chitin. Polysaccharides of galactose or/and mannose are also present in significant amounts, whereas compounds containing glucosamine, galactosamine, glucuronic acid, rhamnose, and/or xylose, are present in lower amounts. Arabinose, fucose, and ribose exist in trace amounts in some specific fungal groups (reviewed in Ruiz-Herrera, 1992).

Some of the wall polysaccharides have mostly a structural role, while others serve as an amorphous, outer layer. These polysaccharides are extremely variable and can be classified according to their composition as homopolymers, composed of a single type of sugar, or heteropolysaccharides, containing more than one type of sugar. Some polysaccharides are neutral, others are acidic, and some are basic. The presence of certain kind of polysaccharides in the cell wall of fungi was shown to have chemotaxonomic value. Bartnicki-Garcia (1968) has demonstrated that this analysis agrees with other characteristics that are used for the fungal classification.

The most important structural polysaccharide, chitin, is composed of N-acetylglucosamine (GlcNAc) units joined by β-1,4-linkages; cellulose, which is very rare in the cell walls of true fungi, is also composed of glucose units in by β-1,4-bonds. β–glucans, which are the most abundant structural polysaccharides, are also composed of glucose residues in either β-1,3- and/or β-1,6-linkages.

The number of amorphous polysaccharides of fungi is extensive. α-Glucans are abundant in some fungi, and are composed glucose units in α-1,3-bonds. Mannan, the most abundant amorphous polysaccharide, is the generic name given to the carbohydrate moiety of the glycoproteins that contain mostly mannose. Polyuronides, acidic polysaccharides that contain uronic acid and some neutral sugars, were discovered in the walls of *Mucor rouxii* (Bartnicki-García, 1968) and other Zygomycetes. Related in structure is mucoran, a heteropolymer containing D-glucuronic acid, D-mannose and L-fucose, with a smaller proportion of galactose and glucose (Bartnicki-García and Lindberg, 1972). On the other hand, chitosan is a basic polysaccharide of glucosamine, originally discovered in *Phycomyces blakesleeanus* (Kreger, 1954), but more recently has been found in many fungi. Nigeran, a polysaccharide composed of D-glucopyranose units in α-1,3- and α-1,4-bonds, was originally described in several species of *Aspergillus* and *Penicillium*. Also noticeable in the walls of different fungi are glucomannogalactans.

The structural polysaccharides, chitin and β-glucans accumulate in the internal layer of the cell wall, which appears less electron-dense, whereas the external layer, which is more electron-dense, is enriched in highly glycosylated mannoproteins (Baba and Osumi, 1987; Cappellaro *et al.,* 1994) that interact with substrates and other cells (Lipke and Kurjan, 1992; Teunissen and Steetsma, 1995; Reynolds and Fink, 2001). The carbohydrate moiety of the glycoproteins contains abundant phosphodiester linkages that confer a negative charge to cells at physiological pH (Jigami and Odani, 1999). These carbohydrate chains also give the cell surface its hydrophilic properties and may be involved in water retention. The external layer represents about one third of the wall weight and is bound to the β-glucan/chitin layer through the lateral chains of β-1,6-glucan.

It is important to stress that the designation of glucans as structural polysaccharides and their location within the cell wall are not absolute properties. Thus, it has been demonstrated that cell wall β-glucans from different fungi have other functions. Accordingly, the β-glucans from some fungi, such as *Pneumocystis carinii*, activate alveolar macrophages to release pro-inflammatory cytokines and chemokines (Lebron *et al.*, 2003).

The cell wall of fungi may not have a constant chemical composition, since changes occur during the life cycle and can be different in the various growth forms of a specific fungus, as, for example, in the spore, in the vegetative and reproductive mycelia, or in the yeast or mycelial forms of dimorphic fungi. Environmental conditions may drastically affect the composition of the cell wall. A classic example of not only quantitative, but also qualitative changes in the chemical composition of the cell wall during cell differentiation, is *Mucor rouxii*, a system exhaustively analyzed in the past (Bartnicki-Garcia, 1968). Other example is *Paracoccidioides brasiliensis* in which the yeast morphology, which is the invasive phase of the fungus, contains α-1,3-glucan in the wall, while the mycelial phase contains β-glucans instead (San-Blas *et al.,* 1977). A thorough discussion of this aspect is beyond the scope of the present review, but interested readers may consult Ruiz-Herrera (1992).

Regarding cell wall proteins, as has already been described, most often, they are glycoproteins. Except for the polysaccharides, proteins constitute the most abundant components in the cell wall. Proteins have extremely important roles in the wall. Some wall proteins have enzymatic activity. They may be involved in the degradation of large molecules that are not taken into the cell, in the synthesis of wall components, or in the degradation of the cell wall. Other proteins are important in the interaction of fungi with the environment. For example, in *Candida albicans,* important antigenic differences due to glycoproteins have been noticed among the yeast-like and mycelial forms (Sundstrom and Kenny, 1985; Pontón and Jones, 1986a; 1986b; Sundstrom *et al.,* 1988; Casanove *et al.*, 1992), and probably are involved in the establishment of the pathogen in the host, or in the response of the latter to invasion. Other glycoproteins may be important from a structural point of view. The existence of different glycoproteins in either phase of dimorphic fungi suggests that they participate in the architecture of the cell wall, and therefore in its morphology. These results have led to suggest that some mannoproteins possess a so-called "morphogenetic code," responsible for the molecular architecture of the wall (Sentandreu *et al.,* 1993).

Lipids constitute minor components of the cell wall. Some lipids may be readily extractable with solvents, whereas others require hydrolysis prior to extraction, suggesting that they are bound to other wall components by means of covalent linkages. The hydrophobic characteristics of the spores of several fungal species are due to the presence of lipids or wax on their surface (reviewed by Ruiz-Herrera, 1992) and may protect the cells from desiccation. In other fungi, surface lipids may be important for their antigenic properties. *C. albicans* cells contain a phospholipomannan (PLM), which reacts with antibodies specific for β-1,2-oligomannosides (Trinel *et al.*, 1999). This component is a novel phosphoinositolmannoside, which might have some pathophysiological relevance to host recognition, protection and signaling properties.

Finally, it is necessary to mention the presence of pigments in the cell wall of some fungi, especially for those organisms that produce spores (see review by Ruiz-Herrera, 1992). Of these, the most abundant pigments are the melanins. Melanins are polymeric compounds, formed by the oxidative polymerization of different phenols (Nicolaus, 1969). Melanins are considered to be important in the protection of fungal spores against UV light. In human pathogens, the production of melanin has been associated with virulence (Gomez and Nosanchuk, 2003). Melanins produced by different pathogenic fungi such as *Cryptococcus neoformans* and *Histoplasma capsulatum*, can bind antifungals, reducing their inhibitory activity, while in others, such as *Aspergillus niger*, melanin may affect the immune response, and even protect the fungus from toxic metals.

3. Chitin

Chitin is one of the most important structural compounds of the fungal cell wall. The chitin content in fungal cell wall ranges from as low as 2 % of the dry weight in *S. cerevisiae*, to *ca.* 60 % in *Allomyces macrogynus* and *Sclerotium rolfsii*, although in most fungi chitin accounts for about 20% of the cell wall weight (for a review see Ruiz-Herrera, 1992).

Chitin is a linear polysaccharide composed of more than 2000 units of *N*-acetylglucosamine (2-acetamido-2-deoxy-D-glucose, GlcNAc) (Calvo-Mendez and Ruiz-Herrera, 1987) joined by β-1,4-linkages. Chains of this polysaccharide are associated through hydrogen bonds to form microfibrils composed of about 20-400 chains. Although three crystalline forms of the polysaccharide exist in nature, as far as is known, fungal cell walls contain the α form only, where chains associate in an antiparallel fashion. Due to this crystalline arrangement, chitin is one of the most insoluble natural products, even more than cellulose.

4. Chitosan

Chitosan is a linear polysaccharide of glucosamine (GlcN) units in β-1,4-linkages. Originally described in the walls of Zygomycetes (Kreger, 1954), it is now known to be present in different Ascomycetes as well (review by Ruiz-Herrera, 1992). In contrast to chitin, chitosan is poorly crystalline and soluble in dilute acids. Chitosan is considered to be a product of the deacetylation of chitin. Analysis of chitosan synthesized *in vitro* by crude extracts from *M. rouxii* revealed that treatment with HNO_2 (known to convert GlcN into anhydromannose), liberated anhydromannose, and short oligosaccharides composed of different proportions of this saccharide and GlcNAc. These results revealed that fungal wall chitosan was not a homopolymer, but a partially deacetylated polysaccharide containing different proportions of GlcN and GlcNAc. Chitosan is present in the walls of hyphae, yeasts and spores of Zygomycetes and in hyphal walls of most Ascomycetes; in *S. cerevisiae* it is only found in the wall of ascospores.

Apparently chitosan does not play a structural role in the wall. It has been suggested that it may protect chitin from the action of chitinases present in the environment. An extracellular chitin deacetylase, suggested to play a role in autolysis, was described in *Aspergillus nidulans* (Alfonso *et al.*, 1995). That it may represent an enzyme originally located in the wall, and involved in chitosan synthesis cannot be disproved. Double mutants of *S. cerevisiae* in the genes *CDA1* and *CDA2,* which together account for its total capacity to synthesize chitosan (see below), form ascospores that are affected in cell wall assembly. The chitosan lacks dityrosine residues and is more sensitive to hydrolytic enzymes, as compared to wild-type cells (Christodoulidou *et al.*, 1996; Mishra *et al.*, 1997; Christodoulidou *et al.*, 1999).

5. Cell Wall Glucans

Glucan polysaccharides are composed of glucose residues joined through either α or β linkages. Glucans are the most abundant polysaccharides of the fungal cell wall, displaying extremely variable structures as well as differences in solubility and size. Their distribution in fungi is universal, but they are absent in the vegetative walls of Zygomycetes (Bartnicki-Garcia, 1968). α-Glucans are present in a large number of fungi that lack phylogenetic relationships. Cell wall-bound α-1,3-glucans are linear homopolymers, water insoluble, but soluble in cold alkali (Sanchez-Hernandez *et al.*, 1993). The α-1,3-glucans from *P. brasiliensis, A. niger, Polyporus betulinus* and *Histoplasma farciminosum* display a structure of 10-14 nm thick fibers when observed by electron microscopy. Nigerans, originally detected in the walls of *Aspergillus* and *Penicillum* species, are composed of D-glucopyranosyl moieties joined through alternating α-1,3- and α-1,4-linkages (Barker *et al.*, 1957). Surprisingly, a recent report described the presence of glycogen covalently bound to β-1,3-glucans in the yeast cell wall (Arvindekar and Patil, 2002).

The most abundant glucans in the fungal walls are β-glucans containing variable proportions of β-1,3- and β-1,6-linkages. Unbranched β-1,3-glucans have a microfibrillar structure as revealed by electron microscopy of material synthesized by protoplasts (Kreger and Kopecka, 1975) or cell-free extracts (Larriba *et al.*, 1981) from *S. cerevisiae*. The glucans crystallize poorly and are similar to the so-called "hydroglucans" obtained after boiling the branched β-glucan from yeast with mineral acids, a treatment that apparently eliminates the β-1,6-bound branches. In agreement with the well-studied β-1,3-glucan, curdlan, it was suggested (Kopecka and Kreger, 1986) that these microfibrils are composed of three intertwined, helical chains that associate through hydrogen bonding, while containing six glucose moieties per turn, similar to lentinan, a β-1,3-glucan from *Lentinus edodes* (Bluhm and Sarko, 1977).

In the cell wall, glucans form microfibrils with a diameter of 7.5-10 nm that associate into a mesh of 20-60 nm in thickness (Kreger and Kopecka, 1975). Selective localization of different glucans has been described for the cell wall of *Schizosaccharomyces pombe* (Humbel *et al.*, 2001). By immuno-electron microscopy, β-1,6-branched/β-1,3-glucan was located in the less electron-dense layer of the cell wall. On the other hand, linear β-1,3-glucan was found to be located exclusively in the primary septum of dividing cells, whereas β-1,6-glucan was present in the secondary septum.

In general, the most abundant β-glucans of the fungal cell wall contain both β-1,3- and β-1,6-linkages; the latter occurring at branching points, although some may form inter-residue linkages. A notorious exception is the *A. fumigatus* cell wall, where β-1,6-glucans have not been detected (Fontaine *et al.* 2000). Some β-1,3-, β-1,6-glucans appear loosely associated with the cell wall in the form of a mucilagenous layer, but most are indispensable to the structure of the wall. Some are alkali-insoluble, whereas others are soluble in alkali. The different behaviour of β-glucans may be related not only by their molecular size, but also by their degree of branching; thus, a mutant of yeast containing larger amounts of alkali-soluble glucan in its wall, was found to be enriched in glucans with a higher degree of β-1,6-linkages (Ha *et al.*, 2002).

The number and length of β-1,6-branches in fungal wall glucans are extremely variable. For example, the wall-bound alkali-soluble glucan from *Coprinus lagopus* contains β-1,6-branches of a single glucose unit (Schaeffer, 1977). On the other hand, glucans from the wall of *S. cerevisiae* (which has been studied the most) show a much higher number of β-1,6-branches composed of a larger number of glucose units (Manners *et al.*, 1973), although the observation that they are dialysable reveals that they do not exceed an Mr of *ca.* 10 kDa. Other glucans containing not only β-1,3- and β-1,6-linkages but also β-1,4-bonds have been isolated from *Monilinia fructigena* (Feather and Malek, 1972) and *Armillaria mellea* (Sanchez-Hernandez *et al.*, 1993). In the cell wall of *A. nidulans*, a novel linear β-1,3-/β-1,4-glucan has been characterized (Fontaine *et al.* 2000).

Determination of the molecular size of glucans has been a difficult task due to the extreme diversity of molecules that fall under this general designation, and because they form complexes in the wall with other macromolecules (see below). A water-soluble glucan from *M. fructigena* was found to have an Mr of 70 kDa (Santamaria *et al.*, 1978). In contrast, the size of the wall-bound β-glucan from yeast was calculated to be 240 kDa, a probable underestimation because of the harsh methods used to extract it. The β-1,3-glucan synthesized by membrane fractions obtained from *S. cerevisiae* protoplasts contained less than 50 sugar residues (Shematek *et al.*, 1980), a result apparently due to β-glucanase contamination. On the other hand, β-1,3-glucan synthesized by yeast membrane fractions obtained by ballistic disruption was greater than 20 x 10^6 Da (Ruiz-Herrera and Larriba, 1995). Agreeing with this value, β-glucan extracted from the wall of *C. albicans* yeasts had a mean Mr of over 10^6 Da, the value being higher for the β-glucan isolated from mycelial walls of the fungus (Miura *et al.*, 2003).

6. Cell Wall Proteins

6.1. General Aspects

Previously, the presence of proteins in the fungal cell walls was a debatable subject. However, several seminal papers have clearly established the existence of covalently-linked proteins in the cell walls of fungi (Pastor *et al.*, 1984; Valentin *et al.*, 1984; Elorza *et al.*, 1985). Our knowledge of the role of these proteins has grown rapidly and steadily in the recent years, and their study has become one of the most exciting subjects in the field. An analysis of wall proteins was preceded by the development of gentle methods for their extraction. One of the new approaches to their study is that of proteomics, which focuses upon their identification. This discipline is expected to increase even further our knowledge of this subject. Early studies described the existence of peptides bound to glucans and mannans (Kessler and Nickerson, 1959; Korn and Northcote, 1960). These results suggested the possibility that wall proteins could be extracted from purified walls or whole cells by digestion of the wall polysaccharides with chemical reagents (Sentandreu and Northcote, 1969) and lytic enzymes (β-1,3-glucanases, chitinases and others) (Elorza *et al.*, 1985; Casanova *et al.*, 1992; Kaptein *et al.*, 1994; Lopez-Ribot *et al.*, 1996; Mrsa *et al.*, 1997; Mormeneo *et al.*, 1996). Preferential radioactive labeling of these proteins facilitated their analysis, and demonstrated that proteins resisting extraction by hot ionic detergent extraction accounted for less than 10 % of the total cell wall-bound protein. Labelling by non-permeable reagents (e.g. sulfo-NHS-LC-biotin) has been widely used to identify those proteins located outside of the permeability barrier of the cell, and has fostered a debate as to whether we consider them to be "wall proteins" *sensu stricto* (Casanova *et al.*, 1992; Mrsa *et al.*, 1997).

The development of extremely sensitive and powerful methods has demonstrated the presence of cell wall proteins from different fungi that are normally considered to be cytoplasmic, such as enolase, glyceraldehyde-3-phosphate dehydrogenase, alcohol dehydrogenase and phosphoglycerate kinase, which are involved in the glycolytic pathway, as well as other cytoplasmic proteins. These results established the interesting question about how to define what a "fungal cell wall protein" is. This is not a mere semantic problem. It involves additional questions such as their role (if any) in the wall, and how they are externalized. For the sake of simplicity, we will avoid this term and instead refer to them as "cell wall-associated proteins". Our definition of a wall-associated protein includes those proteins that have a secretory motif, that are *N*- and/or *O*-glycosylated, and have other specific characteristics, such as a GPI-binding motif or specific inner repeats. These aspects are discussed below.

6.2. Methodology for the Isolation and Identification of Wall Proteins

Sequential extraction is used for the isolation of cell wall proteins. After mechanical breakage of the cells, the cell wall is isolated by differential centrifugation, washed with water or saline solutions, and extracted with a hot ionic detergent, normally SDS. Proteins linked in the wall by disulfide bridges are then extracted with a reducing agent such as β-mercaptoethanol, and finally, the wall is digested with lytic enzymes, such as β-glucanases and chitinases. The extracted proteins are separated by SDS-PAGE in the presence of β-mercaptoethanol. Nevertheless, the number of wall glycoproteins in the wall has been difficult to be determine because SDS-PAGE analysis results in diffuse, polydispersed bands. This fact is due to the *N*- and *O*-linked carbohydrate moieties. Microsequencing has permitted the identification of several protein bands (e. g. Mrsa *et al.*, 1993; 1997), but a similar problem with separation of proteins has limited the technique. The use of tunicamycin, an inhibitor of *N*-glycosylation (Orlean *et al.*, 1991), or mutants affected in protein glycosylation, specifically *mnn9* from *S. cerevisiae* (Tsai *et al.*, 1984), *C. albicans* (Southard *et al.*, 1999), *Yarrowia lipolytica* (Jaafar *et al.*, 2003) and *Hansenula polymorpha* (Kim *et al.*, 2001) (see below) facilitated this analysis (Frevert and Ballou, 1985). Identification of other specific proteins was achieved by use of antibodies (Valentin *et al.*, 1986; Moukadiri *et al.*, 1997; Pardo *et al.*, 1999; 2000; Pitarch *et al.*, 2002). Screening by antibodies against expression libraries made in *E. coli* demonstrated the presence of several glycolytic and other enzymes in the cell wall of *C. albicans* (Sentandreu *et al.*, 1995; Roig *et al.*, 2000). A gene fusion encoding an intracellular invertase has been used to identify secreted proteins (including cell wall-associated ones) in *C. albicans* and *S. cerevisiae* (Monteoliva *et al.*, 2002).

Genes involved in the construction of the cell wall have been identified by systematic analysis of mutants obtained by use of chemicals or by transposon mutagenesis (Ram *et al.*, 1994; Chun and Goebl, 1996; Lussier *et al.*, 1997; Neuveglise *et al.*, 1998). Their phenotypic analyses have included sensitivity to lytic enzymes (Ram *et al.*, 1995), detergents (Kanik-Ennulat *et al.*, 1995), or calcofluor white (see Elorza *et al.*, 1983; Ram *et al.*, 1994).

Introduction of mass spectrometry for peptide sequencing, and the determination of molecular mass, assisted by genome sequencing and annotation of various fungal species, has resulted in a new approach to the identification of cell wall proteins. In *S. cerevisiae* and *C. albicans* a whole array of wall-associated proteins have thus been identified (Pardo *et al.*, 2000; Pitarch *et al.*, 2002; Garcerá *et al.*, 2003).

6.3. Role of Proteins in the Cell Wall and Biology of Fungi; Their Abundance and Distribution

The cell wall of fungi is essential for growth, morphogenesis, adhesion, flocculation, mating and biofilm formation. In pathogenic fungi, wall proteins frequently constitute virulence factors, and important immunogens (Calderone and Fonzi, 2001). Accordingly, it is expected that both structural and catalytic proteins are required to fulfil these functions at specific points during the life cycle of a fungus and in its response to environmental conditions (Ram *et al.*, 1995; Caro *et al.*, 1998; Moukadirii *et al.*, 1999; Rodriguez-Pena *et al.*, 2000). Some of these functions, either catalytic or structural, depend on polypeptide folding, whereas others are affected by the degree of glycosylation; *e. g.*, the proteins that are located at the outer surface of the wall and responsible in part for the permeability properties of the cell, become more permeable when the cell is less glycosylated (Valentin *et al.*, 1987; De Nobel *et al.*, 1991).

It is important to point out that whereas structural wall proteins (see above) are strictly bound to the wall, some of the catalytic proteins are either firmly or loosely-bound, and even secreted into the medium. On the outer surface of the wall are glycoproteins of the fimbriae, filamentous structures that provide the initial contact of fungi such as *C. albicans* with host cells (Tokunaga *et al.*, 1990). Fimbriae are mannoproteins composed of 80 to 85% D-mannose and 10 to 15% protein (Yu *et al.*, 1994; Vitkov *et al.*, 2002).

Hydrophobic properties of the wall surface of a large number of fungi are due to the presence of a group of proteins collectively known as hydrophobins (Wessels, 1999). Due to their hydrophobic nature, they lower the surface tension of cells resulting in the growth of aerial hyphae, or serving as adhesive agents to facilitate attachment to surfaces (Paananen *et al.,* 2003). In the case of pathogenic fungi, the hydrophobins play an important role in overcoming host defences (Paris *et al.*, 2003). One of the properties of hydrophobins is their tendency to form fibrillar or rod-like aggregates at interfaces (Torkkeli *et al.* 2002). Analysis of their amino acid composition has revealed the presence of about 50% hydrophobic residues and at least six cysteine residues that form disulfide bonds. Interestingly, these proteins are not glycosylated (Calonje *et al.*, 2002).

A large number of wall proteins have been described in fungi; they are either catalytic or structural. A partial list of mannoproteins includes the following: the so-called Pir proteins [proteins with internal repeats], glycosylphosphatidylinositol-proteins (GPI-proteins), Cwp1, Bgl2, chitinases, Bar1 (a protease released by βME; Moukadiri *et al.*, 1999), agglutinins (Cappellaro *et al.*, 1994), glycosidases and potential glucanases active in wall metabolism, growth and morphogenetic processes (Cappellaro *et al.*, 1998), adhesins , such as Hwp1, Ala1p/Als5p, Als1p, from *C. albicans* and Epa1p from *Candida glabrata*, which are GPI-proteins (Sundstrom, 2002), flocculins such as Flo1 (Bidard *et al.*, 1995; Lo *et al.*, 1998), lectins such as the flo protein responsible for the formation of large flocculent aggregates in yeast (Stratford 1992; 1996; Bony *et al.*, 1998), and proteins that facilitate the uptake of iron in yeast (FIT1, FIT2, and FIT3) (Protchenko *et al.*, 2001). It seems odd that, not withstanding our great advances, we remain ignorant of the functions of more than half of the predicted cell wall mannoproteins in *S. cerevisiae* and *C. albicans*, the best known fungal models.

Electron microscopic studies have revealed that proteins are far from being homogeneously distributed within or at the surface of the cell wall (see below). Glycoproteins are enriched in the outer surface of the wall and are less abundant in the inner, more electron-transparent layer that is richer in glucans and chitin. Nevertheless their presence in this layer can be demonstrated by different techniques, including labelling with ConA and antibodies (Marcilla *et al.*, 1991; Kapteyn *et al.*, 2000).

6.4. Types of Proteins in the Cell Wall of Fungi

The proteins in the cell wall can be classified according to different criteria, including, molecular structure, function, or by the mechanism through which they are bound to the wall, etc. Traditionally, all wall proteins were thought to be glycoprotein in nature, based upon our knowledge of the secretory pathway whereby proteins are glycosylated. Nevertheless, as mentioned above, the introduction of novel techniques (Sentandreu *et al.*, 1995; 1997; Angiolella *et al.*, 1996), particularly mass spectrometry, has disclosed the presence in the fungal cell walls of non-glycosylated proteins, such as glycolytic enzymes, Hsp40, and others (see above) (Pitarch *et al.*, 2002). As indicated above, whether these proteins can be considered as *bona fide* cell wall proteins, and even more important, their significance, remains to be determined. The different types and structures of glycoproteins are described below.

In regard to their retention in the wall, some proteins remain associated with the wall through their covalent linkages to structural polysaccharides (Pastor *et al.*, 1984; Valentin *et al.*, 1984; Elorza *et al.*, 1985). These proteins were classified separately from those that are bound by non-covalent linkages. Nevertheless, taking into consideration the above described "unusual" proteins in the wall [glycolytic proteins, for example], whose association with the wall remains unknown, we have introduced an operative criterion that separates all proteins into those extractable by ionic detergents or chaotropic agents [the typical non-covalently bound proteins], and those which resist this treatment, and are solubilized only after digestion of the structural polysaccharides. The latter group includes four types of glycoproteins. Three of them are covalently linked to the cell wall, and the fourth group corresponds to proteins devoid of a carbohydrate moiety that are retained by the wall by unknown mechanisms. The covalently-bound glycoproteins are the following: 1) Proteins bound to β-1,6-glucan through a glycosyl phosphatidylinositol (GPI) moiety (De Nobel and Lipke, 1994; Kapteyn *et al.*, 1995; van Der Vaart *et al.*, 1996). 2) Pir proteins (*p*roteins with *i*nternal *r*epeats), characterized as containing repetitive sequences (Russo *et al.*, 1992; Toh-e *et al.*, 1993) and being highly *O*- glycosylated. These are attached to the β-1,3-glucan by unknown alkali-labile bonds (*O*-glycosydic linkage?) (Mrsa *et al.*, 1997; Kollar *et al.*, 1997; Kapteyn *et al.*, 1999a; 1999b; Castillo *et al.*, 2003). 3) Other mannoproteins retained by disulfide bridges which are released by reducing agents such as β-mercaptoethanol (β-ME) or dithiothreitol (DTT; Orlean *et al.*, 1986; Moukadiri and Zueco, 2001). Some Pir proteins are included in this group.

6.5. *In Silico* Study of Wall Proteins

The rapid growth in the number of complete and annotated genome sequences of fungi that are used as model organisms has resulted in a new approach to the study of cell wall proteins. Even if only a partial sequence is obtained, it is possible to use that to "mine" the genomic and proteomic data bases to determine the entire sequence. Oppositely, knowledge of the special characteristics of *bona fide* wall proteins permits *in silico* analysis of the whole genomes for those ORFs that fulfil those characteristics. In the case of *S. cerevisiae*, use of the Von Heijne's algorithm allows the identification of 686 ORFs encoding proteins with potential N-terminal signal sequences for targeting to the secretory pathway. Of these, 51 contained potential GPI-attachment signals, including 20 that were putatively membrane-bound and 37 probable cell wall proteins (Caro *et al.*, 1997). A complementary approach to confirm this *in silico* data has been protein tagging with hemagglutinin (HA), *myc* or flag, followed by determination of the cellular location of the tag (Hamada *et al.*, 1999; Terashima *et al.*, 2002). These analyses have been made more selective through use of the following criteria: (i) a signal sequence for secretion at the N-terminus, (ii) a Ser / Thr-rich sequence which could provide sites for glycosylation, (iii) a C-terminal GPI-attachment site and; (iv) a stretch of hydrophobic amino acid residues. Using these criteria, more than one hundred ORFs encoding putative wall proteins have been identified in *C. albicans* (Garcerá *et al.*, 2003). Using this same approach, similar results were reported by De Groot and Klis (http://www.pasteur.fr/recherche/unites/Galar_Fungail/data.html).

6.6. Chemical Structure of Glycoproteins

The structure of the protein moiety of cell wall glycoproteins is basically the same as all proteins, except that nascent wall proteins, like all extracellular proteins, possess a signal peptide, and as described above, have either a specific motif involved in the attachment of a GPI moiety, or several repeats characteristic of Pir proteins. Normally these proteins are rich in Ser and Thr residues and contain the so-called sequons, tripeptides of *N*-Asn-Xaa-Ser/Thr (Roitsch and Lehle, 1989), where *N*-glycosylation takes place.

The level of glycosylation of wall proteins is variable. *Bona fide* wall glycoproteins have the highest amount of carbohydrate, often as high as 50 to 95% by weight, and are sometimes referred to as proteoglycans (Orlean, 1997). Different forms of glycosylation can be found in wall proteins: *N*-glycosylation, *O*-glycosylation, attachment of a glycosylphosphatidylinositol (GPI) anchor, linkage to β-1,3- or β-1,6-glucan, and probably other additional yet unknown linkages.

Most of what we know about the *N*-glycosidically bound carbohydrate moiety of glycoproteins comes from the studies by Ballou and his co-workers. A rather large polysaccharide composed of mannose residues is linked to the protein via a β-1,4-diacetylchitobiose moiety that involves an asparagine residue of the protein (Ballou, 1990). *N*-linked sugars have two moieties, a core composed of 10-14 mannose residues (Asn-$GlcNAc_2$-Man_{10-14}), and a large outer chain containing a backbone of α-1,6-linked mannose residues, and a variable number of α-1,2-linked mannose side chains. The side chains in turn contain a variable number of α-1,2- and α-1,3-bound mannosyl moieties, and occasionally, phosphomannosyl units that confer a net negative charge on the cell wall, which contributes to the properties of the cell surface (Jigami and Odani, 1999). This residue appears to be the target for osmotin (a plant-defence protein; Ibeas *et al.*, 2000; Veronese *et al.*, 2003). The level of mannosyl phosphorylation increases at late-logarithmic and stationary phases of cell growth, and levels of mannosylphosphate transfer appear to be involved in the cellular response to a variety of stresses (Odani *et al.*, 1997).

Interestingly, the structure of the inner core is conserved in all eukaryotic cells from fungi to plants and animals (Kornfeld and Kornfeld, 1976; Chapman and Kornfeld, 1979). Mutations affecting its synthesis are lethal, contrary to mutants in the outer chain. Mannose chains in *mnn1* mutants lack the α-1,3-bound external mannose units, whereas *mnn2* mutants are devoid of all side chains.

Regarding *O*-glycosylated proteins, chains of up to 5 α-1,2- and α-1,3-bound mannose units are linked to Ser or Thr residues (Sentandreu and Norcothe, 1969; Strahl-Bolsinger *et al.*, 1999; Ernst and Prill, 2001; Girrbach and Strahl, 2003). This type of bond is sensitive

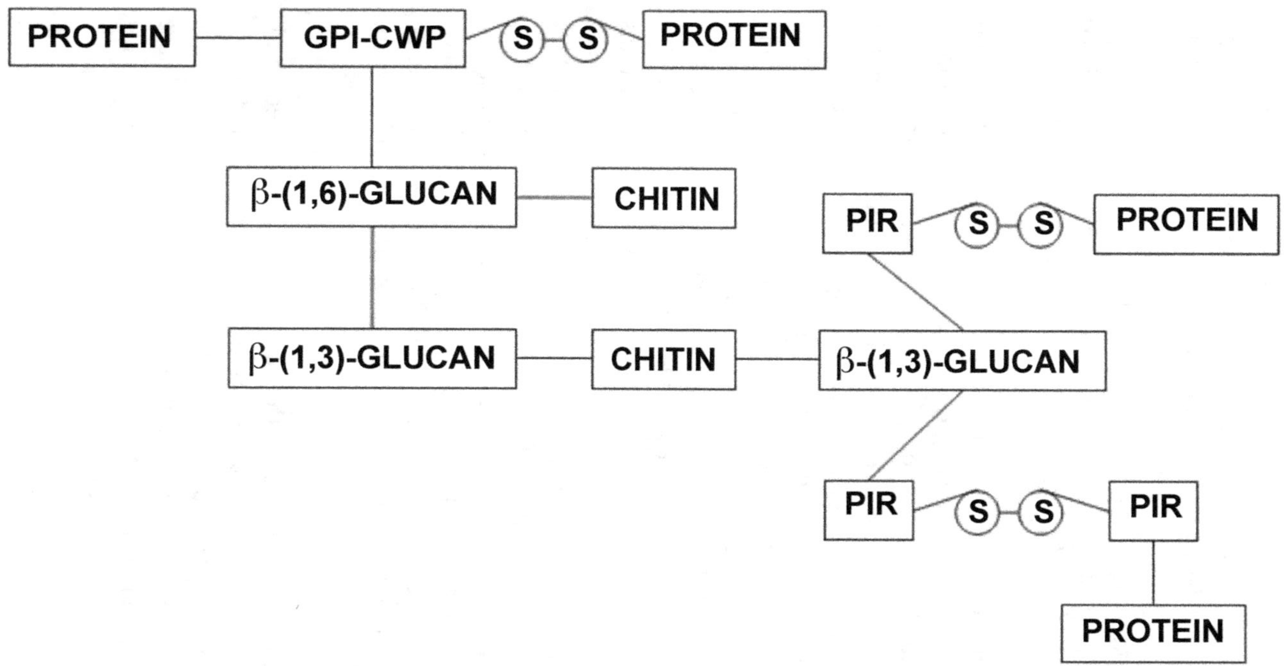

Figure 1. A model for the molecular architecture of yeast cell wall. Pir proteins are linked directly to the β-1,3-glucan-chitin internal skeleton layer whereas the GPI-proteins are indirectly linked through interconnecting β-1,6-glucan moieties. Other mannoproteins are retained only by disulfide bridges to other cell wall proteins. (European Consortium "Novel approaches for the control of fungal diseases", BMC 2000-1449, Unpublished results).

to weak alkali treatment (β-elimination; Sentandreu and Norcothe, 1969; Tanner and Lehle, 1987). These proteins are characterized by their rigid structure that protrudes from the wall surface analogous to antennae (Jentoft, 1990).

Attachment of a glycosylphosphatidylinositol (GPI) anchor to the carboxy terminus of wall proteins was first demonstrated by Lu *et al.* (1994). The core structure of the GPI anchor has been conserved and has the structure Man-α-1,2-Man-α-1,6-Man-α-1,4-GlcN-α-1,6-inositol. It is linked to a phospholipid that binds to the protein through a residue of ethanolamine-phosphate. The structure of the core is branched displaying α-1,3- and -α-1,2-linked mannose (Sipos *et al.*, 1995). In the wall, this structure appears truncated (see below) and bound to β-1,6-glucans (Kapteyn *et al.*, 1996; Kollar *et al.*, 1997; Lipke and Ovalle, 1998) (Figure 1).

Pir proteins are highly *O*-glycosylated proteins. Pir proteins from *S. cerevisiae* contain one (Pir4), or more characteristic repetitive sequences (Pir1, Pir2 and Pir3) (Russo *et al.*, 1992; Toh-e *et al.*, 1993). They have a similar organization that includes the presence of a signal peptide, a Kex2 sensitive site, a domain with 2 to 11 repetitive sequences, and a *C*-terminal sequence with four Cys residues at identical positions (repeat (s)-Cys –66aa–Cys –16aa–Cys –12aa–Cys –COOH). We lack information regarding the structure of Pir proteins from most fungal species, but it is probable that they do not differ significantly from those of *S. cerevisiae* or *C. albicans* cells. Pir proteins do not contain a GPI anchor motif and are attached to the cell wall by unknown alkali-labile bonds, possibly *O*-glycosydic linkages with β-1,3-glucan (Mrsa *et al.*, 1997; 1999; Kapteyn *et al.*, 1999a). At least some of the Pir proteins are retained in the wall exclusively by disulfide bridges, given that some is released by reducing agents such as β-mercaptoethanol (β-ME) or dithiothreitol (DTT) (Orlean *et al.*, 1986; Moukadiri and Zueco, 2001; Castillo *et al.*, 2003).

6.7. Linkages Formed During the Assembly of Wall Components

The oligosaccharide fractions of wall glycoproteins play different roles, some of which may still be unknown (for a review see Dwek, 1995). Among its known functions is the facilitation of peptide folding and recognition, both processes occurring during glycoprotein synthesis and secretion. Once the glycoprotein reaches the cell wall, the recognition of sites for its localization and interaction with external factors occurs. It is also known that the carbohydrate moiety protects glycoproteins against denaturation and proteolysis, and modifies its quaternary structure. The rod-like structure of some *O*-glycosylated proteins depends on the carbohydrate components (Jentoft, 1990), and the amount of the mannose content of the surface proteins determines cell wall porosity (Zlotnik *et al.*, 1984, De Nobel *et al.*, 1990). Also, the role of some wall proteins in virulence and as immunogens may depend to a large extent on their carbohydrate moiety.

Even though the described characteristics refer to the specific attachment of glycoproteins and wall polysaccharides, some authors have advanced the hypothesis that these basic cell wall units may constitute the so-called "flexible building blocks" with a solid structure (Kollar *et al.*, 1997) or "modules" with a lattice structure (Lipke and Ovalle, 1998). According to the latter study, each module would consist of a β-1,3-glucan chain with about 40-50 branches, including one or two β-1,6-glucan chains that are bound to mannoprotein. Chitin chains would form an association with some of these modules through β-1,3-/β-1,6-glucan linkages. In turn, these modules would associate among themselves through non-covalent linkages to the chitin-glucan layer, and by covalent links, including disulfide bonds, to the mannoprotein-rich layer. Although interesting, this model is only descriptive and is based on a hypothesis that lacks experimental evidence and does not take into consideration that the relative composition of the wall is not constant during the life cycle of fungi and that the wall is not a self-assembling structure.

7. Structural Organization of the Cell Wall

Microscopy has been an important tool to advance our understanding of the cell wall structure, including light, electron, and even atomic force microscopy. For these several techniques, specific staining procedures have been very useful, such as chemical methods that recognize different polymers, lectins that are specific for different oligosaccharide residues, and antibodies, each bound to fluorescent compounds or heavy metals. The use of proteins containing metal centers, such as ferritin, has been also a powerful technique.

Studies of the fine structure of the fungal cell walls by electron microscopy were initiated over 45 years ago, and procedures have been refined steadily since then (for reviews see Wright, 2000; Osumi, 1998). More recently, the introduction of the confocal microscopy that allows the study of virtual fine sections (Brakenhoff *et al.*, 1989) has widened our vision of the fungal cell wall.

When thin sections of cells stained by standard techniques are observed under the electron microscope, the cell wall appears to have several layers that are distinguished by their different electron densities. Initial studies showed the cell wall of *S. cerevisiae* was composed of three layers, an external, electron dense, a less dense inner layer, and an internal layer, which was the most electron dense (Darling *et al.*, 1972; Linnemans *et al.*, 1977). In *C. albicans,* depending on the method used, four to eight layers have been visualized (Chattaway *et al.*, 1976; Poulain *et al.*, 1978; Cassone *et al.*, 1978; Tronchin *et al.*, 1979; 1981; 1984). Also, a variable number of layers have been observed in the cell wall of *S. pombe* (Poole and Lloyd, 1973; Johnson *et al.*, 1982; Walther *et al.*, 1984; Humbel *et al.*, 2001), *C. neoformans* (Cassone *et al.*, 1974), *H. farciminosum* (San-Blas and Carbonell, 1974), *Dictyostelium discoideum* (Hemmes *et al.*, 1972), *Agaricus bisporus* (Rast and Hollenstein, 1977) and *Trichophyton mentagrophytes* (Wu-Yuan and Hashimoto, 1977). A useful staining technique in these studies was the PATAg method (Thiery, 1967). This technique employs oxidation with periodate and identifies polysaccharides with vicinal OH groups.

It is difficult to correlate the chemical composition of the wall with the number of layers observed, because of the artifacts that may be introduced during the preparation of cells, the type of cells observed [yeasts versus filaments], and the phase of the cell cycle (see Odds, 1985). It is also important to take into consideration the staining method; cell walls appear different if stained with osmium tetroxide or with $KMnO_4$. Accordingly, to construct a unified model of the fungal cell wall based on these studies is almost impossible because of the extreme variability in chemical composition. Nevertheless, some generalities can be reached. Layers of the wall richer in structural polysaccharides appear less electron dense, because they do not take up osmium; thus, chitin and β-1,3-glucans are not stained by the method of Thiery. On the other hand, electron dense layers contain more glycoproteins that are stained by osmium; thus, mannans, and other polysaccharides are stained by the method of Thiery. On this basis, the data suggest that glycoproteins are accumulated in the outer layer of the wall, where they fulfill important functions in the so-called "social activities" of the cell, such as adhesion, agglutination, cell recognition, etc. Instead, structural polysaccharides are accumulated in the middle layer of the wall. This differential distribution is confirmed by the use of selective staining with Fe-labelled Concanavalin (CoA) (that binds to mannans), Au-labelled specific antibodies against proteins or glucans, or Au-labelled wheat germ agglutinin (that binds to chitin) (Horisberger and Vonlanthen, 1977; Horisberger *et al.*, 1978; Cassone *et al.*, 1979; Valentin *et al.*, 1987). The presence of glucans in the middle layer was also demonstrated indirectly by its reduction in thickness after treatment with glucanases (Valentin *et al.*, 1987; Kopeka *et al.*, 1995). Other mannoproteins, including invertase (Bielefeld and Hollenberg, 1989) or alkaline phosphatase (Linnemans *et al.*, 1977; Arnold and Garrison, 1979; Haguenauer-Tsapis, 1986) are present in the innermost layer of the wall (periplasmic space?).

It is important to stress that location of certain components within the wall proper is not absolute. Accordingly, chitin appears to be distributed in different layers of the wall as shown by labeling with Au-wheat germ lectin (Marcilla *et al.*, 1991), and mannoproteins were found to be present also in the middle [inner] layer of the wall. This was confirmed in *S. cerevisiae, C. albicans* and other fungi, by partial digestion of the glucans, a treatment that permitted access of ferritin-labelled ConA, and resulted in a heavy labelling of the median layer (Valentin *et al.*, 1987).

Scanning electron microscopy (SEM) has resulted in the demonstration of specific features on the wall surface. Thus, the yeast cell wall displays a smooth surface, except for the bud scars (Watson and Arthur, 1977; Pringle *et al.*, 1979; Koch and Rademacher, 1980). The surface becomes eroded when the glycoprotein layer is removed (Koch and Rademacher, 1980). These results have been confirmed by use of the atomic force microscopy (AFM) (Ahimou *et al.*, 2003), which has also shown that the wall surface is far from being uniform (Gad *et al.*, 1997). Last, but not least, it is important to mention the use of laser confocal microscopy that permitted the location of specific compounds in the walls if combined with fluorescent antibodies or lectins (Ramon *et al.*, 1996; Navarro-Garcia *et al.*, 1998; Poulain *et al.*, 2002).

In summary, these results suggest that the fungal cell wall is organized from the inner to the outer surfaces as follows. Glycoproteins, some with enzymatic activity, accumulate in an inner layer that has been referred to as the "periplasmic space". A middle layer is then seen that is enriched in the structural polysaccharides, glucans and chitin, which in turn are bound to glycoproteins and other proteins that may be involved in wall maturation. This layer would provide resistance for the cell wall to stress. Finally, an external layer enriched primarily in glycoproteins, is bound to the structural polysaccharides. These proteins would be involved, as mentioned above, in the interaction of the cell with the environment, regulate permeability, participate in adhesion, flocculation, sexual recognition, antigenicity, etc. Fimbriae and hydrophobins would be in the most external surface of this outer layer that facilitate attachment, influence surface tension, and in pathogenic fungi, overcome host defences (Paananen *et al.*, 2003; Paris *et al.*, 2003).

8. Acknowledgements

Work in the laboratory of Valencia was supported in part by grants from the European Union (QLK2CT 2000-00795), Spanish Ministerio de Ciencia y Tecnología (BMC2000-1449) and Agencia Valenciana de Ciencia i Tecnologia de la Generalitat Valenciana (Grupos 03/187). Original work from the Irapuato laboratory was partially supported by Consejo Nacional de Ciencia y Tecnología, México.

References

Ahimou, F., Touhami, A., and Dufrene, Y.F. 2003. Real-time imaging of the surface topography of living yeast cells by atomic force microscopy. Yeast 20: 25-30.

Alfonso, C., Nuero, O.M., Santamaria, F., and Reyes, F. 1995. Purification of a heat-stable chitin deacetylase from *Aspergillus nidulans* and its role in cell wall degradation. Curr. Microbiol. 30: 49-54.

Angiolella, L, Facchin, M., Stringaro, A., Maras, B., Simonetti, N., and Cassone, A. 1996. Identification of a glucan-associated enolase as a main cell wall protein of *Candida albicans* and an indirect target of lipopeptide antimycotics. J. Infect. Dis. 173: 684-690.

Arnold, W.N., and Garrison, R.G. 1979. An Fe^{3+}-activated acid phosphatase in *Saccharomyces rouxii*. J. Biol. Chem. 254: 4919-4924.

Arvindekar, A.U., and Patil, N.B. 2002. Glycogen, a covalently linked component of the cell wall in *Saccharomyces cerevisiae*. Yeast 19: 131-139.

Baba, M., and Osumi, M. 1987. Transmission and scanning electron microscopic examination of intracellular organelles in freeze-substituted *Kloeckera* and *Saccharomyces cerevisiae* yeast cells. J. Electron Microsc. 5: 249-261.

Ballou, C. E. 1990. Isolation, characterization, and properties of *Saccharomyces cerevisiae mnn* mutants with nonconditional protein glycosylation defects. Meth. Enzymol. 185: 440-470.

Barker, S.A., Bourne, E.J., O'Mantz, D.M., and Stacey, M. 1957. Studies of *A. niger*. Part VI. The separation and structures of oligosaccharides from nigeran. J. Chem. Soc. 2448-2454.

Bartnicki-Garcia, S. 1968. Cell wall chemistry, morphogenesis and taxonomy of fungi. Annu. Rev. Microbiol. 22: 87-108.

Bartnicki-Garcia, S., and Lindberg, B. 1972. Partial characterization of mucoran, the glucoromannan component. Carbohyd. Res. 23: 75-85.

Bidard, F., Bony, M., Blondin, B., Dequin, S., and Barre, P. 1995. The *Saccharomyces cerevisiae* FLO1 flocculation gene encodes for a cell surface protein. Yeast 11: 809-822.

Bielefeld, M., and Hollenberg, C.P. 1989. Mutant invertase proteins accumulate in the yeast endoplasmic reticulum. Mol. Gen. Genet. 215: 401-406.

Bluhm, T.L., and Sarko, A. 1977. Packing analysis of carbohydrates and polysaccharides. V. Crystal structures of two polymorphs of pachyman triacetate. Biopolymers 16: 2067-2089.

Bony, M., Barre, P., and Blondin, B. 1998. Distribution of the flocculation protein, flop, at the cell surface during yeast growth: the availability of flop determines the flocculation level. Yeast 14: 25-35.

Brakenhoff, G.J, van der Voort, H.T, van Spronsen, E.A., and Nanninga, N. 1989.Three-dimensional imaging in fluorescence by confocal scanning microscopy. J. Microsc. 153: 151-159.

Calderone, R.A., and Fonzi, W.A. 2001.Virulence factors of *Candida albicans*. Trends Microbiol. 9: 327-335.

Calonje, M., Bernardo, D., Novaes-Ledieu, M., and Garcia-Mendoza, C. 2002. Properties of a hydrophobin isolated from the mycoparasitic fungus *Verticillium fungicola*. Can. J. Microbiol. 48: 1030-1034.

Calvo-Mendez, C., and Ruiz-Herrera, J. 1987. Biosynthesis of chitosan in membrane fractions from *Mucor rouxii* by the concerted action of chitin synthetase and a particulate deacetylase. Exp. Mycol. 11: 128-140.

Cappellaro, C., Balderman, C., Rachel, R., and Tanner, W. 1994. Mating type–specific cell-cell recognition of *Saccharomyces cerevisiae*: cell wall attachment and active sites of α- and a-agglutinin. EMBO J. 13: 4737-4744.

Cappellaro, C., Mrsa, V., and Taner, W. 1998. New potential cell wall glucanases of *Saccharomyces cerevisiae* and their involvement in mating. J. Bacteriol. 180: 5030-5037.

Caro, L.H., Smits, G.J., van Egmond, P., Chapman, J.W., and Klis, F.M. 1998. Transcription of multiple cell wall protein-encoding genes in *Saccharomyces cerevisiae* is differentially regulated during the cell cycle. FEMS Microbiol. Lett. 161: 345-349.

Caro, L.H., Tettelin, H., Vossen, J.H., Ram, A.F., van den Ende, H., and Klis, F.M. 1997. *In silico* identification of glycosyl-phosphatidylinositol-anchored plasma-membrane and cell wall proteins of *Saccharomyces cerevisiae*. Yeast 13: 1477-1489.

Casanova, M., Lopez-Ribot, J.L., Martinez, J.P., and Sentandreu R. 1992. Characterization of cell wall proteins from yeast and mycelial cells of *Candida albicans* by labelling with biotin: comparison with other techniques. Infect. Immun. 60: 4898-4906.

Cassone, A., Kerridge, D., and Gale, E.F. 1979. Ultrastructural changes in the cell wall of *Candida albicans* following cessation of growth and their possible relationship to the development of polyene resistance. J. Gen. Microbiol. 110: 339-349.

Cassone, A., Mattia, E., and Boldrini, L. 1978. Agglutination of blastospores of *Candida albicans* by concanavalin A and its relationship with the distribution of mannan polymers and the ultrastructure of the cell wall. J. Gen. Microbiol. 105: 263-273.

Cassone, A., Simonetti, N., and Strippoli, V. 1974. Wall structure and bud formation in *Cryptococcus neoformans*. Arch. Microbiol 95: 205- 212.

Castillo, L Martinez, A.I., Garcerá, A., Elorza, M.V., Valentin, E., and Sentandreu, R. 2003. Functional analysis of cysteine residues and the repetitive sequences of Sc Pir4: the first repetitive sequence is needed to binding the cell wall β-1,3-glucan. Yeast 20: 973-983.

Chapman, A., and Kornfeld, R. 1979. Structure of the high mannose oligosaccharides of a human IgM myeloma protein. I. The major oligosaccharides of the two high mannose glycopeptides. J. Biol. Chem. 254: 816-823.

Chattaway, F.W., Shenolikar, S., O'Reilly, J., and Barlow, A.J. 1976. Changes in the cell surface of the dimorphic forms of *Candida albicans* by treatment with hydrolytic enzymes. J. Gen. Microbiol. 96: 335-337.

Chun, K.T., and Goebl, M.G. 1996. The identification of transposon-tagged mutations in essential genes that affect cell morphology in *Saccharomyces cerevisiae*. Genetics 142: 39-50.

Christodoulidou, A., Bouriotis, V., and Thireos, G. 1996. Two sporulation-specific chitin deacetylase-encoding genes are required for the ascospore wall rigidity of *Saccharomyces cerevisiae*. J. Biol. Chem. 271: 31420-31425.

Christodoulidou, A., Briza, P., Ellinger, A., and Bouriotis V. 1999. Yeast ascospore wall assembly requires two chitin deacetylase isozymes. FEBS Lett. 460: 275-279.

Darling, S., Theilade. J., and Birch-Andersen, A. 1972. Structure and chemical composition of prospheroplast envelopes of *Saccharomyces cerevisiae* and *Hansenula anomala*. J. Bacteriol. 110: 336-450.

De Groot, P.W.J., Ruiz, C., Vazquez de Aldana, C.R., Andel, A., Caubin, J., Arroyo, J., Garcia, J.C., Gil, C., Molina, M., Garcia, L.J., Nombela, C., and Klis, F.M. 2001. A genomic approach for the identification and classification of gene involved in cell wall formation and its regulation in *Saccharomyces cerevisiae*. Comp. Funct. Genomics 2: 124-142.

De Nobel, H., and Lipke, P.N. 1994. Is there a role for GPIs in cell wall assembly in yeast? Trends Cell Biol. 4: 42-45.

De Nobel, J.G., Klis, F.M., Priem, J., Munnik, T., and van den Ende, H. 1990.The glucanase-soluble mannoproteins limit cell wall porosity in *Saccharomyces cerevisiae*. Yeast 6: 491-499.

De Nobel, J.G., Klis, F.M., Ram, A., Van Unen, H., Priem, J., Munnik, T., and Van Den Ende, H. 1991. Cyclic variations in the permeability of the cell wall of *Saccharomyces cerevisiae*. Yeast 7: 589-598.

Dwek, R.A. 1995. Glycobiology: more functions for oligosaccharides. Science 269: 1234-1235.

Elorza, M.V., Murgui, A., and Sentandreu, R. 1985. Dimorphism in *Candida albicans*: contribution of mannoproteins to the architecture of yeast and mycelial cell walls. J. Gen. Microbiol. 131: 2209-2216.

Elorza, M.V., Rico, H., and Sentandreu, R. 1983. Calcofluor white alters the assembly of chitin fibrils in *Saccharomyces cerevisiae* and *Candida albicans* cells. J. Gen. Microbiol. 129: 1577-1582.

Ernst, J.F., and Prill, S.K. 2001. O-glycosylation. Med. Mycol. 39 Suppl 1: 67-74.

Feather, M.S., and Malek, A. 1972. A highly branched exocellular D-glucan from *Monolinia fructicola*. Biochim. Biophys. Acta 264: 103-105.

Fleet, G.H. 1991. Cell walls. In: The Yeast, Vol 4. A.H Rose and J.S. Harrison, eds. Academic Press, New York. p. 199-277.

Fontaine, T., Simenel, C., Dubreucq, G., Adam, O., Delepierre, M., Lemoine, J., Vorgias, C.E., Diaquin, M., and Latge, J.P. 2000. Molecular organization of the alkali-insoluble fraction of *Aspergillus fumigatus* cell wall. J. Biol. Chem. 275: 27594-607.

Frevert, J., and Ballou, C.E. 1985. *Saccharomyces cerevisiae* structural cell wall mannoprotein. Biochemistry 24: 753-759.

Gad, M., Itoh, A., and Ikai, A. 1997. Mapping cell wall polysaccharides of living microbial cells using atomic force microscopy. Cell Biol. Int. 21: 697-706.

Garcerá, A., Martinez, A.I., Castillo, L., Elorza, M. V., Sentandreu, R., and Valentín, E. 2003. Identification and study of a *Candida albicans* protein homologous to *Saccharomyces cerevisiae* Ssr1p, an internal cell-wall protein. Microbiology 149: 2243-2250.

Girrbach, V., and Strahl, S. 2003. Members of the evolutionarily conserved PMT family of protein O-mannosyltransferases form distinct protein complexes among themselves. J. Biol. Chem. 278: 12554-12562.

Gomez, B.L., and Nosanchuk, J. 2003. Melanin and fungi. Curr. Opin. Infect. Dis. 16: 91-99.

Ha, C.H., Lim, K.H., Kim, Y.T., Lim, S.T., Kim, C.W., and Chang, H.I. 2002. Analysis of alkali-soluble glucan produced by *Saccharomyces cerevisiae* wild-type and mutants. Appl. Microbiol. Biotechnol. 58: 370-377.

Haguenauer-Tsapis, R., Nagy, M., and Ryter, A.A. 1986. Deletion that includes the segment coding for the signal peptidase cleavage site delays release of *Saccharomyces cerevisiae* acid phosphatase from the endoplasmic reticulum. Mol. Cell. Biol. 6: 723-729.

Hamada, K., Terashima, H., Arisawa, M., Yabuki, N., and Kitada, K. 1999. Amino acid residues in the omega-minus region participate in cellular localization of yeast glycosylphosphatidylinositol-attached proteins. J. Bacteriol. 181: 3886-3889.

Hemmes, D.E., Kojima-Buddenhagen, E.S., and Hohl, H. R. 1972. Structural and enzymatic analysis of the spore wall layers in *Dictyostelium discoideum*. J. Ultrastruct. Res. 41: 406-417.

Horisberger, M., and Vonlanthen, M. 1977. Location of mannan and chitin on thin sections of budding yeasts with gold markers. Arch. Microbiol. 115: 1-7

Horisberger, M., Vonlanthen, M., and Rosset, J. 1978. Localization of α-galactomannan and of wheat germ agglutinin receptors in *Schizosaccharomyces pombe*. Arch. Microbiol. 119: 107-111.

Humbel, B.M., Konomi, M., Takagi, T., Kamasawa, N., Ishijima, S.A., and Osumi, M. 2001. *In situ* localization of beta-glucans in the cell wall of *Schizosaccharomyces pombe*. Yeast 18: 433-444.

Ibeas, J.I., Lee, H., Damsz, B., Prasad, D.T., Pardo, J.M., Hasegawa, P.M., Bressan, R.A., and Narasimhan, M.L. 2000. Fungal cell wall phosphomannans facilitate the toxic activity of a plant PR-5 protein. Plant J. 23: 375-383.

Jaafar, L., Leon M., and Zueco, J. 2003. Isolation of the *MNN9* gene of *Yarrowia lipolytica* (*YlMNN9*) and phenotype analysis of a mutant mnn9Delta strain. Yeast 20: 633-644.

Jentoft, N. 1990.Why are proteins O-glycosylated? Trends Biochem. Sci. 15:291-294.

Jigami, Y., and Odani, T. 1999. Mannosylphosphate transfer to yeast mannan. Biochim. Biophys. Acta 1426: 335-345.

Johnson, B.F., Calleja, G.B., Yoo, B.Y., Zuker, M., and McDonald, I.J. 1982. Cell-division-key to cellular morphogenesis in the fission yeast, *Schizosaccharomyces*. Int. Rev. Cytol. 75: 167-208

Kanik-Ennulat, C., Montalvo, E., and Neff, N. 1995. Sodium orthovanadate-resistant mutants of *Saccharomyces cerevisiae* show defects in Golgi-mediated protein glycosylation, sporulation and detergent resistance. Genetics 140: 933-943.

Kapteyn, J.C., Hoyer, L.L., Hecht, J.E., Muller, W.H., Andel, A., Verkleij, A.J., Makarow, M., Van Den Ende. H., and Klis, F.M. 2000. The cell wall architecture of *Candida albicans* wild-type cells and cell wall-defective mutants. Mol. Microbiol. 35: 601-611.

Kapteyn, J.C., Montijn, R.C., Dijkgraaf, G.J., and Klis, F.M. 1994. Identification of β-1,6-glucosylated cell wall proteins in yeast and hyphal forms of *Candida albicans*. Eur. J. Cell Biol. 65: 402-407.

Kapteyn, J.C., Montijn, R.C., Dijkgraaf, G.J., Van den Ende, H., and Klis, F.M. 1995. Covalent association of β-1,3-glucan with beta-1,6-glucosylated mannoproteins in cell walls of *Candida albicans*. J. Bacteriol. 177: 3788-3792.

Kapteyn, J.C., Montijn, R.C., Vink, E., de la Cruz, J., Llobell, A., Douwes, J.E., Shimoi, H., Lipke, P.N., and Klis, F.M. 1996. Retention of *Saccharomyces cerevisiae* cell wall proteins through a phosphodiester-linked β-1,3-/β-1,6-glucan heteropolymer. Glycobiology 6: 337-345.

Kapteyn, J.C., Van Egmond, P., Sievi, E., Van Den Ende, H., Makarow, M., and Klis, F.M. 1999a. The contribution of the O-glycosylated protein Pir2p/Hsp150 to the construction of the yeast cell wall in wild-type cells and β-1,6-glucan-deficient mutants. Mol. Microbiol. 31:1835-1844.

Kapteyn, J.C., Van Den Ende, H., and Klis, F.M. 1999b. The contribution of cell wall proteins to the organization of the yeast cell wall. Biochim. Biophys. Acta 1426: 373-383.

Kessler, G., and Nickerson,W.J. 1959. Glucomannan-protein complexes from cell walls of yeast. J. Biol. Chem. 234: 2281-2285.

Kim, S.Y., Sohn, J.H., Kang, H.A., Yoo, S.K., Pyun, Y.R., and Choi, E.S. 2001. Cloning and characterization of the *Hansenula polymorpha* homologue of the *Saccharomyces cerevisiae MNN9* gene. Yeast 18: 455-461.

Koch, Y., and Rademacher, K.H. 1980. Chemical and enzymatic changes in the cell walls of *Candida albicans* and *Saccharomyces cerevisiae* by scanning electron microscopy. Can. J. Microbiol. 26: 965-970.

Kollar, R., Reinhold, B.B., Petrakova, E., Yeh, H.J., Ashwell, G., Drgonova, J., Kapteyn, J.C., Klis, F.M., and Cabib, E. 1997. Architecture of the yeast cell wall. β(1→6)-glucan interconnects mannoprotein, β(1→3)-glucan, and chitin. J. Biol. Chem. 272: 17762-17775.

Korn, E.D., and Northcote, D.H. 1960. Physical and chemical properties of polysaccharides and glycoproteins of the yeast-cell wall. Biochem. J. 75:12-17.

Kopecka, M., Fleet, G.H., and Phaff, H.J. 1995. Ultrastructure of the cell wall of *Schizosaccharomyces pombe* following treatment with various glucanases. J. Struct. Biol. 114: 140-152.

Kopecka, M.and Kreger, D.R. 1986. Assembly of microfibrils *in vivo* and *in vitro* from (1→3)-β–D-glucan synthesized by protoplasts of *Saccharomyces cerevisiae*. Arch. Microbiol. 143: 387-395.

Kornfeld, R., and Kornfeld, S. 1976. Comparative aspects of glycoprotein structure. Annu. Rev. Biochem. 45: 217-237.

Kreger, D.R. 1954. Observations on cell wall of yeast and some other fungi by X-ray diffraction and solubility test. Biochim. Biophys. Acta. 13: 1-9.

Kreger, D.R., and Kopecká, M. 1975. On the nature and formation of the fibrillar nets produced by protoplasts of *Saccharomyces cerevisiae* in liquid media: an electronmicroscopic, X-ray diffraction and chemical study. J. Gen. Microbiol. 92: 207-220.

Larriba, G., Morales, M., and Ruiz-Herrera, J. 1981. Biosynthesis of β-glucan microfibrills by cell-free extracts from *Saccharomyces cerevisiae*. J. Gen. Microbiol. 124: 375-383.

Lebron, F., Vassallo, R., Puri, V., and Limper, A.H. 2003. *Pneumocystis carinii* cell wall β-glucans initiate macrophage inflammatory responses through NF-kB activation. J. Biol. Chem. 278: 25001-25008.

Linnemans, W.A., Boer, P., and Elbers, P.F. 1977. Localization of acid phosphatase in *Saccharomyces cerevisiae*: a clue to cell wall formation. J. Bacteriol. 131: 638-644.

Lipke, P.N., and Ovalle, R. 1998. Cell wall architecture in yeast: new structure and new challenges. J. Bacteriol. 180: 3735-3740.

Lipke, P.N. and Kurjan, J. 1992. Sexual agglutination in budding yeasts: structure, function, and regulation of adhesion glycoproteins. Microbiol. Rev. 56: 180-194.

Lo, W.S., and Dranginis, A.M. 1998. The cell surface flocculin Flo11 is required for pseudohyphae formation and invasion by *Saccharomyces cerevisiae*. Mol. Biol. Cell 9: 161-171.

Lopez-Ribot, J.L., Alloush, H.M., Masten, B.J., and Chaffin, W.L. 1996. Evidence for presence in the cell wall of *Candida albicans* of a protein related to the hsp70 family. Infect. Immun. 64: 3333-3340.

Lu, C.F., Kurjan, J., and Lipke, P.N. 1994. A pathway for cell wall anchorage of *Saccharomyces cerevisiae* α-agglutinin. Mol. Cell. Biol. 14: 4825-4833.

Lussier, M., Sdicu, A.M., Winnett, E., Vo, D.H., Sheraton, J., Dusterhoft, A., Storms, R.K., and Bussey, H. 1997.Completion of the *Saccharomyces cerevisiae* genome sequence allows identification of *KTR5*, *KTR6* and KTR7 and definition of the nine-membered *KRE2/MNT1* mannosyltransferase gene family in this organism. Yeast 13: 267-274.

Manners, D.J., Masson, A.J., and Patterson, J.C. 1973. The structure of a β–(1 →3)-D-glucan from yeast cell walls. Biochem. J. 135: 19-30.

Marcilla, A., Elorza, M.V., Mormeneo, S., Rico, H., and Sentandreu, R. 1991. *Candida albicans* mycelial wall structure: supramolecular complexes released by zymolyase, chitinase and β-mercaptoethanol. Arch. Microbiol. 155: 312-319.

Mishra, C., Semino, C.E., McCreath, K.J., de la Vega, H., Jones, B.J., Specht, C.A., and Robbins, P.W. 1997. Cloning and expression of two chitin deacetylase genes of *Saccharomyces cerevisiae*. Yeast 13: 327-336.

Miura, N.N., Adachi, Y., Yadomae, T., Tamura, H., Tanaka, S., and Ohno, N. 2003. Structure and biological activities of β–glucans from yeast and mycelial forms of *Candida albicans*. Microbiol. Immunol. 47: 173-182.

Monteoliva, L., Matas, M.L., Gil, C., Nombela, C., and Pla, J. 2002. Large-scale identification of putative exported proteins in *Candida albicans* by genetic selection. Eukaryotic Cell 1: 514-525.

Mormeneo, S., Rico, H., Iranzo, M., Aguado, C., and Sentandreu, R. 1996. Study of supramolecular structures released from the cell wall of *Candida albicans* by ethylenediamine treatment. Arch. Microbiol. 166: 327-335.

Moukadiri, I., Armero, J., Abad, A., Sentandreu, R., and Zueco, J. 1997. Identification of a mannoprotein present in the inner layer of the cell wall of *Saccharomyces cerevisiae*. J. Bacteriol. 179: 2154-262.

Moukadiri, I., Jaafar, L., and Zueco, J. 1999. Identification of two mannoproteins released from cell walls of a *Saccharomyces cerevisiae mnn1 mnn9* double mutant by reducing agents. J. Bacteriol. 181: 4741-4745.

Moukadiri, I., and Zueco, J. 2001. Evidence for the attachment of Hsp150/Pir2 to the cell wall of *Saccharomyces cerevisiae* through disulfide bridges. FEMS Yeast Research 1: 241-245.

Mrsa, V., Ecker, M., Strahl-Bolsinger, S., Nimtz, M., Lehle, L., and Tanner, W. 1999. Deletion of new covalently linked cell wall glycoproteins alters the electrophoretic mobility of phosphorylated wall components of *Saccharomyces cerevisiae*. J. Bacteriol. 181: 3076-3086.

Mrsa, V., Klebl, F., and Tanner, W. 1993. Purification and characterization of the *Saccharomyces cerevisiae* BGL2 gene product, a cell wall endo-beta-1,3-glucanase. J. Bacteriol. 175: 2102-2106.

Mrsa, V., Seidl, T., Gentzsch, M., and Tanner, W. 1997.Specific labelling of cell wall proteins by biotinylation. Identification of four covalently linked O-mannosylated proteins of *Saccharomyces cerevisiae*. Yeast 13: 1145-1154.

Navarro-Garcia, F., Monge, A.R., Rico, H., Pla, J., Sentandreu, R., and Nombela, C. 1998. A role for the MAP kinase gene MKC1 in cell wall construction and morphological transitions in *Candida albicans*. Microbiology 144: 411-424.

Neuveglise, C., Nicauda, J.M., Ross-Macdonald, P., and Gaillardin, C. 1998. A shuttle mutagenesis system for tagging genes in the yeast *Yarrowia lipolytica*. Gene 213: 37-46.

Nicolaus, R. A. 1969. Melanins. Herman, Paris.

Odani, T., Shimma, Y., Wang, X.H., and Jigami,Y. 1997. Mannosylphosphate transfer to cell wall mannan is regulated by the transcriptional level of the *MNN4* gene in *Saccharomyces cerevisiae*. FEBS Lett. 420: 186-190.

Odds, F. C. 1985. Morphogenesis in *Candida albicans*. CRC Crit. Rev. Microbiol. 12: 45-93.

Orlean, P. 1997. Biogenesis of yeast wall and surface components. In: The Molecular and Cellular Biology of the Yeast *Saccharomyces*. J. R. Pringle, J. R. Broach and E. W. Jones, eds. Cold Spring Harbor Laboratory Press, Cold Spring Harbor. p. 229-362.

Orlean, P., Ammer, A., Watzele, M., and Tanner, W. 1986. Synthesis of an O-glycosylated cell surface protein induced in yeast by alpha-factor. Proc. Natl. Acad. Sci. U.S.A. 83: 6263-6266.

Orlean, P., Kuranda, M.J., and Albright, C.F. 1991. Analysis of glycoproteins from *Saccharomyces cerevisiae*. Methods Enzymol. 194: 682-697.

Osumi, M. 1998. The ultrastructure of yeast: cell wall structure and formation. Micron 29: 207-233.

Paananen, A., Vuorimaa, E., Torkkeli, M., Penttila, M., Kauranen, M., Ikkala, O., Lemmetyinen, H., Serimaa, R., and Linder, M.B. 2003. Structural hierarchy in molecular films of two class II hydrophobins. Biochemistry 42: 5253-5258.

Pardo, M., Monteoliva, L., Pla, J., Sanchez, M., Gil, C., and Nombela, C. 1999. Two-dimensional analysis of proteins secreted by *Saccharomyces cerevisiae* regenerating protoplasts: a novel approach to study the cell wall. Yeast 15: 459-472.

Pardo, M., Ward, M., Bains, S., Molina, M., Blackstock, W., Gil, C., and Nombela, C.A. 2000. A proteomic approach for the study of *Saccharomyces cerevisiae* cell wall biogenesis. Electrophoresis 21: 3396-3410.

Paris, S., Debeaupuis, J.P., Crameri, R., Carey, M., Charles, F., Prevost, M.C., Schmitt, C., Philippe, B., and Latge, J.P. 2003. Conidial hydrophobins of *Aspergillus fumigatus*. Appl. Environ. Microbiol. 69: 1581-1588.

Pastor, F.I.J., Valentin, E., Herrero, E., and Sentandreu, R. 1984. Structure of the *Saccharomyces cerevisiae* cell wall: Mannoproteins released by zymolyase and their contribution to wall architecture. Biochim. Biophys. Acta 802: 292-300.

Pitarch, A., Sanchez, M., Nombela, C., and Gil, C. 2002. Sequential fractionation and two-dimensional gel analysis unravels the complexity of the dimorphic fungus *Candida albicans* cell wall proteome. Mol. Cell. Proteomics 1: 967-982.

Ponton, J., and Jones, J.M. 1986a. Identification of two germ tube-specific cell wall antigens of *Candida albicans*. Infect. Immun. 54: 864-868.

Ponton, J., and Jones, J.M. 1986b. Analysis of cell wall extracts of *Candida albicans* by sodium dodecyl sulfate-polyacrilamide gel electrophoresis and western blot techniques. Infect. Immun. 53: 565-572.

Poole, R.K., and Lloyd, D. 1973. Effect of 2-deoxy-D-glucose on growth and cell walls of *Schizosaccharomyces pombe* 972h. Arch. Mikrobiol. 88: 257-272.

Poulain, D., Slomianny, C., Jouault, T., Gomez, J.M., and Trinel, P.A. 2002. Contribution of phospholipomannan to the surface expression of β-1,2-oligomannosides in *Candida albicans* and its presence in cell wall extracts. Infect. Immun. 8: 4323-4328.

Poulain, D., Tronchin, G., Dubremetz, J.F., and Biguet, J. 1978. Ultrastructure of the cell wall of *Candida albicans* blastospores: study of its constitutive layers by the use of a cytochemical technique revealing polysaccharides. Ann. Microbiol. 129: 141-145.

Pringle, A.T., Forsdyke, J., and Rose, A.H. 1979. Scanning electron microscope study of *Saccharomyces cerevisiae* spheroplast formation. J. Bacteriol. 140: 289-293.

Protchenko, O., Ferea, T., Rashford, J., Tiedeman, J., Brown, P.O., Botstein, D., and Philpott, C.C. 2001. Three cell wall mannoproteins facilitate the uptake of iron in *Saccharomyces cerevisiae*. J. Biol. Chem. 276: 49244-49250.

Ram, A.F., Brekelmans, S.S., Oehlen, L.J., and Klis, F.M. 1995. Identification of two cell cycle regulated genes affecting the β 1,3-glucan content of cell walls in *Saccharomyces cerevisiae*. FEBS Lett. 358: 165-170.

Ram, A.F., Wolters, A., Ten Hoopen, R., and Klis, F.M. 1994. A new approach for isolating cell wall mutants in *Saccharomyces cerevisiae* by screening for hypersensitivity to calcofluor white. Yeast 10: 1019-1030.

Ramon, A.M., Gil, R., Burgal, M., Sentandreu, R., and Valentin, E. 1996. A novel cell wall protein specific to the mycelial form of *Yarrowia lipolytica*. Yeast 12: 1535-1548.

Rast, D., and Hollenstein, G.O. 1977. Architecture of the *Agaricus bisporus* spore cell wall. Can. J. Bot. 55: 2251.

Reynolds, T.B., and Fink, G.R. 2001. Bakers' yeast, a model for fungal biofilm formation. Science 291: 979-881.

Rodriguez-Pena, J.M., Cid, V.J., Arroyo, J., and Nombela, C. 2000. A novel family of cell wall-related proteins regulated differently during the yeast life cycle. Mol. Cell. Biol. 20: 3245-3255.

Roig, P., Martinez, J.P., Gil, M.L., and Gozalbo, D. 2000. Molecular cloning and characterization of the *Candida albicans* UBI3 gene coding for a ubiquitin-hybrid protein. Yeast 16: 1413-1419.

Roitsch, T., and Lehle, L. 1989. Structural requirements for protein N-glycosylation. Influence of acceptor peptides on cotranslational glycosylation of yeast invertase and site-directed mutagenesis around a sequon sequence. Eur. J. Biochem. 181: 525-529.

Ruiz-Herrera, J. 1992. Fungal Cell Wall: Structure, Synthesis, and Assembly. CRC Press, Boca Raton.

Ruiz-Herrera, J., and Larriba, G. 1995. High molecular weight precursors of glucans in *Saccharomyces cerevisiae*. Antonie van Leeuwenhoek 68: 231-235.

Russo, P., Kalkkinen, N., Sareneva, H., Paakkola, J., and Makarow, M. 1992. A heat shock gene from *Saccharomyces cerevisiae* encoding a secretory glycoprotein. Proc. Natl. Acad. Sci. U.S.A. 89: 3671-3675.

San-Blas, G., and Carbonell, L.M. 1974. Chemical and ultrastructural studies on the cell walls of the yeastlike and mycelial forms of *Histoplasma farciminosum*. J. Bacteriol. 119: 602-611.

San-Blas, G., San-Blas, F., and Serrano, L.E. 1977. Host-parasite relationship in the yeast like form of *Paracoccidioides brasiliensis* strain IVICPb9. Infect. Immun. 15: 343-346.

Sanchez-Hernandez, M.E., Garcia Mendoza, C., and Novaes-Ledieu M. 1993. Two different alkali-soluble α-glucans in hyphal walls of the basidiomycete *Armillaria mellea* Microbiologia 9: 34-42.

Santamaria, F., Reyes, F., and Lahoz, R. 1978. Extracellular glucan containing (1→3)-β and (1→6)-β linkages isolated from *Monilinia fructigena*. J. Gen. Microbiol. 109: 287-293.

Schaefer, H.P. 1977. An alkali-soluble polysaccharide from the cell walls of *Coprinus lagopus*. Arch. Microbiol. 113: 79-82.

Sentandreu, M., Elorza, M.V., Valentin, E., Sentandreu, R., and Gozalbo, D. 1995. Cloning of cDNAs coding for *Candida albicans* cell surface proteins. J. Med. Vet. Mycol. 33: 105-111.

Sentandreu, R., Elorza, M.V., Mormeneo, S., Sanjuán, R., and Iranzo, M. 1993. Possible roles of mannoproteins in the construction of

Candida albicans. In: *Candida albicans*, Cellular and Molecular Biology. R. Prasad, ed. Springer, Berlin. p. 72-88.

Sentandreu, M., Nieto, A., Iborra, A., Elorza, M.V., Ponton, J., Fonzi, W.A., and Sentandreu, R. 1997.Cloning and characterization of CSP37, a novel gene encoding a putative membrane protein of *Candida albicans*. J. Bacteriol. 179: 4654-4663.

Sentandreu, R., and Northcote, D.H. 1969. The characterization of oligosaccharides attached to threonine and serine in a mannan glycopeptide obtained from the cell wall of yeast. Carbohyd. Res. 10: 584-585.

Shematek, E.M., Braatz, J.A., and Cabib, E. 1980. Biosynthesis of the yeast cell wall. I. Preparation and properties of β-(1→3)glucan synthetase. J. Biol. Chem. 255: 888-894.

Sipos, G., Puoti, A., and Conzelmann, A. 1995. Biosynthesis of the side chain of yeast glycosylphosphatidylinositol anchors is operated by novel mannosyltransferases located in the endoplasmic reticulum and the Golgi apparatus. J. Biol. Chem. 270: 19709-19715.

Smith, A.E., Zhang, Z., Thomas, C.R., Moxham, K.E., and Middelberg, A.P. 2000.The mechanical properties of *Saccharomyces cerevisiae*. Proc. Natl. Acad. Sci. U.S.A. 97: 9871-9874.

Southard, S.B., Specht, C.A., Mishra, C., Chen-Weiner, J., and Robbins, P.W. 1999. Molecular analysis of the *Candida albicans* homolog of *Saccharomyces cerevisiae MNN9*, required for glycosylation of cell wall mannoproteins. J. Bacteriol. 181: 7439-7448.

Strahl-Bolsinger, S., Gentzsch, M., and Tanner, W. 1999. Protein O-mannosylation. Biochim. Biophys. Acta 1426: 297-307.

Stratford, M. 1992. Yeast flocculation: receptor definition by *mnn* mutants and concanavalin A. Yeast 8: 635-645.

Stratford, M. 1996. Induction of flocculation in brewing yeasts by change in pH value. FEMS Microbiol. Lett. 136: 13-18.

Sundstrom, P. 2002. Adhesion in *Candida* spp. Cell Microbiol. 4: 461-469.

Sundstrom, P.M., and Kenny, G.E. 1985. Enzymatic release of germ-tube specific antigens from cell walls of *Candida albicans*. Infect. Immun. 49: 609-614.

Sundstrom, P.M., Tam, M.R., Nichols, E.J., and Kenny, G.E. 1988. Antigenic differences in the surface mannoproteins of *Candida albicans* as revealed by monoclonal antibodies. Infect. Immun. 56: 601-620.

Tanner, W., and Lehle, L. 1987. Protein glycosylation in yeast. Biochim. Biophys. Acta 906: 81-99.

Terashima, H., Fukuchi, S., Nakai, K., Arisawa, M., Hamada, K., Yabuki, N., and Kitada, K. 2002. Sequence-based approach for identification of cell wall proteins in *Saccharomyces cerevisiae*. Curr. Genet. 40: 311-316.

Teunissen, A.W., and Steensma H.Y. 1995. The dominant flocculation of *Saccharomyces cerevisiae* constitutes a new subtelomeric gene family. Yeast 11: 1001-1013.

Thiery, J.P. 1967. Mise en evidence des polysaccharides sur coupes fines en microscopie électronique. J. Microsc. Paris 6: 987-1018.

Toh-e, A., Yasunaga, S., Nisogi, H., Tanaka, K., Oguchi, T., and Matsui, Y. 1993. Three yeast genes, *PIR1*, *PIR2* and *PIR3*, containing internal tandem repeats, are related to each other, and *PIR1* and *PIR2* are required for tolerance to heat shock. Yeast 9: 481-494.

Tokunaga, M., Niimi, M., Kusamichi, M., and Koike, H. 1990. Initial attachment of *Candida albicans* cells to buccal epithelial cells. Demonstration of ultrastructure with the rapid-freezing technique. Mycopathologia 111: 61-68.

Torkkeli, M., Serimaa, R., Ikkala, O., and Linder, M. 2002. Aggregation and self-assembly of hydrophobins from *Trichoderma reesei*: low-resolution structural models. Biophys. J. 83: 2240-2247.

Trinel, P.A., Plancke, Y., Gerold, P., Jouault, T., Delplace, F., Schwarz, R.T., Strecker, G., and Poulain, D. 1999. The *Candida albicans* phospholipomannan is a family of glycolipids presenting phos phoinositolmannosides with long linear chains of β-1,2-linked mannose residues. J. Biol. Chem. 274: 30520-30526.

Tronchin, G., Poulain, D., and Biguet, J. 1979. Cytochemical and ultrastructural studies of the cell wall of *Candida albicans*. Arch. Microbiol. 123: 245-249.

Tronchin, G., Poulain, D., Herbaut, J., and Biguet, J. 1981. Localization of chitin in the cell wall of *Candida albicans* by means of wheat germ agglutinin. Fluorescence and ultrastructural studies. Eur. J. Cell. Biol. 26: 121-128.

Tronchin, G., Poulain, D., and Vernes, A. 1984. Cytochemical and ultrastructural studies of *Candida albicans*. III. Evidence for modifications of the cell wall coat during adherence to human buccal epithelial cells. Arch. Microbiol. 39: 221-224.

Tsai, P.K., Frevert, J., and Ballou, C.E. 1984. Carbohydrate structure of *Saccharomyces cerevisiae* mnn9 mannoprotein. J. Biol. Chem. 259: 3805-3811.

Valentin, E., Herrero, E., Pastor, F.I.J., and Sentandreu, R. 1984. Solubilization and analysis of mannoprotein molecules from the cell wall of *Saccharomyces cerevisiae*. J. Gen. Microbiol. 130: 1419-1428.

Valentin, E., Herrero, E., Rico, H., Miragall, F., and Sentandreu, R. 1987. Cell wall mannoproteins during the population growth phases in *Saccharomyces cerevisiae*. Arch. Microbiol. 148: 88-94.

Valentin, E., Herrero, E., and Sentandreu, R. 1986. Incorporation of mannoproteins into the walls of aculeacin A-treated yeast cells. Arch. Microbiol. 146: 214-220.

van der Vaart, J.M., Biesebeke, R., Chapman, J.W., Klis, F.M., and Verrips, C.T. 1996. The β-1,6-glucan containing side-chain of cell wall proteins of *Saccharomyces cerevisiae* is bound to the glycan core of the GPI moiety. FEMS Microbiol. Lett. 145: 401-407.

Veronese, P., Ruiz, M.T., Coca, M.A., Hernandez-Lopez, A., Lee, H., Ibeas, J.I., Damsz, B., Pardo, J.M., Hasegawa, P.M., Bressan, R.A., and Narasimhan, M.L. 2003. In defence against pathogens. Both plant sentinels and foot soldiers need to know the enemy. Plant Physiol. 131: 1580-1590.

Vitkov, L., Krautgartner, W.D., Hannig, M., Weitgasser, R., and Stoiber, W. 2002. *Candida* attachment to oral epithelium. Oral Microbiol. Immunol. 17: 60-64.

Walther, P., Muller, M., and Schweingruber, M.E. 1984. The ultrastructure of the cell surface and plasma membrane of exponential and stationary phase cells of *Schizosaccharomyces pombe*, grown in different media. Arch. Microbiol. 137: 128-134.

Watson, K., and Arthur, H. 1977. Cell surface topography of *Candida* and *Leucosporidium* yeasts as revealed by scanning electron microscopy. J. Bacteriol.130: 312-317.

Wessels, J.G. 1999. Fungi in their own right. Fungal Genet. Biol. 27: 134-145.

Wright, R. 2000. Transmission electron microscopy of yeast. Microsc. Res. Tech. 51: 496-510.

Wu-Yuan, C.D., and Hashimoto, T. 1977. Architecture and chemistry of microconidial walls of *Trichophyton mentagrophytes*. J. Bacteriol. 129: 1584-1592.

Yu, L., Lee, K.K., Ens, K., Doig, P.C., Carpenter, M.R., Staddon, W., Hodges, R.S., Paranchych, W., and Irvin, R.T. 1994. Partial characterization of a *Candida albicans* fimbrial adhesin. Infect. Immun. 62: 2834-2842.

Zlotnik, H., Fernandez, M.P., Bowers, B., and Cabib, E. 1984. *Saccharomyces cerevisiae* mannoproteins form an external cell wall layer that determines wall porosity. J. Bacteriol. 159: 1018-1026.

From: Pathogenic Fungi: Structural Biology and Taxonomy
Edited by: G. San-Blas and R.A. Calderone

Chapter 2

Biosynthesis of the Fungal Cell Wall

José Ruiz-Herrera,
M. Victoria Elorza, Peggy E. Alvarez,
and Rafael Sentandreu

Abstract

The biosynthesis of the fungal cell wall is an extremely complex phenomenon. The wall is made of different components, some of which are synthesized intracellularly, while others are polymerized outside the permeability barrier, but all of them must come together at a precise time and place, and in the precise amounts, to react and be organized into a coherent structure. Accordingly, there must exist precise mechanisms that regulate these processes. We know that synthesis of the cell wall is a polarized process that depends on the transport of components and biosynthetic enzymes to the growing sites. This is a function that depends on the cytoskeleton that carries all these components inside secretory vesicles and microvesicles to sites where wall expansion takes place. This process is tightly associated with the functioning of the secretory pathway. During its operation, proteins are synthesized, and placed into the lumen of the endomembrane system, where they become glycosylated and associated in part with some other polysaccharides. This pathway is also operative for the enzymes responsible for the biosynthesis of structural polysaccharides. By means of specific vesicles, the chitosomes, the enzymes responsible for chitin synthesis are carried to the cell surface where synthesis of the polysaccharide takes place. Similar mechanisms must be operative for glucan synthesis. Once at the cell surface, association among the different components takes place to produce a three-dimensional structure maintained and organized by non-covalent and covalent bonds that join them together. These processes are themselves subjected to regulatory mechanisms that permit changes in the wall, both qualitative and quantitative, in response to internal or external effectors. These changes permit the fungus to alter its morphology during its cell and life cycles, when interacting with other organisms, to better exploit the environment, compensate reduction in the synthesis of wall components, or adapt to different types of stress.

1. Introduction

Growth of a fungus is strictly dependent on the synthesis of the cell wall, the outermost structure that is responsible for cell containment and shape. One important aspect to understand how the growth of the wall occurs is that whereas this structure is located outside of the permeability barrier of the cell, all of its components must come from the protoplast itself. Accordingly, complex phenomena must occur to bring altogether the different components that make up the wall. Moreover, the delivery of these components must be coordinated in time and space, and their proportions carefully maintained. As a final step, the wall components must be associated among themselves to give rise to a coherent structure.

We know now that growth of the fungal cell wall represents a manifestation of the activity of the secretory pathway of the cell. Wall components and biosynthetic enzymes are synthesized through the secretory route, and finally packaged into secretory vesicles and microvesicles that are targeted to and fused with the plasma membrane. These vesicular elements along with their membranes contribute to the extension of the plasmalemma and discharge components that are incorporated into the wall along with enzymes that synthesize, modify or degrade the wall constituents. Controlled degradation seems to be important to soften the otherwise rigid wall, and thus permit its expansion (reviewed by Ruiz-Herrera, 1992; Bartnicki-Garcia, 2003). The fungal wall is made of two different kinds of components, structural polysaccharides (chitin and β-glucans) that exist in the form of long microfibrils, and amorphous compounds, mostly glycoproteins, that act as cementing material for the structural polysaccharides. Synthesis of chitin occurs by a vectorial process that involves the transport of chitin synthases by chitosomes to the cell surface where chitin microfibrils are formed and deployed in the outer surface of the plasmalemma (Ruiz-Herrera, 1992). Synthesis of β-1,3-glucan occurs at the level of the plasmalemma as result of the activity of a catalytic subunit and a GTPase activator (Douglas *et al.,* 1994; Mazur *et al.,* 1995), but is initiated by binding to a protein acceptor (Andaluz *et al.*, 1986; 1988). β-1,6-Glucan synthesis is initiated in the endomembrane system and is terminated at the cell surface (Shahinian and Bussey, 2000). Cell wall glycoproteins are synthesized by complex mechanisms during their transfer through the secretory pathway, and finally delivered to the cell surface by secretory vesicles (Novick and Schekman, 1979; Novick *et al.,* 1980).

2. Synthesis of Chitin

2.1 Chitin Synthases

A number of reviews on chitin synthesis have been published recently (Munro and Gow, 2001; Ruiz-Herrera *et al.*, 2001; 2002; Ruiz-Herrera and Ruiz-Medrano, 2003; Ruiz-Herrera and San-Blas, 2003). Chitin is synthesized by enzymes known with the generic name of "chitin synthases" (Chsp), none of which, unfortunately, has been purified to homogeneity yet. One significant observation was the realization that fungi contain not one, but several chitin synthases. This conclusion arose from the original observation that disruption of *CHS1* from *S. cerevisiae* did not lead to any significant phenotypic alteration (Bulawa *et al.*, 1986). This result, and the observation that mutants in a second gene (*CHS2*) also did not display phenotypic alterations (Bulawa and Osmond, 1990), led to the isolation of a third gene (*CHS3*) whose mutation reduced significantly chitin levels in the wall (Valdivieso *et al.*, 1991; Bulawa, 1992).

This multiplicity of chitin synthase-encoding genes (*CHS*s) is not restricted to *S. cerevisiae*, but to fungi in general, and probably constitutes a mechanism to protect the cell from the loss of one chitin synthase. Accordingly, most single, and even double mutants display almost no phenotypic alterations (reviewed by Ruiz-Herrera *et al.*,

2001). This result is probably due to compensatory mechanisms as reported for *Wangiella* (*Exophiala*) *dermatitidis* (Wang *et al.*, 2002). This hypothesis is reinforced by the observation that some single mutants do not display any significant phenotype, whereas mutation in a second gene brings about significant phenotypic alterations e.g. *A. nidulans chsb*/*chsD* double mutants (Ichnomiya *et al.*, 2002). *chs2*/*chs3* double mutants of *W. dermatitidis*, in contrast to single mutants have low virulence for mice (Wang *et al.*, 2001), and *chsG-*/*chsE* double mutants from *A. fumigatus* contrasting with single mutants, display morphological alterations and reduced radial growth (Mellado *et al.*, 2003).

With a few organisms specific roles have been suggested for some chitin synthases. In *S. cerevisiae*, Chs3p is responsible for the synthesis of most of the wall chitin, the chitin ring in the septum that separates the bud from the mother cell, and chitosan from the spore wall; Chs2p is involved in the formation of septum chitin, whereas Chs1p plays a mere repair function after cell separation (Shaw *et al.*, 1991). Nevertheless, it must be indicated that neither the single or double, but only the triple mutation is lethal. Other examples are ChsDp from *A. nidulans* that appears to contribute to wall rigidity in those structures growing isodiametrically (Specht *et al.*, 1996), Chs1p from *C. albicans* whose null mutants fail to make septa, and eventually lyse (Munro *et al.*, 2001), and *Colletotrichum graminicola chsA* mutants that produce conidia that burst and hyphal tips that swell in the absence of osmotic support (Amnuaykanjanasin and Epstein, 2003). The clearest example of an essential *CHS* gene is *CHS3* from *N. crassa* (Beth-Din and Yarden, 2000), whose mutated spores fail to germinate.

In other organisms, mutation of *CHS* genes has identified chitin synthases of importance, either from a quantitative point of view, or under certain conditions such as in media of low osmotic strength, under stress, or during host invasion. Examples are the yeast Chs3p (see above), *Glomerella graminicola* ChsAp (Amnuaykanjanasin and Epstein, 2003) whose mutant spores lyse under hypotonic conditions, chsA of *Aspergillus oryzae* whose mutants show morphological alterations (Muller *et al.*, 2002), a class V chitin synthase from *Fusarium oxysporum* (Madrid *et al.*, 2003) and Umchs6 of *Ustilago maydis* (Garcerá-Teruel *et al.*, 2004)

2.2. Taxonomy and Evolution of Chitin Synthases

Analysis of similarities among fungal *CHS* gene fragments obtained by PCR led to their separation into three classes (Bowen *et al.*, 1992). As more genes or their fragments were isolated, new analyses led to the establishment of five classes (Specht *et al.*, 1996; Xoconostle-Cazares *et al.*, 1997). Considering only the complete gene sequences, it has been observed that fungal chitin synthases fall into two divisions (1 and 2), each one radiating into three (I, II and III), and two (IV and V) classes respectively (Two new classes, VI and VII, have been recently proposed in Division 2; see San-Blas and Niño-Vega's Chapter 5, this book). Class I is subdivided into two subclasses, one that includes enzymes from Ascomycetes and Basidiomycetes, whereas the other groups contain the Zygomycete enzymes. These analyses also revealed that fungal chitin synthases separated from the animal enzymes at early stages during evolution (Ruiz-Herrera *et al.*, 2002).

Considering that chitin is present in the eukaryote kingdoms only, but absent in higher plants, it may be suggested that chitin synthase is probably an ancient enzyme that may have appeared once plants had diverged from the eukaryotic groups, *i. e.* approximately one billion years ago. Its precursor may have been the ancestor of a group of modern β-glycosyl-transferases to which cellulose synthases also belong, and which share important motifs involved in catalysis (see Ruiz-Herrera *et al.*, 2001).

The lack of correlation between the phylogenetic relationships among chitin synthases and fungi should be noticed. Accordingly all fungal groups contain enzymes from both chitin synthase divisions and five classes (reviewed by Ruiz-Herrera *et al.*, 2001). The only feasible hypothesis to this lack of correlation is that diversification of chitin synthases preceded the separation of the fungal groups, an event that apparently occurred approximately half a billion years ago (Walker, 1985; Berbee and Taylor 1992; Redecker *et al.* 2000).

Comparison of the amino acid sequence of chitin synthases belonging to Divisions 1 and 2 revealed that their conserved motifs were restricted to a short stretch differentially located in the polypeptide chain (Ruiz-Herrera *et al.*, 2001). It is therefore likely that both divisions originated from a gene duplication event, with the probable loss of a dispensable genomic fragment. New duplications would have given rise to the corresponding enzymatic classes. The presence of two or more enzymes of the same class in a certain fungal species probably represents late gene duplication events, whereas the opposite phenomenon, absence of enzymes from a certain class, is probably due to loss of the corresponding parental gene during evolution.

2.3. Structure of Chitin Synthases

Since no chitin synthase has been purified yet, the information we have on the structure of the enzyme comes from indirect procedures. Hydrophobic cluster analysis (HCA) of several β-glycosyl-transferases including *S. cerevisiae* Chs1, identified three aspartic acid residues plus the universal motif for β-glycosyl-transferases, QXXRW, to be involved in catalysis (Saxena *et al.*, 1995). The role of this last motif was confirmed by the use of the QSLAVE program (Horsch and Sowdhamini, 1996), also demonstrating that chitin synthases belong to the $(\beta/\alpha)_8$-barrel fold family of proteins. In agreement with these results was the observation that the three Asp (or Asn) residues, and the QXRRW motif were absolutely conserved in all the chitin synthases analyzed thus far (Ruiz-Herrera *et al.*, 2002). The role of residues Asp441, Asp562, and the amino acids Gln601, Arg604, and Trp605 from the QRRRW motif, in catalysis by the *S. cerevisiae* Chs2 were confirmed by site-directed mutagenesis (Nagahashi *et al.*, 1995).

Comparative analysis of the amino acid sequences of chitin synthases revealed the existence of three conserved motifs besides QXRRW: QXXEY, LP(G/A) and LXEDRXL (Ruiz-Herrera *et al.*, 2002). Use of the Psi-Pred program revealed that the conserved motifs and amino acids K662, E665, G669, L714, L720, T744 and P747 were invariant in the structure of the enzyme core region (Ruiz-Herrera *et al.*, 2002).

2.4. The Mechanism of Chitin Synthesis

Unfortunately, in recent years important advances in our understanding of the mechanisms involved in chitin synthesis have not occurred. No further efforts to purify the enzyme have been published, and the status of the field remains as described in previous reviews on the subject (Ruiz-Herrera, 1992; Merz *et al.*, 1999; for more recent reviews with emphasis on inhibition of chitin synthesis see Cohen, 2001; Ruiz-Herrera and San-Blas, 2003).

Our knowledge of chitin synthesis *in vitro*, comes from the use of crude membrane preparations, chitosomes, and preparations obtained by extraction with digitonin. Digitonin-extracted enzyme occurs in the form of 16 S aggregates with a Mr of 500 kDa. Highly purified 16 S aggregates from *M. rouxii* were made by several polypeptides whose Mr did not exceed 50 kDa (Lending *et al.*, 1991). This result was surprising, since all *CHS* genes encode proteins of much larger Mr.

We must pay homage to Glaser and Brown (1957) who using crude membrane preparations from *N. crassa* determined most of the kinetic properties of fungal chitin synthases. Accordingly, we know that the substrate for the enzymes is UDP-N-acetyl glucosamine (UDPGlcNAc) that by an apparently simple transglycosylase reaction donates glycosyl residues for chain growth, liberating UDP. A divalent metal, usually Mg^{2+}, is required for activity. The enzymes display a sigmoidal dependence on substrate, probably due to a cooperative behavior; Km is in the mM range, optimal temperatures are around 15-40 °C, and pH around 5-8. The enzyme is activated by GlcNAc that binds to an allosteric site (Horsch *et al.*, 1996). The involvement of a lipid intermediate for GlcNac transfer, as occurs during glycoprotein synthesis has been excluded, and the existence of a high energy intermediate dismissed (Ruiz-Herrera, 1982).

The interesting suggestion that chitin synthesis, in a similar way as occurs for glycogen synthesis (reviewed in Alonso *et al.*, 1995), is a two-step reaction has been put forward by Merz *et al.* (1999) who described that a priming reaction was responsible for initial glycosylation of a protein acceptor, followed by a step involving further addition of glycosyl units to this acceptor to produce chitin chains. By use of specific inhibitors and by the differential effect of metal ions, results that agree with this hypothesis have been obtained with 16 S subunits of *M. rouxii* (Ruiz-Herrera and Ruiz-Medrano, 2003). This mechanism is not rare; similar ones occur in the synthesis of glycogen, starch, and cellulose.

2.5. Regulation of Chitin Synthesis

Regulation of chitin synthesis is the result of complex interactions between mechanisms operating at transcriptional, and post-translational levels, that depend on the specific enzymes and the environmental conditions. Analysis of the upstream regulatory regions of different chitin synthase genes (Ruiz-Herrera *et al.*, 2002) revealed the existence of cannonical sites for the binding of different transcription factors responding to heat shock, stress, pH, nitrogen deprivation and phosphate levels, among others. Agreeing with these theoretical results, a transcript corresponding to *CHS3* from *Exophiala dermatitidis* whose 5′ upstream region displays Stre01, PacC, and Nit2 (AreA) sites, was up-regulated by high temperature, acidic pH, or nitrogen deprivation. Examples of transcriptional regulation during fungal differentiation are known. Transcripts from *CaCHS2* of *C. albicans* (Chen-Wu *et al.*, 1992), from several *CHS* genes of *U. maydis*, (Xoconostle-Cazares *et al.*, 1996; 1997), and from *Mucor circinelloides CHS1* (Lopez-Matas *et al.*, 2000) increased during the yeast-mycelium dimorphic transition. Additionally, transcriptional regulation of *CHS* genes during sporulation of *A. nidulans* by the transcriptional regulator AbaA (Park *et al.*, 2003), and during fruiting body formation in *Tuber borchii* (Balestrini *et al.*, 2000) have been described.

Examples exist of post-transcriptional regulation of chitin synthases. Chsp from *P. blakesleeanus* was activated by light *in vitro* and *in vivo* (Herrera-Estrella and Ruiz-Herrera, 1983). Osmotic stress increased chitin synthase activity in *Benjaminiella poitrasii* (Deshpande *et al.*, 1997). Alterations in cell wall synthesis may affect Chs regulation, probably as the result of compensatory mechanisms. Changes may also result of alterations in the cellular location, or accumulation of chitosomes (Martin *et al.*, 1999).

Most fungal chitin synthases can be activated by limited proteolysis, although the physiological importance of this phenomenon is still obscure (Ruiz-Herrera *et al.*, 1992). The main obstacle in reaching a final conclusion on this matter is the absence of purified enzymes that would allow polypeptide analysis before and after proteolysis. Activation of chitin synthases by calcium-calmodulin has been reported for *P. blakesleeanus* (Martinez-Cadena and Ruiz-Herrera, 1987), and *N. crassa* (Suresh and Subramanyam, 1997). It is feasible that this process is related to a cascade of reactions, not yet identified.

3. Synthesis of Chitosan

Chitosan is not synthesized by *de novo* transglycosylation as chitin is, but by the partial deacetylation of chitin by specific enzymes. A soluble chitin deacetylase was originally described in *M. rouxii* (Araki and Ito, 1974; Davis and Bartnicki-Garcia 1984a; 1984b), although later data suggested a particulate location (Calvo-Mendez and Ruiz-Herrera, 1987). In agreement with this last result, a chitin deacetylase from *Absidia coerulea* was found associated with the cell wall (Gao *et al.*, 1995). Also an extracellular location of the enzyme was found in *A. nidulans* (Alfonso *et al.*, 1995). It is likely that nascent chitin that reaches the wall is partially deacetylated by the wall-associated or extracelular deacetylases, giving rise to chitosan that associates and covers nascently crystallized chitin microfibrils. The observation that HNO_2-sensitive aminopolysaccharide (chitosan) was absent at the apical region of *P. blakesleeanus* hyphae (Herrera-Estrella and Ruiz-Herrera, 1983) is in agreement with this hypothesis.

CDA1 and *CDA2,* the genes coding for the two deacetylases of yeast, are regulated during development, being expressed only at distinct time periods during sporulation (Christodoulidou *et al.*, 1996). Cda2p has been purified and found to be a 43 kDa glycoprotein with a carbohydrate content of approximately 18% by weight. Deglycosylation of the enzyme causes a total loss of activity, but this can be restored by addition of $CoCl_2$ (Martinou *et al.*, 2002).

4. Synthesis of β-Glucans

4.1. General Properties of β-1,3-Glucan Synthases

Early studies described activity of β-1,3-glucan synthase in fungal crude extracts, mostly membrane fractions (reviewed by Ruiz-Herrera, 1991; 1992; Douglas, 2001). These studies demonstrated that UDP-glucose (UDPG) was the universal glycosyl donor with Km values around 1-5 mM (Lopez-Romero and Ruiz-Herrera, 1978). Maximal activity was found at pH 6.7 to 8, and the addition of Mg^{++} or Mn^{++} stimulated activity. The enzyme was activated by nucleoside triphospates (Shematek and Cabib, 1980) and competitively inhibited by UDP, and glucono-D-γ-lactone (Lopez-Romero and Ruiz-Herrera, 1978).

The synthesized polysaccharide was water-insoluble and soluble in alkali, resistant to α-amylase, and completely solubilized by β-1,3-endo-glucanase. UDP was the byproduct of the reaction. No evidence was found for a lipid-linked intermediate (Shematek *et al.*, 1980). Nascent β-1,3-glucan synthesized by mixed membrane fractions from *S. cerevisiae* had Mr above 1,500 kDa but below 20,000 kDa, and this nascent material served as an acceptor for further glucose transfer, giving rise to glucan molecules over 20,000 kDa. (Ruiz-Herrera and Larriba, 1995). Cell free extracts of *S. cerevisiae* incubated with UDPG produced microfibrils insoluble in water, ethanol and acid, but soluble in alkali with an average length of about 0.5 μm. The microfibrils were digested by both endo and exo β-1,3-glucanases giving rise to glucose as the only product, which indicated that they consisted of β-1,3-glucan molecules with no detectable branches (Larriba *et al.*, 1981). Glucan synthase activity was then extracted from membrane fractions. The highest level of purification was obtained from *N. crassa* with a specific activity of 14,000 nmoles glucose incorporated per min per mg protein (Awald *et al.*, 1994).

It is generally accepted that β-1,3-glucan synthesis occurs in the exterior of the cell. Nevertheless, convincing evidence exists that synthesis of the polymer requires a protein acceptor (Andaluz, 1986; 1988; Ruiz-Herrera and Larriba, 1995). Accordingly, it has been shown that *in vitro* synthesized glucan had no free reducing end, and that during synthesis two different glucan fractions were made, one soluble and one insoluble in SDS. The SDS-soluble fraction was sensitive to digestion with β-1,3-glucanase or a protease, and chase experiments showed transfer of the soluble fraction into the insoluble one.

Purification of the catalytic protein has been unsuccessful, although several polypeptides in samples of different fungi were labeled with the UDPG photosensitive analog 5-azido-(β-(^{32}P)-UDPG (reviewed by Douglas, 2001). Use of a highly purified sample from *N. crassa* resulted in the unequivocal identification of one of the labelled peptides as Fks1p (Schimoler-O'Rourke *et al.*, 2003). β-1,3-Glucan synthase activity was stimulated by what originally was suggested to be a regulatory subunit (Kang and Cabib, 1986), and then found to be the small monomeric GTPase Rho1p (Drgonova *et al.*, 1996; Qadota *et al.*, 1996), which is involved in other reactions required for cell growth (see below).

4.2. Genes Encoding β-1, 3-Glucan Synthases

The search for genes involved in β-1,3-glucan synthase activity originated from the study of different *S. cerevisiae* mutants: mutants resistant to inhibitors of β-1,3-glucan synthesis such as papulacandin B (papulacandin B resistance, *pbr* mutants) (Castro *et al.*, 1995) and echinocandin (echinocandin target gene, *etg* mutants) (Douglas *et al.* 1994; El-Shebeini and Clemas, 1995), Calcofluor white hypersensitivity mutants (*cwh53* mutants) (Ram *et al.*, 1995), *gsc1* mutants (Abe *et al.*, 2001), mutants with calcineurin dependent growth (*cnd1* mutants) (Garret-Engele *et al.*, 1995) or hypersensitive to tracolimus (FK506-supersensitive, *fks1* mutants) (Parent *et al.*, 1993). The genes affected in these mutants: *CND1*, *CWH53, ETG1, FKS1, GSC1* and *PBR1* were the same, and all are designated as *FKS*.

In *S. cerevisiae* two homologous genes involved in β-1,3-glucan synthesis, *FKS1* and *FKS2*, have been identified and correspond to the catalytic subunits of different β-1,3-glucan synthase complexes. This concept is supported by the observation that the *fks1-fks2* double mutation is lethal. Both genes differ primarily in their regulation. Under optimal growth conditions, *FKS1* is predominantly expressed, and its mRNA levels fluctuate periodically throughout the cell cycle. Expression of *FKS2* is low under optimal growth conditions, but expression is induced in response to treatment with the mating pheromone or $CaCl_2$, by growth in poor carbon sources, or in the absence of *FKS1*. Fks2 is essential for sporulation (Mazur *et al.*, *1*995). Induction of Fks2 by growth with poor carbon sources seems to respond to glucose depletion and is under the control of the Snf1 protein kinase and the MigI transcriptional repressor (Zhao *et al.*, 1998).

Fks proteins are fairly conserved in fungi, and as occurs in yeast, mutations affecting their synthesis are lethal in *Cryptococcus neoformans* (Thomson *et al.*, 1999) and *Yarrowia lipolytica* (Leon *et al.*, 2002). Expression of the antisense gene in *N. crassa* reduced enzyme levels, altered cell-wall morphology and inhibited growth (Tentler *et al.*, 1997). Homologous genes to yeast *FKS* have been also described in other fungi, for example *Paracoccidioides brasiliensis* (Pereira *et al.*, 2000), *A. nidulans* (Kelly *et al.*, 1996) and *S. pombe* (Ishiguro *et al.*, 1997). In *C. albicans* three *FKS* homologs with high similarity to yeast *FKS1* have been described: *CaGSC1, CaGSL1* and *CaGSL2*, although activity seemed to be restricted to Gsc1p

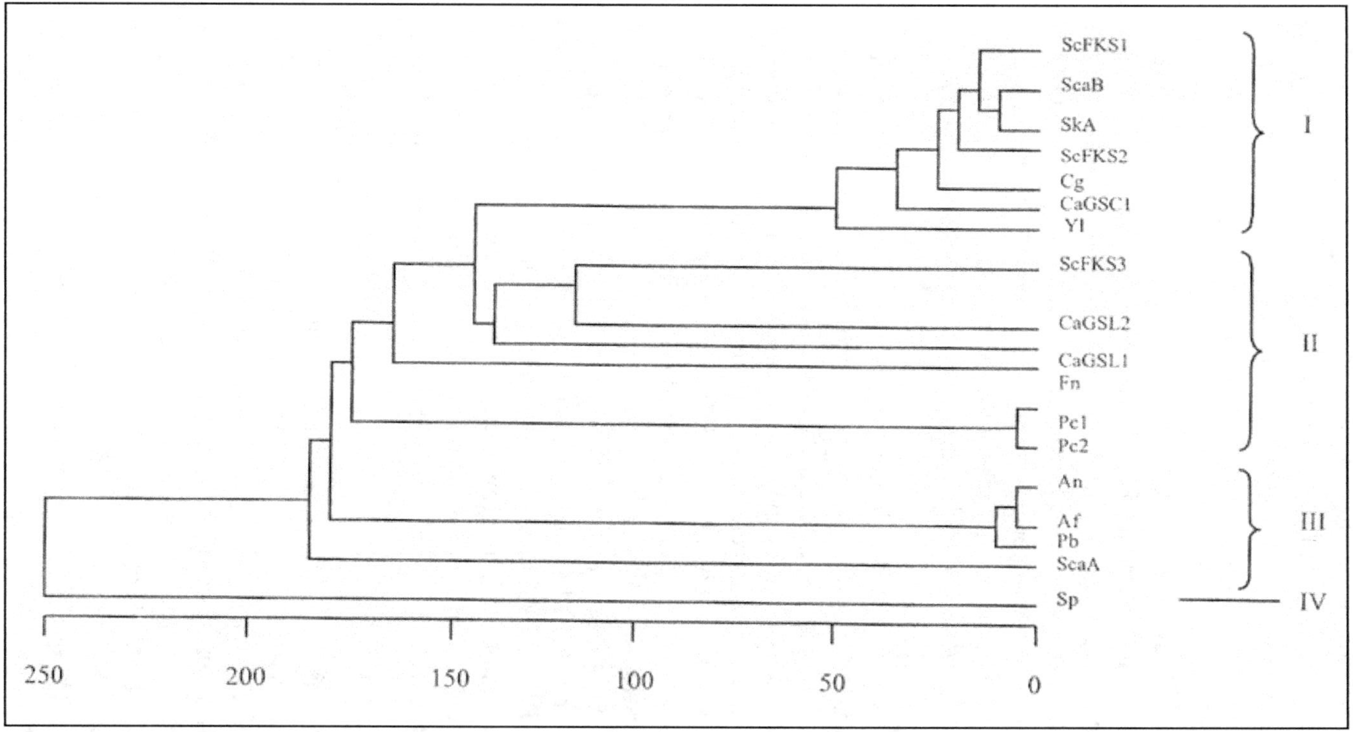

Figure 1. Dendrogram of fungal β-1,3-glucan synthases. Only fully known sequences were included. Af, *Aspergillus flavus*; An, *Aspergillus nidulans*; Ca, *Candida albicans*, Cg, *Candida glabrata*; Fn, *Filobasidiella neoformans*; Pb, *Paracoccidioides brasiliensis*; Pc, *Pneumocystis carinii*; Sc, *Saccharomyces cerevisiae*; Sca, *Saccharomyces castelli*; Sk, *Saccharomyces kluyveri*; Sp, *Schizosaccharomyces pombe*; Yl, *Yarrowia lipolytica*.

(Mio *et al.*, 1997), which also appeared to be the target for anti-glucan synthase inhibitors (Douglas *et al.*, 1997).

Hydrophobicity analyses reveal that Fks are membrane proteins with 16 transmembrane helices and a central hydrophilic domain (Douglas *et al.*, 1994) that is probably the catalytic site. Interestingly, all β-1,3-glucan synthases, including those from plants, have important structural differences compared to the rest of the glucosyltransferase families, including the cellulose and chitin synthases. Accordingly, they lack the characteristic conserved amino acids (three Asp residues) and the QXXRW motif, associated with the catalytic activity of the latter (Campbell *et al.*, 1997; reviewed by Douglas, 2001), and the proposed UDP glucose-binding site (R/K) XGG found in glycogen synthases (Farkas *et al.*, 1990). This difference has suggested the existence of two different families of synthetic glucosyltransferases of probable distinct phylogenetic origin.

Characterization of a group of mutants resistant to glucan synthesis inhibitors in *S. cerevisiae,* defined a critical "inhibitor resistance cluster". This was similar among Fks proteins from different fungi, and showed the consensus sequence FL(V/I/T/)LS(L/I/F)(R/K)D (Douglas, 2001). Whether this cluster corresponds or is associated with the catalytic site remains to be determined.

We have made a comparison of the sequences of the proteins encoded by different genes described as β-1,3-glucan synthases from fungi. Their similarity is very high, with the exception of Cps1 from *S. pombe*, which has lower similarity, because of its larger size. A dendrogram of the full cloned sequences plus a fragment from the *U. maydis* homologue is shown in Figure 1. It may be observed that proteins separate into four well-defined classes. Class I appears as a homogeneous group, with similarities varying from 52.1 to 86.9 %. All these proteins belong to Ascomycetes including the two *S. cerevisiae* Fksps, and the *C. albicans* and *C. glabrata* proteins. The proteins from *S. castelli* B and *S. kluyveri* are the smallest ones in the class, about 600 amino acids in length, while the rest are made of about 1900 amino acid residues (size average 1904±32). Enzymes of this group have 76 conserved motifs in common (a motif was defined as a sequence of at least four or more identical amino acid residues), with the exception of *S. castelli* and *S. kluyveri* that are smaller and have only 14 motifs in common with the rest of proteins. The second group includes enzymes from Ascomycetes and Basidiomycetes. These

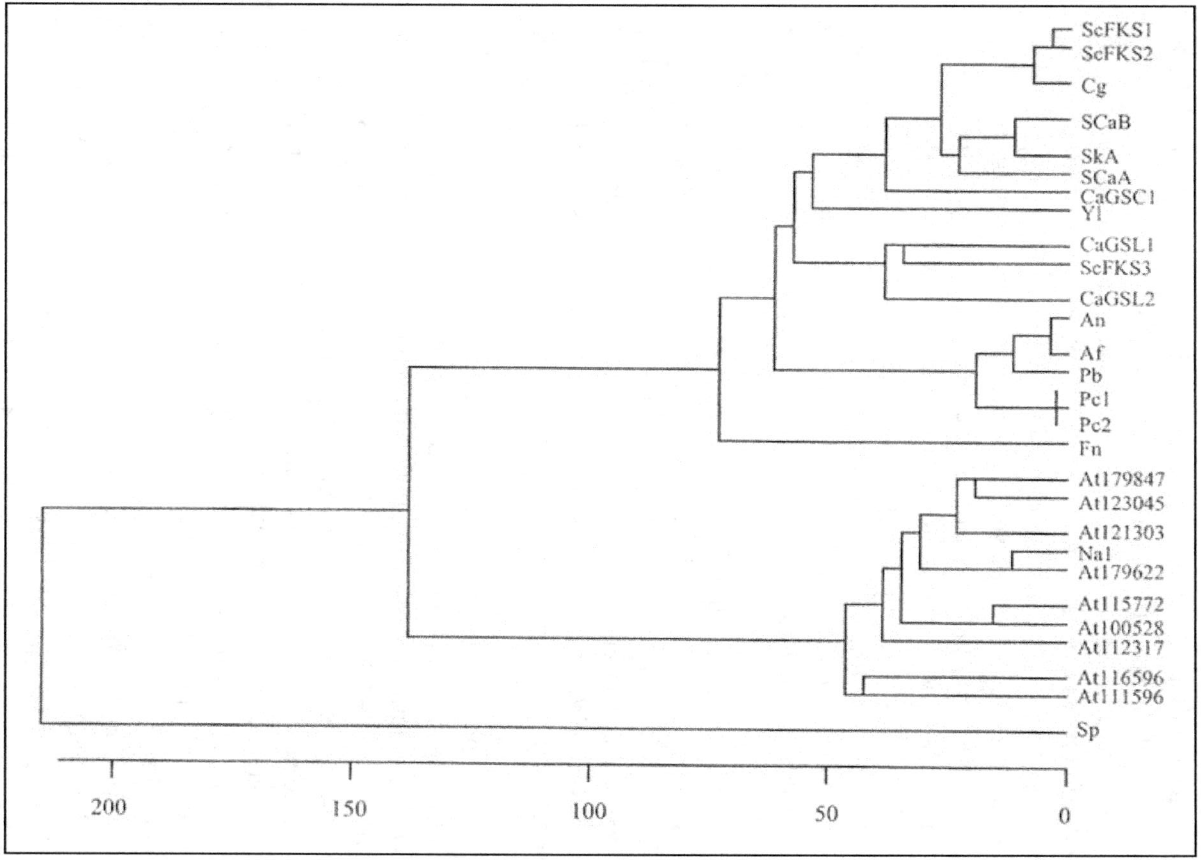

Figure 2. Comparative dendrogram of β-1,3-glucan synthases from fungi and plants. Only fully known sequences were included. Abbreviations for fungal enzymes as in Figure 1. At, *Arabidopsis thaliana*; Na, *Nicotiana aldata*. Numbers for *A. thaliana* enzymes correspond to Gene Bank accesion numbers.

proteins share 22 motifs, 16 of them identified also in the *U. maydis* Umfks1 fragment. Similarities ranged between 41.4 and 97.6 %. The size of the proteins vary from 1000 to approximately 1900 (size average 1698±352). The third group is the most homogeneous one. Their similarities are about 77.4 to 87.6 %, and they have 90 common motifs, except for the *S. castelli* enzyme that shares only 28 motifs. The proteins of this group have about 1900 amino acid residues (size average 1911±13) except for the *S. castelli* protein. Fksp csp1 from *S. pombe* is the only member of the fourth group. This protein has 2353 amino acid residues, thus being the largest β-1,3-glucan synthase described thus far. It has motifs in common with the rest of the glucan synthases along its full length.

In plants there is a group of genes similar to fungal *FKS*. These genes encode callose synthases, responsible for the synthesis of the characteristic plant β-1,3-glucans, collectively referred to as callose. When the sequences of plant callose synthases were compared with fungal Fksp, and included in a dendrogram, all of them separated into a compact group isolated from the fungal enzymes, suggestive of their early evolutionary departure (Figure 2).

The transglycosidases are different kinds of enzymes that have been classified into a new group of glycoside hydrolases, Family 72, that appears to be important for the structural organization of the cell wall (reviewed by Popolo *et al.*, 2001). All these proteins are bound to the plasmalemma through a GPI anchor and are widely distributed in fungi. Initially described in *C. albicans* as involved in pH-regulated morphogenesis, it was later suggested that they played a role in the linking of β-1,6-and β-1,3-glucans in the wall of the fungus (Fonzi, 1999). A similar enzyme was described in *A. fumigatus* as a novel transglycosidase encoded by the *GEL1* gene (Mouyna *et al.*, 2000). The enzyme splits internally a β-1,3-glucan molecule and transfers the newly generated reducing end to the non-reducing end of another β-1,3-glucan molecule forming a β-1,3-linkage. This enzyme is homologous in structure and activity to Gas proteins from yeast (Gasp1 to 5), Phr1p, and Phr2p from *C. albicans,* Phr1p, and Phr2p from, *C. dubliniensis*, and Phr1 of *Pneumocystis carinii* (reviewed by Popolo *et al.*, 2001). Mutation in the genes encoding Gas1 in *S. cerevisiae* (Popolo *et al.*, 1993), Phr1 (Saporito-Irwin *et al.*, 1995) or Phr2 in *C. albicans* (Muhlschlegel and Fonzi, 1997) induces abnormal morphology, and affects cell wall properties and structure. These alterations are related to a reduction in the incorporation of β-glucans into the cell wall, and release of GPI proteins to the medium, evidence for their importance in cell wall organization.

4.3. The Regulatory Role of the Rho GTPase

Rho1 was identified as a regulatory subunit of β-1,3-glucan synthase in *S. cerevisiae*. (Drgonova *et al.*, 1996; Qadota *et al.*, 1996). The Rho1 homologue in *C. neoformans* complemented the yeast *rho1* mutation, although it manifested distinct features of the yeast protein in certain aspects (Chang and Penoyer, 2000). In *A. fumigatus* four Rho proteins were identified, Afrho1 having the highest similarity to Rho1 from *S. cerevisiae* (Beauvais, 2001). In *S. pombe* initially two genes called *RHO1* and *RHO2* were identified. The encoded Rho1 protein showed 72.7 % identity to *S. cerevisiae* Rho1, whereas the encoded Rho2 protein showed 53.2 % identity to yeast Rho2 (Nakano and Mabuchi, 1995). Interestingly, it has been suggested that Rho1 regulates the synthesis of β-1,3-glucan by activation of the β-1,3-glucan synthase, whereas Rho2 regulates the synthesis of α-glucan through its effect on the α-1,3-glucan synthase Mok1p (Calonge *et al.*, 2000). Later results have described the existence of more *RHO* genes, and one of them (*RHO4*) was suggested to be involved in the control of cell shape and septum formation (Nakano *et al.*, 2003). Mutation of Rho1p is lethal in *S. cerevisiae* and *S. pombe* (Nakano *et al.*, 1997), but this may not occur in other fungi. Thus, the homolog gene from *Y. lipolytica* (*YlRHO1*) complemented *rho1* lethality in *S. cerevisiae*, but its disruption was not lethal, suggesting the existence of other homologous genes (Leon *et al.*, 2003). In general it has been concluded that the role of the small GTPase Rho is related to cell polarity (Arellano *et al.*, 1999; Guo, 2001).

It has been found that Rho1p interacts with Cdc42 in pathways involving, besides wall synthesis, exocytosis, septin organization, actin assembly, microtubule organization, autophagy and cytokinesis (Drees *et al.*, 2001). The mechanisms of regulation of Rho, and its interactions with different proteins involved in polarity, secretion, and wall synthesis are extremely complex. Accordingly, it was demonstrated that in yeast, distinct networks regulated two effector proteins of Rho1p, Fks1p and Pkc1p (Sekiya-Kawasaki *et al.*, 2002). Bem2 and Sac7, but not Bag7, were involved in the control of the Rho1-mediated activation of Mpk1, whereas Bag7 and Sac7, but not Bem2, affected other Rho1 functions, ensuring their individual regulation at their appropriate place and time (Schmidt *et al.*, 2002). Rho1p is regulated in *S. pombe* through the activity of a GAP specific for Rho1p named Rga5p. Rho1p activity is increased in *rga5* mutants and decreased in cells over expressing *RGA5*. These changes were reflected not only on β-1,3-glucan synthase activity, but also on other known Rho1p effectors, including, Pck1p and Pck2p (Calonge *et al.*, 2003).

4.4. Synthesis of β-1,6-Glucans

β-1,6-Glucan is a highly branched glucose polymer with an average of *ca.* 350 monomers, corresponding to about 10% of the total glucans in the *S. cerevisiae* cell wall (Shahinian and Bussey, 2000). In contrast to our knowledge of β-1,3-glucan synthesis, the mechanism for β-1,6-glucan formation is mostly unknown, although important advances in the subject have been made in recent years (reviewed by Shahinian and Bussey, 2000). *In vitro* synthesis of β-1,6-glucans has been described in only a few fungal systems. Cell-free extracts from *S. cerevisiae* synthesized mostly β-1,3-glucan, but small amounts of β-1,6-glucans were also detected (Lopez-Romero and Ruiz-Herrera, 1978).

A search for the genes encoding enzymes involved in β-1,6-glucan synthesis has been developed in *S. cerevisiae* with the use of killer toxin K1 resistant (*kre*) mutants (Brown *et al.*, 1993), taking into consideration that the receptor for this toxin is β-1,6-glucan (Bussey *et al.*, 1979; Hutchins and Bussey, 1983). Another strategy used for this search has been the complementation of Calcofluor white hypersensitive mutants (Ram *et al.*, 1994). By these procedures at least 10 genes involved in β-1,6-glucan biosynthesis were identified: *KRE1, KRE5, KRE6, KRE9, KRE11, CNE1, CWH41/GLS1, KNH1, ROT2/GLS2* and *SKN1* (reviewed by Shahinian and Bussey, 2000).

Genetic and structural studies suggest that β-1,6-glucan synthesis may occur at the cell surface, but several intracellular events involving the secretory pathway appear to be crucial for the synthesis of the polymer. Accordingly, several gene products involved in β-1,6-synthesis have an intracellular location. Kre1p is a cytoplasmic protein that plays a role in the trafficking of ER-derived vesicles. The products encoded by *KRE5* and *CWH41* are located in the endoplasmic reticulum, whereas *KRE6* and *SKN1* encode transmembrane Golgi proteins (Roemer *et al.*, 1994). *KRE5,* whose mutation gives rise to

K1 resistance and a reduction in the β-1,6-glucan levels in the wall (Boone *et al.*, 1990; Meaden *et al.*, 1990), has significant similarity to a glucosyltransferase (Parker *et al.*, 1995), although a possible role in the binding of GPI proteins to β-1,6-glucans was suggested (Shahinian and Bussey, 2000). Cwh41 is a transmembrane protein, and mutation of the encoding gene produces a phenotype similar to *kre5*. *CWH41* and *ROT2* encode glucosidases I and II, respectively, both of which are involved in trimming during GPI-bound protein assembly of glucans. Another important Golgi-protein is Big1p. Deletion of *BIG1* reduces about 95% of the cell wall β-1,6-glucan, and mutants grow slowly requiring osmotic support (Azuma *et al.*, 2002). On the other hand, *KRE1* and *KRE9* encode exocellular products suggested to be involved in the enlargement of the β-1,6-glucan chains. Mutation in either gene confers K1 resistance and reduces the level of β-1,6-glucan in the wall, apparently by shortening the glucan chains (Boone *et al.*, 1990; Roemer and Bussey, 1995). Kre1p is a GPI-bound protein located in the plasmalemma that is enriched in the yeast mother cells (Roemer and Bussey, 1995), whereas Kre9p is located at the cell surface (Brown and Bussey, 1993).

Thus, there is substantial information on the genes and proteins involved in the synthesis of β-1,6-glucans, but we still lack a model to explain how the polymer is finally assembled. Failure to identify the transglycosidases responsible for chain growth is possibly one of the most important missing links for understanding the synthesis of this important cell wall polymer.

5. Synthesis of α-1,3-Glucans

Despite the abundance of α–glucans in the cell walls of numerous fungi, little is known about their synthesis, and numerous attempts to synthesize them *in vitro* have failed. α–1,3-Glucan is an important component of the cell wall of *S. pombe* that accounts for 28-32 % of the cell wall dry weight. The enzymes responsible for α–1,3-glucan synthesis are encoded by the family of *MOK* genes, *MOK1* being the most important one (Katayama *et al.*, 1999). Mok1p is involved in the first step of glucan bundle formation, an event that it is required for new cell wall synthesis during vegetative growth (Konomi *et al.*, 2003). Calonge *et al.* (2000) have suggested that, similar to β-1,3-glucan synthase that is activated by Rho1p (see above), Rho2p is involved in Mok1p activation through the action of the protein kinase Psk2p.

6. Synthesis of Wall Glycoproteins

6.1. General Concepts

All proteins are synthesized at the ribosomes where information conveyed by mRNAs is translated. The difference between cytoplasmic, membrane-bound, extracellular and wall proteins, is that the last three classes are synthesized by ribosomes bound to the endoplasmic reticulum (ER) by a vectorial mechanism, through which the proteins are transferred into the ER lumen.

True wall proteins are *N*- and/or *O*-glycosylated. Also, some proteins are bound to β-1,6-glucans through a GPI bridge, and others are bound to β-1,3-glucans, and the possibility exists of other kinds of additions and linkages. Binding of some substituents occurs during the transfer of the proteins through the secretion pathway, whereas others occur in the wall itself (for older and recent reviews consult Tanner and Lehle, 1987; Kukuruzinska *et al.*, 1987; Ballou, 1990; Hercovics and Orlean, 1993; Orlean, 1997; Kukuruzinska and Lennon, 1998; Gemmill and Trimble, 1999; Cutler, 2001; Ernst and Prill, 2001; Klis *et al.*, 2002).

6.2. Initiation of Synthesis of Wall Glycoproteins

N-glycosylation of proteins has been studied more deeply in *S. cerevisiae* and *C. albicans*. Synthesis of proteins is initiated at the ribosomes by the translation of the so-called "signal peptide", that brings the ribosome to the ER where transfer of the polypeptide to the lumen is initiated (Walter and Blobel, 1980). The signal peptide, which may be processed later on, is normally at the N-terminus of the proteins, and is characterized by the presence of one or more charged amino acids, followed by 6 to 12 hydrophobic residues. Transfer of long polypeptides to the ER lumen occurs by a co-translational process (Walter and Blobel, 1980), whereas transfer of small polypeptides is normally a post-translational process (Rothblatt and Meyer, 1986). These polypeptides maintain their dissociated state through the activity of chaperones (Matlack *et al.*, 1999).

6.3. Synthesis of The Glycosidic Inner Core of Wall Proteins

Synthesis of the inner core ($GlcNAc_2Man_9Glc_3$) initiates at the ER and is completed in the Golgi. This process is analogous in all eukaryotic cells and occurs while the oligosaccharide is bound to an isoprenoid lipid that plays the role of transporter. The first seven sugars are transferred from the corresponding sugar nucleotides UDPGlcNAc and GDPMan, with formation of the molecule Dol-PP-$GlcNAc_2Man_5$. This unit is transferred to the lumen of the ER, where the remaining sugars are transferred from the corresponding sugar isoprenoid derivatives Dol-PMann and Dol-PGlc (Burda and Aebi, 1999). Mutants affected in late genes are viable, but mutations affecting early genes, in which the transfer of the inner core is affected, are lethal.

6.4. Transfer of the Inner Core to the Proteins

An oligosaccharyl transferase (OT) catalyzes the transfer of the inner core from Dol-PP-$GlcNAc_2Man_9Glc_3$ to the amido group of an asparagine residue of the acceptor protein. It is composed of several subunits, nine transmembrane polypeptides in the case of *S. cerevisiae* (Karaoglu *et al.*,1995; Silberstein and Gilmore, 1996; Knauer and Lehle, 1999). The genes encoding these proteins have been cloned, and six of the proteins (Wbp1, Swp1, Ost1; Ost2, Ost3 and Ost5) have been purified (Kelleher and Gilmore, 1994; Knauer and Lehle, 1994; Pathak *et al.*, 1995; Silberstein and Gilmore, 1996; Reiss *et al.*, 1997). Some show synthetic interactions between them (Knauer and Lehle, 1999). Five genes are essential, while mutation of the others only reduces efficiency of transfer *in vitro*. The enzyme must recognize a tripeptide (the sequon) made of Ans-X-Ser/Thr (Lehle and Bause, 1984). Not all sequons are glycosylated, and the resulting structures may be different. Invertase from *S. cerevisiae* contains 14 sequons, but only 8 or 9 are normally glycosylated (Reddy *et al.* 1988). This selection is probably due to steric hindrance. *Alg* mutants with shorter inner cores transfer them to a minor number of sequons; they transfer inefficiently $GlcNAc_2Man_5$ to invertase (Verostek *et al.*, 1993a; 1993b). N-glycosylation is reduced in mutants that stop synthesis at the $GlcNAc_2Man_5$ stage (Muñoz *et al.*, 1994). Probably glucose residues in Dol-PP-oligosaccharides favour the recognition by the transferase (Alvarado *et al.*, 1991).

6.5. Processing and Maturation of the *N*-Linked Oligosaccharides

After transfer to a protein, the oligosaccharide is trimmed. The terminal α-1,2-bound glucose is removed, followed by the two α-1,3-bound sugar moieties that are eliminated by glucosidase II. Nevertheless, the *gls1* mutant that retains the 3 glucosyl residues grows normally. In several systems the trimmed oligosaccharide may be glycosylated again by a UDPGlc:glycoprotein glucosyl tranferase, an enzyme that forms part of the "quality control" system that regulates protein folding in the ER (Herbert *et al.*, 1995). This system is involved in the retention of misfolded proteins in the "periplasmic space" by the membrane-associated chaperonine calnexin (Boisrame *et al.*, 2002). Correctly trimmed proteins that reach their normal folding are not substrates for UDPGlc:glycoprotein glucosyl transferase and may leave the ER.

6.6. Glycosylation of Proteins along the Secretory Pathway

In fungi, proteins bearing the trimmed oligosaccharide are further glycosylated to produce large glycoprotein molecules (reviewed by Ballou, 1982; 1990). The external chains are variable and depend on the protein. Accordingly, three of the four carbohydrate moieties of *S. cerevisiae* Y carboxypeptidase retain inner cores composed of 9-13 mannose residues, whereas 8 to 12 mannose units are added to the fourth (Trimble *et al.*, 1983). In yeast invertase some chains are composed of 8 to 13 sugar residues, whereas others contain as many as 200 mannose units. Apparently these differences are due to the accessibility of the chains to the mannosyl transferases that catalyse the addition of other mannose units. Studies with mutants affected at different steps of glycosylation have revealed that glycosylation takes place in succesive Golgi compartments (Franzussof and Schekman, 1989; Graham and Emr, 1991; Franzussof, 1992; Lussier *et al.*, 1995; Burda *et al.*, 2002).

Synthesis of the outer chain is initiated by the Och1p α-1,6-mannosyl transferase (Nakanishi *et al.*, 1993; Romero *et al.*, 1994). Och1p is a type II membrane protein located at the early Golgi compartment (Chapman and Munro, 1994; Gaynor *et al.*, 1999; Harris and Walters, 1996). Och1p can glycosylate the inner core of late *alg* mutants, such as *alg3*, but not the *alg2* mutants. *Och1* mutants are viable, but their glycoproteins lack the outer chains. Addition of mannose in α-1,2-position to the α-1,6-chains apparently serves as signal to stop further addition of mannosyl units (Ballou, 1990). This α-1,2-transfer activity requires the product of the *MNN2* gene, but surprisingly, *mnn2* null mutants or Mnn2p overexpressing mutants do not show any change in transferase activity (Devlin and Ballou, 1990). *mnn5* mutants are affected in the transfer of the second α-1,2-bound mannosyl unit, and *mnn1* mutants are blocked in the transfer of the α-1,3-mannosyl units to the chain, and contrary to *mnn2* mutants, both lack the corresponding enzymatic activities. These enzymes are located at the median *trans*-Golgi (Graham *et al.*, 1994). The outer chain, as well as the inner core can be further modified by the addition of mannosylphosphate from GDPMan.

Two models have been suggested to explain the transport of proteins through the Golgi apparatus. The model of the stable compartment suggests that non-transported proteins remained anchored in the Golgi, whereas those destined to be secreted are transported by anterograde vesicles. The second model (cisternal maturation) suggests that *cis*-Golgi is formed by incoming vesicles from the ER, and that the proteins to be secreted are transferred through maturation of cisternae, while resident proteins are retained by the action of retrograde vesicles (Allan and Balch, 1999). The time the proteins spend in the Golgi alters their glycosylation levels. Hyperglycosylated forms observed in some *sec* mutants are aberrant molecules that have spent abnormally long times in the secretory pathway.

6.7. Phenotypic Alterations Due to Defects in Protein Glycosylation

Alterations in protein glycosylation may affect the entire secretion process, or give rise to cells with phenotypic alterations, some of them unexpected. For example, *mnn7/8*, *mnn9* and *mnn10* mutants with reduced glycosylation show an increase in resistance to orthovanadate, and sensitivity to hygromycin B (Ballou *et al.*, 1991). The phenotype may be corrected in some cases by increased levels of GDP-Man (Dean, 1995; Kanik-Ennulat *et al.*, 1995). *mnn10/bed1* mutants present a delay in bud formation, which become larger and rounder, and show misslocalization of the chitin ring (Mondésert and Reed, 1996).

6.8. Regulation of Protein Glycosylation

Despite its importance, little is known regarding regulation of protein glycosylation. We know of the existence of a "quality control" of glycosylation associated with the correct folding of proteins (see above). Misfolded proteins that accumulate in the secretory pathway are transferred to the cytoplasm where they become deglycosylated and degraded by the proteosome (Susuki and Lennarz, 2000). *O*-mannosylation of proteins (see below) may inhibit *N*-glycosylation, because it blocks the hydroxyamino acid in the sequon (Ecker *et al.*, 2003), suggesting a regulation between both processes. Microarray analyses of the differential gene expression in the *agl7* (lipid-linked oligosaccharide precursor) mutant showed an up regulation in 29 genes, and the repression of 23 more (Schleip *et al.*, 2001). These changes involved the following routes: MAPK cascades, metabolism of phosphate and amino acids, mitochondrial activity, and ATP synthesis (Klebl *et al.*, 2001). The regulation of all these processes is unknown.

6.9. *O*-Glycosylation of Proteins

Many proteins, including many *N*-glycosylated ones, and those carrying a GPI anchor are *O*-glycosylated, with one to 5 mannosyl units, bound to Ser or Thr residues (Sentandreu and Northcote, 1968; Tanner and Lehle, 1987). Contrary to *N*-mannosylation, *O*-mannosylation does not require a consensus amino acid motif. The process is initiated at the ER by transfer of mannosyl units from DolP-Man by the *O*-mannosyltransferases Pmt1, Pmt2 and Pmt4. Completion of glycosylation occurs at the Golgi by the activity of mannosyl transferases (Mnt-proteins) that form distinct protein complexes among themselves (Girrbach and Strahl, 2003). In *C. albicans* 5 Pmts have been described, of which only Pmt1 and Pmt6 have been studied in detail. Glycosylation may occur in vicinal Ser and Thr residues (Ecker *et al.*, 2003) resulting in a protein with a rigid structure (Jentoft, 1990). *O*-glycosylation is required for different essential cellular processes, such as maintenance of cell wall integrity and budding (Strahl-Bolsinger *et al.*, 1999), dimorphic transition, and resistance to diverse antifungals. Protein glycosylation by Pmt1 and Pmt6 leads to a significant increase in the capacity of *C. albicans* to adhere to the host cells, as well as in its virulence in model systems (Ernst and Prill, 2001).

6.10. Synthesis of GPI-Glycoproteins

De Nobel and Lipke (1994) suggested that similar to α-aglutinin involved in adhesion, some membrane and wall proteins might contain a glycosyl phosphatidylinositol (GPI) anchor at their carboxy termini to bind to the corresponding structures. Use of the Von Heijne algorithm permitted the identification in the *S. cerevisiae* genome of 686 putative ORFs containing the N-terminal signal, from which 51 contained the potential GPI-binding motif (Caro *et al.*, 1997; see below). From these, about 30 may be located in the wall (Kollar *et al.*, 1997, Hamada *et al.*, 1999). These *in silico* studies have been extended to *C. albicans*, *S. pombe* and the filamentous fungus *N. crassa* for the search of putative GPI proteins, and have led to the definition of the GPI-binding motif as follows: [NSGDAC]-[GASVIETKDLF]-[GASV]-X(4,19)-[FILMVAGPSTCYWN](10)-C-terminal end of the protein (De Groot *et al.*, 2003). The results obtained with *C. albicans* were similar to those previously described by Garcerá *et al.* (2003).

GPI-proteins were initially described in *S. cerevisiae* (Conzelmann *et al.*, 1988). Their general structure is the following: protein-CO-NH-CH_2-CH_2-PO_4-6-Man-α-1,2-Man-α-1,6-Man-α-1,4-GlcN-α-1,6-myo-inositol-1-PO_4-lipid (Reggiori *et al.*, 1999). In cell wall proteins, the anchor is trimmed, and only a part is retained at the C-terminus, which participates in binding of the protein to β-1,6-glucan (Kollar *et al.* 1997; Fujii *et al.*, 1999). The conserved fraction is made by ethanolamine-PO_4-$(Man)_4$-, and is substituted by new ethanolamine phosphate groups (Gaynor *et al.*,1999; Imhof *et al.*, 2000; Richard *et al.*, 2002). GPI-proteins are rich in Ser and Thr residues, and are highly glycosylated.

The GPI anchor is transferred at the C-terminal residue of the proteins by a transpeptidase reaction (Orlean, 1997). This reaction involves a GPI:protein transamidase that recognizes the above described motif, cleaves a peptide bond, and binds the GPI moiety to the protein through an amide linkage (Hamburger *et al.*, 1995). This process occurs during the transit of the proteins in the ER (a process dependent on sphingolipids) (Skrzypek *et al.*, 1997; Meyer *et al.*, 2000), suggesting that GPI-proteins must possess a signal peptide. The N-terminal third of the GPI-proteins contains the ligand-binding domain, the GPI anchor being essential for cross-linking in the wall (Frieman *et al.*, 2002). The GPI-proteins are linked covalently through the GPI remnant, [a mannose reducing end (Fujii *et al.*, 1999)] to a non-reducing end of the β-1,6-glucan, and then to a non-reducing end of β-1,3-glucan (Klis *et al.*, 2002).

Wall GPI proteins are secreted at the growing region of the cell (Roberg *et al.*, 1997), suggesting that they are transferred in secretory vesicles. Synthesis of GPI-proteins is essential for all eukaryotes (Gaynor *et al.*, 1999), including fungi (Orlean, 1997). Under limited conditions of synthesis, growth is reduced with formation of aberrant cells.

7. Alterations in the Coordinate Synthesis of the Cell Wall

Synthesis of the cell wall can be affected by different kinds of inhibitors that block several processes involved in its formation, including, synthesis of structural polysaccharides, synthesis of glycoproteins, lipids, or other minor components, operation of the secretory pathway, as well as other processes. The effect of these inhibitors is therefore extremely variable. These inhibitors can be important on one hand as tools for the analysis of specific steps in the synthesis of the wall, and to establish the role of certain components, and, on the other, to control fungal diseases.

Cycloheximide inhibits the synthesis of the protein moiety of glycoproteins but does not stop incorporation of glucans into the wall (Elorza *et al.,* 1985). Tunicamycin, an inhibitor of N-acetylglucosaminyl transferase, as well as the metabolic analogue 2-deoxy-D-glucose, inhibit the formation of the glycosidic moiety of glycoproteins.

Aculeacins (Mizoguchi *et al.,* 1977), echinocandins and papulacandins (Baguley *et al.,* 1979) are groups of natural substances that inhibit synthesis of β-1,3-glucans (reviewed by Douglas, 2001), but also have some collateral effects on the cell, such as a quantitative alterations in membrane sterols, and wall chitin. Chitin biosynthesis is inhibited by a large number of substances, of which polyoxins and nikkomycins are the most active ones (reviewed by Ruiz-Herrera and San-Blas, 2003).

A different way by which synthesis of wall components may be affected is by mutation of different genes. There are large numbers of genes involved in the formation of the cell wall subject to extremely complex regulation pathways, and compensatory mechanisms exist to cope with their mutations. In *S. cerevisiae* a decrease in the synthesis of any wall component gives rise to a significant increase in the levels of chitin (Popolo *et al.,* 1997). This is accompanied by an increased bonding of the polymer to β-1,6-glucan and GPI-proteins up to 20 times their normal levels (Kapteyn *et al.,* 1997; Lipke and Ovalle, 1998; Ram *et al.,* 1998). In turn, a defect in the amounts of β-1,6-glucan induced an increase in the amounts of β-1,3-glucan-bound Pir proteins and chitin (Kapteyn *et al.,* 1999). Mutants deficient in *O*-glycosylation (*pmt* mutants) or *N*-glycosylation (*mnn* mutants) display twice the normal levels of chitin (Gaughran *et al.,* 1994; Gentzsch and Tanner, 1996). Such mutants also displayed an elevation in the relative amounts of Pir and GPI proteins (Kapteyn *et al.,* 2000). In *C. glabrata* (Nagahashi *et al.,* 1998), and *C. albicans* (Popolo and Vai, 1998) mutants affected in β-1,6-glucan synthesis contained levels of chitin higher than normal. Thermosensitive *S. cerevisiae rho1* mutants displayed lytic defects at the bud apex at the non-permissive temperature (Kamada *et al.,* 1996; Qadota *et al.,* 1996), and decreased amounts of β-1,3-glucan, β-1,6-glucan and manoproteins (Roh *et al.,* 2002).

The composition and structure of the fungal wall can also change in response to either internal or external stimuli. Noticeable are the changes observed when fungi enter into the stationary phase. The wall becomes thicker (Northcote and Horne, 1952; Valentin *et al.,* 1987; De Nobel *et al.,* 1990), more resistant to β-glucanases (De Nobel *et al.,* 1990; 2000), and less permeable to macromolecules (De Nobel *et al.,* 1990). Apparently these phenomena are due to deposition of new mannoproteins, and their almost negligible turnover (Valentín *et al.,* 1987). Other changes are an increase in the levels of mannoprotein phosphorylation (Odani *et al.,* 1997), and a 6-7 fold increase in the amount of disulfide-bound proteins. Apparently these changes are not accompanied by an increase in the cross-linking of the wall polymers (Smith *et al.,* 2000). Similarly chitin content increases in older hyphae (Kopecek and Raclavsky, 1999).

8. Post-Synthetic Reactions for the Maturation of the Cell Wall

8.1. Establishment of Associations Among Wall Components

Organization of the cell wall requires the connections among its different components. Cell wall components associate among themselves not only by hydrogen, hydrophobic or polar bonding, but also by covalent bonds. In filamentous fungi, the newly synthesized and the mature walls have different mechanical and structural properties. Thus, the apical wall is plastic whereas the mature wall is viscoelastic, the apical zone is more sensitive to lytic enzymes, nascent chitin is more susceptible to chitinases and deacetylases than the mature one, β-glucans in the newly deposited wall are alkali-soluble, whereas a certain amount becomes alkali-resistant in the mature wall, Calcofluor white and Congo red interfere with chitin crystallization (for a review of the older literature see Ruiz-Herrera, 1992).

Some changes are due to crystallization of the structural polysaccharides, since synthesis and microfibril formation are temporally separated processes (see Ruiz-Herrera, 1992). But other modifications are due to the establishment of covalent bonds among the different wall components. Chitin and glucan are covalently bound in the mature wall, accounting for the alkali-insolubility of a certain amount of β-glucans. In *Schizophyllum commune*, the existence of a bond between glucan and chitin through a short amino acid chain was suggested (Sietsma and Wessels, 1979), but in *C. albicans* (Surarit *et al.,* 1988) and *S. cerevisiae* (Kollar *et al.,* 1995), a glycosidic linkage between β-1,3-glucan and chitin was described.

Taking into consideration the large alterations in the cell wall that occur when its synthesis is affected, it has been suggested that in *S. cerevisiae* β-1,6-glucan is the key molecule to maintain an association with β-1,3-glucan, chitin and glycoproteins (Kollar *et al.*, 1997). By use of inhibitors it has been suggest that the order of association of polymers in the wall is the following: β-1,3-glucan, β-1,6-glucan, manoproteins, and chitin (Roh *et al.,* 2002). Pir and GPI proteins become associated with β-1,3-glucans and β-1,6-glucans respectively, and the family 72 of transglycosidases may function in the cross-linking of β-glucans in the cell wall (Fonzi, 1999; Mouyna *et al.*, 2000; Popolo *et al.*, 2001). Association of proteins among themselves may occur through the formation of disulfide bonds, a reaction that must occur in the wall itself by a redox reaction that is not fully understood.

9. Cell Wall Growth

9.1. Introduction

The growth of a fungal cell depends on the expansion of the wall which "surrounds" the protoplast. Knowledge of several features is necessary in order to explain cell wall, and therefore fungal growth, including: 1) the mechanisms that explain cell polarization; 2) the mechanism through which precursors and enzymes involved in wall formation reach the cell surface; 3) how cell wall expansion takes place; and 4) how the different components of the wall interact to give rise to a coherent structure, and not a mere amorphous aggregate. Established concepts may be consulted in Ruiz-Herrera (1992). We summarize some general conclusions and describe recent information below.

9.2. Polarization of Cell Growth

Hyphal cells characteristically grow in a polarized way, contrary to yeasts, where most of their expansion occurs isodiametrically. Hyphal growth occurs at the apex and depends on special characteristics of the hyphal tip, a zone where the wall is extended (Ortega *et al.*, 1975), where wall biosynthetic enzymes accumulate (Herrera-Estrella and Ruiz-Herrera, 1983), and where wall structural polymers are deposited (Bartnicki-Garcia and Lippman, 1969).

The lack of an ideal model to study all aspects involved in polarization is unfortunate. The most studied model, *S. cerevisiae,* has obvious genetic and genomic advantages, but its polarized growth is extremely limited. Different aspects of cell polarity in this organism have been the subjects of several important reviews (Drubin and Nelson, 1996; Pruyne and Bretscher, 2000a; 2000b; Casamayor and Snyder, 2002). It is known that polarization in yeast is tightly associated with the cell cycle, and that synthesis of 50 % of the wall proteins is cycle-regulated (Spellman *et al.*, 1998). Pir1-3 proteins are incorporated into the wall during G_1, whereas Pir4 is synthesized mostly during G_2 (Moukadiri *et al.*, 1999). On the other hand, proteins involved in cell separation, such as chitinase Cts and the glucanases Scw1, Dse2 and Dsc4 are synthesized at the end of the division process (Capellaro *et al.*, 1998; Colman-Lerner *et al.*, 2001). Two patterns of budding are displayed by *S. cerevisiae* since haploid cells undergo axial budding, whereas diploid cells show a bipolar budding pattern. During most of the G_1 stage, haploid yeast cells grow in a quasi-isodiametric (isotropic) pattern, but at its end, at the START point, a site for budding is selected and growth becomes polarized giving rise to a bud. Growth of the mother cell stops and the bud initially displays a polarized pattern of growth that becomes depolarized when the bud increases in size. Finally, growth becomes concentrated for the formation of the septum that separates mother and daughter cells. Polarity is established and maintained in shmoo formation during cell mating, and also during pseudo-filamentous growth. In contrast to *S. cerevisiae*, hyphal growth in *C. albicans* is independent of the cell cycle (Hazan *et al.*, 2002).

Polarization is a complex phenomenon involving the interplay of large numbers of gene products (Pruyne and Bretscher, 2000a), but it is necessary to point out the outstanding importance of the small GTPases of the Rho class. Accordingly, a large amount of evidence indicates the essential role of the Rho-type GTPase Cdc42 in the establishment of cell polarity (see Pruyne and Bretscher, 2000a; Casamayor and Snyder, 2002, for details). Polarization is initiated by the recruitment of Cdc42, which is more or less randomly distributed on the cell surface during G_1, to the site of bud emergence, and its activation by the guanosine-nucleotide-exchange factor (GEF) Cdc24p. Association of these elements with different components provides specificity to the aggregates involved at each step: initiation, and maintenance of polarity (Pruyne and Bretscher, 2000a; Casamayor and Snyder, 2002). In agreement with these results, data obtained with *A. nidulans* temperature-sensitive mutants suggest that different gene products are involved in the establishment of polarity and its maintenance (Momany *et al.*, 1999).

In the dimorphic human pathogen *Penicillium marneffei*, the Cdc42 homolog cflA is also involved in polarization in the hyphal and yeast-like phases, but not during the yeast-to-hypha transition (Boyce *et al.*, 2001; see San-Blas and Niño-Vega's Chapter 5 of this book for more details on this and other pathogenic fungi). Similarly, the homologues of Cdc42 and Cdc24 are required for polarization in *Ashbya gossypii* (Wendland and Phillippsen, 2001). These data suggest that possibly the polarization mechanisms in mycelial fungi are basically similar to yeast. Interestingly, Cdc42 apparently is also required for polarized secretion in yeast (Adamo *et al.*, 2001).

The polarizing aggregates interact with actin patches and actin bundles to direct the movement of the elements responsible for cell growth to specific points of the cell surface. Myo2p, a class V myosin, is the motor involved in the transport of vesicles in yeast, using as guidance the actin cables of the cytoskeleton (reviewed by Pruyne and Bretscher, 2000b). Similarly, in the plant pathogen *U. maydis*, observation on the behavior of the cytoskeleton during bud morphogenesis revealed that an actin cytoskeleton, constituted as in *S. cerevisiae* by patches and bundles, polarizes the bud site and is responsible for the secretion of wall components (Banuett and Herskowitz, 2002). In *C. albicans* the only existing type myosin I is absolutely required for hypha formation (Oberholzer *et al.*, 2002). In this process myosin I becomes phosphorylated, a reaction necessary for the normal location of actin patches and bundles.

A role for actin in polarization and mobilization of vesicles to the growing zone is well documented also in mycelial fungi (*e. g.* Torralba *et al.*, 1998; Harris *et al.*, 1999; for a discussion see Bartnicki-Garcia, 2003). Interestingly, and in contrast to yeasts, a functional microtubular cytoskeleton appears to be required for normal hyphal growth (Inoue *et al.*, 1998; Riquelme *et al.*, 2000). In hyphal fungi, in common with other apically growing organisms or cells, there exits a tip-high gradient of calcium that is intimately related to apical growth (Jackson and Heath, 1993). Alterations in this gradient by different agents perturb or even stop mycelial growth. Also ionic gradients, driven mainly by Ca^{2+} and H^{+} (McGillivray and Gow, 1986; 1987; reviewed by Ruiz-Herrera, 1992; Gooday, 1994; Gow, 1994), appear to play an important role in the polarization of fungal growth, but the mechanisms by which these gradients originate, and are involved in apical growth are unknown (Torralba and Heath, 2001).

9.3. Mechanism of Externalization of Wall Components

It is accepted that wall components are transported through the exocytic route (Ruiz-Herrera, 1992; Sietsma and Wessels, 1994; Bartnicki-Garcia, 1996; Harold, 1997; Bartnicki-Garcia, 2003). A simple scheme on how this operates in fungi is as follows:

ER → Golgi → Vesicles → Spitzenkörper →Apical vesicles, Microvesicles → Cell Wall

The final step in secretion involves the participation of an organelle characteristic of mycelial fungi, the Spitzenkörper. Originally described as a basophilic body in the apical region of *Coprinus* species by Brunswick in 1924, its role in growth was firmly established in the middle of the 20th century by the studies of Girbardt, who used phase contrast, time-lapse photomicrography, and electron microscopy (Girbardt, 1955; 1957; 1969). These results have been later confirmed, mainly by experiments with video microscopy and image analysis, where the Spitzenkörper was manipulated with the help of laser "tweezers" (Bracker *et al.*, 1997).

The Spitzenkörper is composed of an accumulation of vesicles and microvesicles, microtubules, actin microfilaments and some unidentified components (reviewed by Bartnicki-Garcia, 2003). It is thought to operate as a structure that directs apical vesicles and microvesicles in their final journey to the cell surface. By use of computer simulation studies it was suggested that hyphal shape in fungi was reached through a random fusion with the plasmalemma of vesicles and microvesicles that originated at a specific point, the so-called vesicle supply center (VSC). Mathematical analysis of the data, and comparison with electron micrographs of median sections of different fungi, led to the suggestion that the VSC corresponded to the Spitzenkörper (Bartnicki-Garcia *et al.*, 1989; Bartnicki-Garcia, 2003; see also Chapter 7 of this volume). The origin of the Spitzenkörper remains unknown, although it is hypothesized that it forms *de novo* by the accumulation of vesicles perhaps around nucleation elements. It is important to indicate that the shape of the Spitzenkörper is different in the several fungal groups, and that in some organisms like yeasts, it has not been observed.

Ideas on the precise function of the Spitzenkörper are still hypothetical, because of the lack of information regarding its exact composition and enzymatic activities. It has been suggested that it may be involved in the maturation of vesicles or in their association with actin filaments. The observation that the fluorescent dye FM4-64 is internalized and labels both endosomes and the Spitzenkörper before vacuolar membranes and mitochondria (Fischer-Parton *et al.*, 2000) suggests that the Spitzenkörper is also involved in vesicle recycling.

A role of the secretory pathway in the mobilization of enzymes involved in the synthesis of wall polysaccharides has also been demonstrated. We described above the mechanism of association of glycoproteins to β-1,6-glucan at Golgi, and the probable initiation of the synthesis of this polysaccharide at this stage. In the case of chitin, specialized microvesicles, also referred to as chitosomes, are responsible for chitin synthase mobilization to the cell surface (Bracker *et al.*, 1976; Ruiz-Herrera, 1984). The process, similar to that occurring for other vesicles, has been studied mostly in *S. cerevisiae*. It involves the operation of a large number of gene products, among others those encoded by genes *CHS4* to *CHS7* (see Roncero *et al.*, 2001). Chs5p is a Golgi protein involved in the transport of Chs3p to the cell periphery (Santos and Snyder, 1997). This protein apparently also plays a role in the attachment of Chs3 to the neck of budding cells by an interaction with Bin4 (DeMarini *et al.*, 1997), and the protein phosphatase Glc7 (Kozubowski *et al.*, 2003). Two additional gene products Sbe2p and Sbe22p are probably involved in the transport of cell wall components from the Golgi apparatus to the cell surface periphery in a pathway independent of Chs5p (Santos and Snyder, 2000). Chs6 is also involved in the correct mobilization of Chs3 (Ziman *et al.*, 1998), whereas Chs7p, an ER-resident protein, has been suggested to operate as a chaperone for Chs3 (Trilla *et al.*, 1999).

9.4. Cell Wall Expansion

Cell wall expansion occurs by the constant addition of building materials to a fabric that stretches continuously. In general, we may consider two mechanisms for the incorporation of new materials into the growing wall: apposition, by which new layers of cell wall components are added at the inner surface of the expanding wall; and intussusception through which new material is intercalated into the wall. The ingenious experiments of Castle (1940) with the giant sporangiophore of *P. blakesleeanus* produced convincing evidence that the wall, at least in this structure and organism, grows by apposition.

Expansion of the cell wall in fungi most probably depends on turgor pressure (for a discussion see Bartnicki-Garcia, 2003). The two most striking pieces of evidence in favor of the role of turgor pressure in cell expansion are experiments made with *P. blakesleeanus* sporangiophores (Ortega *et al.*, 1989), and three-dimensional analysis of video imaging of *Rhizoctonia solani* (Bartnicki-Garcia *et al.*, 2000). In the first study, when inert silicon was injected into the vacuole of the sporangiophore, the growth rate was a strict function of turgor pressure. In the second one it was demonstrated that expansion of the wall surface at the hyphal apex followed an orthogonal pattern, a property that agrees only with a uniform driving force.

Different models have been proposed to explain how a rigid wall can expand, but in general they can be reduced to two basic mechanisms: (1) the nascent wall is not rigid, but plastic (see Sietsma and Wessels, 1994); and, (2) lytic enzymes transform the rigid structure into a plastic one (see Bartnicki-Garcia, 2003). Both hypotheses have compelling arguments suggesting that are probably not contradictory or absolute, but quantitative, depending on the particular conditions (see Ruiz-Herrera, 1992 for a discussion). Lytic enzymes would be more important in branching, bud initiation, re-initiation of growth of a hypha, etc, and less or not necessary when a plastic cell wall is growing continuously by the constant deposition of its components at the hyphal apex.

A point not resolved, and which remains an elusive problem, is how the shape of a fungal hypha is attained (independently of the modifications brought about by the Spitzenkörper behavior, see above). The model described above on the role of a VSC that randomly distributes vesicles ("growth units") (Bartnicki-Garcia *et al.*, 1989) is probably the best, but the authors accept the fact that it is based on some aprioristic simplifications. It is possible that the answer to the question is not biological, but physical. Accordingly, it was suggested recently that the shape of filamentous organisms including actinomycetes and fungi can be described in terms of self-similar growth at the apex, driven by pressure and sustained by a constant process of wall formation (Gloriely and Tabor, 2003; see also Chapter 7).

9.5. Regulatory Mechanisms that Affect Cell Wall Structure and Composition

The cell wall is a coherent structure, made by the interaction of a great number of components. Considering the large number of gene products involved in wall synthesis, the existence of an exquisite regulation of their interactions is expected. Regulation may be brought about at the transcriptional or at several post-transcriptional levels, in response to external stimuli, This regulation would be necessary to cope with the changing conditions, or to produce specific morphogenetic responses. Compensation mechanisms for the failure or absence of certain gene product(s) also have been mentioned above. In recent years interest has been developed on the regulatory mechanisms that permit the cells to respond to stress and to conditions demanding alterations in the cell wall composition and structure (see review by Popolo *et al.*, 2001). These mechanisms operate through MAPK cascades, as occurs with other signaling pathways (for a review of MAPK cascades in fungi see Banuett, 1998). Stress affecting cell wall composition depends on at least two MAP kinases: Mpk1 (Slt2), and Kss1. Their role is to phosphorylate different proteins, noticeably transcription factors at the end of the signal transduction chains, and provoke changes in cell wall structure and composition. The exact signal that activates the pathways involved in cell wall integrity or how it is perceived are not clear. In the most often studied Mpk1 pathway, two alternative possibilities have been proposed: (1) that a debilitated cell wall provokes a hypotonic stress, or the activation of specific receptors. It was shown that hypotonic stress gave rise to the activation of a calcium channel susceptible to stretching (Batiza *et al.*, 1996) that could initiate the transduction cascade; (2) alternatively, some data point to a role for the integral transmembrane proteins Mid2 and Slg1 (similar to the integrins which are present in the extracellular matrix of animal cells and are able to activate signalling cascades) (see Ketela *et al.*, 1999). Independent of which is the true receptor mechanism, the resulting effect is the activation of the MAPKKK Pkc1 that initiates a signal transduction cascade. Pkc1 activation is itself regulated by the small GTPase Rho1, which as described before, is involved in different aspects of cell polarization.

At the end of the Pkc1 cascade, activation occurs of the Rlm1 transcription factor, which is responsible in turn for the activation of at least 25 genes involved in the biosynthesis of the cell wall (Jung *et al.*, 2002), including the enzymes responsible for the synthesis of structural polysaccharides, Fks1p and Chs3p (Igual *et al.*, 1996). Cell damage such as observed in *gas1* (Popolo *et al.*, 1997) or *fks1* (Garcia-Rodriguez *et al.* 2000) mutants activates this pathway. In these mutants, *FKS2* and *CHS3* genes become up-regulated, and an increase of chitin levels in the wall occurs, a phenomenon that compensates for the structural alterations. This compensation phenomena apparently utilizes the normal secretion pathway and requires the operation of the *CHSP4-7* gene products (Carotti *et al.*, 2002).

Recent findings suggest that acting in parallel with the Pkc1-dependent pathway, regulation of cell wall integrity also involves a signalling pathway dependent on phosphoinositides. Accordingly, the phosphoinositide phosphatase Sac1 was suggested to play an essential role in lipid-based signal transduction and in secretion, acting as an antagonist of the phosphatidylinositol 4-kinase Pik1p in Golgi trafficking. Mutations in *SAC1* led to an increased transport of chitin synthases and cell wall defects (Schorr *et al.*, 2001). Agreeing with this concept, the yeast protein kinases Pkh1 and Pkh2 (that are functionally redundant homologs of the mammalian protein kinase, phosphoinositide-dependent protein kinase-1) together with their substrate proteins Ypk1 and Ykr2, are involved in this pathway (Roelants *et al.*, 2002).

10. Acknowledgements

Original work from the Irapuato laboratory was partially supported by Consejo Nacional de Ciencia y Tecnología, México. Work in the laboratory of Valencia was supported in part by grants from the European Union (QLK2CT 2000-00795) and Spanish Ministerio de Ciencia y Cultura (BMC2000-1449).

References

Abe, M., Nishida, I., Minemura, M., Qadota, H., Seyama, Y., Watanabe, T., and Ohya, Y. 2001. Yeast β-1,3-glucan synthase activity is inhibited by phytosphingosine localized to the endoplasmic reticulum. J. Biol. Chem. 276: 26923-26930.

Adamo, J.E., Moskow, J.J., Gladfelter, A.S., Viterbo, D., Lew, D.J., and Brenwald, P.J. 2001. Yeast Cdc42 functions as a late step in exocytosis, specifically during polarized growth of the emerging bud. J. Cell. Biol. 155: 581-592.

Alfonso, C., Nuero, O.M., Santamaria, F., and Reyes, F. 1995. Purification of a heat-stable chitin deacetylase from *Aspergillus nidulans* and its role in cell wall degradation. Curr. Microbiol. 30: 49-54.

Allan, B.B., and Balch, W.E. 1999. Protein sorting by directed maturation of Golgi compartments. Science 285: 63-66.

Alonso, M.D., Lomako, J., Lomako, W.M., and Whelan, W.J. 1995. A new look at the biogenesis of glycogen. FASEB J. 9: 1126-1137.

Alvarado, E., Nukada, T., Ogawa, T., and Ballou, C.E. 1991. Conformation of the glucotriose unit in the lipid-linked oligosaccharide precursor for protein glycosylation. Biochemistry 30: 881-886.

Amnuaykanjanasin, A., and Epstein, L. 2003. A class V chitin synthase gene, *chsA* is essential for conidial and hyphal wall strength in the fungus *Colletotrichum graminicola* (*Glomerella graminicola*). Fungal Genet. Biol. 38: 272-285.

Andaluz, E., Guillen, A., and Larriba, G. 1986. Preliminary evidence for a glucan acceptor in the yeast *Candida albicans*. Biochem. J. 240: 495-502.

Andaluz, E., Guillen, A., and Larriba, G. 1986. Preliminary evidence for a glucan acceptor in the yeast *Candida albicans*. Biochem. J. 240: 495-502.

Andaluz, E., Ridruejo, J.C., Ramirez, M. Ruiz-Herrera, J., and Larriba, G. 1988. Initiation of glucan synthesis in yeast. FEMS Microbiol. Lett. 49: 251-255.

Araki, Y., and Ito, E. 1974. A pathway of chitosan formation in *Mucor rouxii*: enzymatic deacetylation of chitin. Biochem. Biophys. Res. Commun. 56: 669-675.

Arellano, M., Coll PM., and Perez, P. 1999. Rho GTPases in the control of cell morphology, cell polarity, and actin localization in fission yeast. Microsc. Res. Tech. 47: 51-60.

Awald, P.D., Frost, D., Drake, R.R., and Selitrennikoff, C.P. 1994. β-1,3-Glucan synthase activity of *Neurospora crassa*: identification of a substrate-binding protein. Biochim. Biophys. Acta. 1201: 312-320.

Azuma, M., Levinson, J.N., Page, N., and Bussey, H. 2002. *Saccharomyces cerevisiae* Big1p, a putative endoplasmic reticulum membrane protein required for normal levels of cell wall β-1,6-glucan. Yeast 19: 783-793.

Baguley, B.C., Rommele, G., Grunner, J., and Wehrli, W. 1979. Papulacandin B: an inhibitor of glucan synthesis in yeast spheroplasts. Eur. J. Biochem. 97: 345-351.

Balestrini, R., Mainieri, D., Soragni, E., Garnero, L., Rollino, S., Viotti, A., Ottonello, S., and Bonfante, P. 2000. Differential expression of chitin synthase III and IV mRNAs in ascomata of *Tuber borchii* Vittad. Fungal Genet. Biol. 31: 219-32.

Ballou, C.E. 1982. Yeast cell wall and cell surface. In: The Molecular Biology of the yeast *Saccharomyces*: Metabolism and Gene Expression. J.N. Strather, ed. Cold Spring Harbor Laboratory. New York. p. 335-360.

Ballou, C.E. 1990. Isolation, characterization, and properties of *Saccharomyces cerevisiae mnn* mutants with nonconditional protein glycosylation defects. Meth. Enzymol.185: 440-470.

Ballou, L., Hitzeman, R.A., Lewis, M.S., and Ballou, C.E. 1991. Vanadate-resistant yeast mutants are defective in protein glycosylation. Proc. Natl. Acad. Sci. U.S.A. 88: 3209-3213.

Banuett, F. 1998. Signalling in the yeasts: an informational cascade with links to the filmentous fungi. Microbiol. Mol. Biol. Rev. 62: 249-274.

Banuett, F., and Herskowitz, I. 2002. Bud morphogenesis and the actin and microtubule cytoskeletons during budding in the corn smut fungus, *Ustilago maydis*. Fungal Genet. Biol. 37: 149-170.

Bartnicki-Garcia, S. 1996. The hypha: unifying thread of the fungal kingdom. In: A Century of Mycology. B. Sutton, ed. Cambridge University Press, Cambridge. p. 105-133.

Bartnicki-Garcia, S. 2003. Hyphal tip growth: outstanding questions. In: Molecular Biology of Fungal Development. H.D. Osiewacz,. ed. Marcel Dekker, New York. p. 29-58.

Bartnicki-Garcia, S., Bracker, C.E., Gierz, R., Lopez-Franco, R. and Lu, H. 2000. Mapping the growth of fungal hyphae: orthogonal cell wall expansion during tip growth and the role of turgor. Biophys. J. 79: 2382-2390.

Bartnicki-Garcia, S., Hergert, F., and Gierz, G. 1989. Computer simulation of fungal morphogenesis and the mathematical basis for hyphal tip growth. Protoplasma 153: 46-57.

Bartnicki-Garcia, S., and Lippman, E. 1969. Fungal morphogenesis: cell wall construction in *Mucor rouxii*. Science 165: 302-304.

Bartnicki-Garcia, S., Ruiz-Herrera, J., and Bracker, C.E. 1979. Chitosomes and chitin synthesis. In: Fungal Walls and Hyphal Growth. J.H. Burnett and A.P.J. Trinci, eds. Cambridge University Press, Cambridge. p. 149-168.

Batiza, A.F., Schulz, T., and Masson, P.H. 1996. Yeast respond to hypotonic shock with a calcium pulse. J. Biol. Chem. 271: 23357-23362.

Beauvais, A., Bruneau, J.M., Mol, P.C., Buitrago, M.J., Legrand, R., and Latge, J.P. 2001. Glucan synthase complex of *Aspergillus fumigatus*. J. Bacteriol. 183: 2273-2279.

Berbee, M.L., and Taylor, J.W. 1992. Detecting morphological convergence in true fungi, using 18S rRNA gene sequence data. Biosystems 28: 117-125.

Beth-Din, A., and Yarden, O. 2000. The *Neurospora crassa chs3* gene encodes an essential Class I chitin synthase. Mycologia 92: 65-73.

Boisrame, A., Chasles, M., Babour, A., Beckerich, J.M., and Gaillardin, C. 2002. Sbh1p, a subunit of the Sec61 translocon, interacts with the chaperone calnexin in the yeast *Yarrowia lipolytica*. Cell Sci. 115: 4947-4956.

Boone, C., Sommer, S.S., Hensel, A., and Bussey, H. 1990. Yeast KRE genes provide evidence for a pathway of cell wall β-glucan assembly. J. Cell. Biol. 110: 1833-1843.

Bowen, A.R., Chen-Wu, J.L., Momany, M., Young, R., Szaniszlo, P.J., and Robbins, P.W. 1992. Classification of fungal chitin synthases. Proc. Natl. Acad. Sci. U.S.A. 89: 519-523.

Boyce, K.J., Hynes, M.J., and Andrianopoulos, A. 2001. The CDC42 homolog of the dimorphic fungus *Penicillium marneffei* is required for correct cell polarization during growth but not development. J. Bacteriol. 183: 3447-3457.

Bracker, C.E., Murphy, D.J., and López-Franco, L. 1997. Laser microbeam manipulation of cell morphogenesisi in growing fungal hyphae. In: Functionl Imaging and Optical Manipulation of Living Cells. D.L. Farkas and D.J.Tromberg, eds. Bellingam W. A.: SPIE-The International Society for Optical Engineering (Proceedings of SPIE) Vol. 2983, p. 67-80.

Bracker, C.E., Ruiz-Herrera, J., and Bartnicki-Garcia, S. 1976. Structure and transformation of chitin synthetase particles (chitosomes) during microfibril synthesis *in vitro*. Proc. Natl. Acad. Sci. U.S.A. 73: 4570-4574.

Brown, J.L., and Bussey, H. 1993. The yeast KRE9 gene encodes an O glycoprotein involved in cell surface β-glucan assembly. Mol. Cell. Biol. 13: 6346-6356.

Brown, J.L., Kossaczka, Z., Jiang, B., and Bussey, H. 1993. A mutational analysis of killer toxin resistance in *Saccharomyces cerevisiae* identifies new genes involved in cell wall β-1,6-glucan synthesis. Genetics 133: 837-849.

Bulawa, C.E. 1992. *CSD2, CSD3*, and *CSD4* genes for chitin synthesis in *Saccharomyces cerevisiae*: the *CSD* gene product is related to chitin synthases and to developmentally regulated proteins of *Rhizobium* species and *Xenopus levis*. Mol. Cell. Biol. 12: 1764-1776.

Bulawa, C.E., and Osmond, B.C. 1990. Chitin synthase I and chitin synthase II are not required for chitin synthesis *in vivo* in *Saccharomyces cerevisiae*. Proc. Natl. Acad. Sci. U.S.A. 87: 7424-7428.

Bulawa, C.E., Slater, M., Cabib, E., Au-Young, J., Sburlati, A., Adair, W.L., and Robbins, P.W. 1986. The *Saccharomyces cerevisiae* structural gene for chitin synthase is not required for chitin synthesis *in vivo*. Cell 46: 213-225.

Burda, P., and Aebi, M. 1999. The dolichol pathway of N-linked glycosylation. Biochim. Biophys. Acta 1426: 239-257.

Burda, P., Padilla, S.M., Sarkar, S., and Emr, S.D. 2002. Retromer function in endosome-to-Golgi retrograde transport is regulated by the yeast Vps34 PtdIns 3-kinase. J. Cell. Sci. 115: 3889-3900.

Bussey, H., Saville, D., Hutchins, K., and Palfree, R.G. 1979. Binding of yeast killer toxin to a cell wall receptor on sensitive *Saccharomyces cerevisiae*. J. Bacteriol. 140: 888-892.

Calonge, T.M., Arellano, M., Coll., P.M., and Perez, P. 2003. PRga5p is a specific Rho1p GTPase-activating protein that regulates cell integrity in *Schizosaccharomyces pombe*. Mol. Microbiol. 47: 507-518.

Calonge, T.M., Nakano, K., Arellano, M., Arai, R., Katayama, S., Toda, T., Mabuchi, I., and Perez, P. 2000. *Schizosaccharomyces pombe* rho2p GTPase regulates cell wall α-glucan biosynthesis through the protein kinase pck2p. Mol. Biol. Cell. 11: 4393-4401.

Calvo-Mendez, C., and Ruiz-Herrera, J. 1987. Biosynthesis of chitosan in membrane fractions from *Mucor rouxii* by the concerted action of chitin synthetase and a particulate deacetylase. Exp. Mycol. 11: 128-140.

Campbell, J.A., Davies, G.J., Bulone, V., and Henrissat, B. 1997. A classification of nucleotide-diphospho-sugar glycosyltransferases based on amino acid sequence similarities. Biochem. J. 326: 929-939.

Cappellaro, C., Mrsa, V., and Taner,W. 1998. New potential cell wall glucanases of *Saccharomyces cerevisiae* and their involvement in mating. J. Bacteriol. 180: 5030-5037.

Caro, L.H., Tettelin, H., Vossen, J.H., Ram, A.F., van den Ende, H., and Klis, F.M. 1997. *In silico* identification of glycosyl-phosphatidylinositol-anchored plasma-membrane and cell wall proteins of *Saccharomyces cerevisiae*. Yeast 13: 1477-1489.

Carotti, C., Ferrario, L., Roncero, C., Valdivieso, M.H., Duran, A., and Popolo, L. 2002. Maintenance of cell integrity in the *gas1* mutant of *Saccharomyces cerevisiae* requires the Chs3p-targeting and activation pathway and involves an unusual Chs3p localization. Yeast 19: 1113-1124.

Casamayor, A., and Snyder, M. 2002. Bud-site selection and cell polarity in budding yeast. Curr. Opin. Microbiol. 5: 179-186.

Castle, E.S. 1940. Discontinuous growth of single plant cells measured at short intervals and the theory of intussusception. J. Cell. Comp. Physiol. 15: 285-297.

Castro, C., Ribas, J.C., Valdivieso, M.H., Varona, R., del Rey, F., and Duran, A. 1995. A papulacandin B resistance in budding and fission yeasts: isolation and characterization of a gene involved in β-1,3-D-glucan synthesis in *Saccharomyces cerevisiae*. J. Bacteriol. 177: 5732-5739.

Chang, Y.C., and Penoyer, L.A. 2000. Properties of various Rho1 mutant alleles of *Cryptococcus neoformans*. J. Bacteriol. 182: 4987-4991.

Chapman, R.E., and Munro, S. 1994. The functioning of the yeast Golgi apparatus requires an ER protein encoded by *ANP1*, a member of a new family of genes affecting the secretory pathway. EMBO J. 13: 4896-4907.

Chen-Wu, J.L., Zwicker, J., Bowen, A.R., and Robbins, P.W. 1992. Expression of chitin synthase genes during yeast and hyphal growth phases of *Candida albicans*. Mol. Microbiol. 6: 497-502.

Christodoulidou, A., Bouriotis, V., and Thireos, G. 1996. Two sporulation-specific chitin deacetylase-encoding genes are required for the ascospore wall rigidity of *Saccharomyces cerevisiae*. J. Biol. Chem. 271: 31420-31425.

Cohen, E. 2001. Chitin synthesis and inhibition: a revisit. Pest Manag. Sci. 57: 946-950.

Colman-Lerner, A., Chin, T.E., and Brent, R. 2001. Yeast Cbk1 and Mob2 activate daughter-specific genetic programs to induce asymmetric cell fates. Cell 107: 739-750.

Conzelmann, A., Riezman, H., Desponds, C., and Bron, C.A. 1988. Major 125-kd membrane glycoprotein of *Saccharomyces cerevisiae* is attached to the lipid bilayer through an inositol-containing phospholipid. EMBO J. 7: 2233-2240.

Cutler, J.E. 2001. N-glycosylation of yeast, with emphasis on *Candida albicans*. Med. Mycol. 39, Suppl. 1: 75-86.

Davis, L.L., and Bartnicki-Garcia, S. 1984a. Chitosan synthesis by the tandem action of chitin synthetase and chitin deacetylase from *Mucor rouxii*. Biochemistry 23: 1065-1073.

Davis, L.L., and Bartnicki-Garcia, S. 1984b. The coordination of chitosan and chitin synthesis in *Mucor rouxii*. J. Gen. Microbiol. 130: 2095-2102.

De Groot, P.W., Hellingwerf, K.J., and Klis, F.M. 2003. Genome-wide identification of fungal GPI proteins. Yeast 20: 781-796.

De Marini, D.J., Adams, A.E.M., Fares, H., De Virgilio, C., Valle, G., Chuang, J.S., and Pringle, J.R. 1997. A septin-base hierarchy of proteins required for localized deposition of chitin in the *Saccharomyces cerevisiae* cell wall. J. Cell Biol. 139: 75-93.

De Nobel, H., and Lipke, P.N. 1994. Is there a role for GPIs in cell wall assembly in yeast? Trends Cell Biol. 4: 42-45.

De Nobel, H., Klis, F.M., Priem, J., Munnik, T., and Van den Ende, H. 1990. The glucanase-soluble mannoproteins limit cell wall porosity in *Saccharomyces cerevisiae*. Yeast 6: 491-499.

De Nobel, H., Ruiz, C., Martín, H., Morris, W., Brul, S., Molina, M., and Klis, F.M. 2000. Cell wall perturbation in yeast results in dual phosphorylation of the Slt2/Mpk1 MAP kinase and in an Slt2-mediated increase in FKS2-lacZ expression, glucanase resistance and thermotolerance. Microbiology 146: 2121-2132.

Dean, N. 1995. Yeast glycosylation mutants are sensitive to aminoglycosides. Proc. Natl. Acad. Sci. U.S.A. 92: 1287-1291.

Deshpande, M.V., O'Donnell, R., and Gooday, G.W. 1997. Regulation of chitin synthase activity in the dimorphic fungus *Benjaminiella poitrasii* by external osmotic pressure. FEMS Microbiol. Lett. 152: 327-332.

Devlin C and Ballou CE. 1990. Identification and characterization of a gene and protein required for glycosylation in the yeast golgi. Mol. Microbiol. 4: 1993-2001.

Doblin, M.S., De Melis, L., Newbigin, E., Bacic, A., and Read, S.M. 2001. Pollen tubes of *Nicotiana alata* express two genes from different β-glucan synthase families. Plant Physiol. 125; 2040-2052.

Douglas, C.M. 2001. Fungal β(1,3)-D-glucan synthesis. Med. Mycol. 39 (Suppl 1): 55-66.

Douglas, C.M., D'Ippolito, J.A., Shei, G.J., Meinz, M., Onishi, J., Marrinan, J.A., Li, W., Abruzzo, G.K., Flattery, A., Bartizal, K., Mitchell, A., and Kurtz, M.B. 1997. Identification of the FKS1 gene of *Candida albicans* as the essential target of β-1,3-D-glucan synthase inhibitors. Antimicrob. Agents Chemother. 41: 2471-2479.

Douglas, C.M., Foor, F., Marrinan, J.A., Morin, N., Nielsen, J.B., Dahl, A.M., Mazur, P., Baginsky, W., Li, W., El-Sherbeini, M., Clemas, J.A., Mandala, S.M., Frommer, B.R., and Kurtz M.B. 1994. The *Saccharomyces cerevisiae* FKS1 (ETG1) gene encodes an integral membrane protein which is a subunit of β-1,3-D-glucan synthase. Proc. Natl. Acad. Sci. U.S.A. 91: 12907-12911.

Drees, B.L., Sundin, B., Brazeau, E., Caviston, J.P., Chen, G.C., Guo, W., Kozminski, K.G., Lau, M.W., Moskow, J.J., Tong, A., Schenkman, L.R., McKenzie, A. 3rd, Brennwald, P., Longtine, M., Bi, E., Chan, C., Novick, P., Boone, C., Pringle, J.R., Davis, T.N., Fields, S., and Drubin, D.G. 2001. A protein interaction map for cell polarity development. J. Cell. Biol. 154: 549-571.

Drgonova, J., Drgon, T., Tanaka, K., Kollar, R., Chen, G.C., Ford, R.A., Chan, C.S., Takai, Y., and Cabib, E. 1996. Rho1p, a yeast protein at the interface between cell polarization and morphogenesis. Science 272: 277-279.

Drubin, D.G., and Nelson, W J. 1996. Origins of cell polarity. Cell 84: 335-344.

Ecker, M., Mrsa, V., Hagen, I., Deutzmann, R., Strahl, S., and Tanner, W. 2003. O-mannosylation precedes and potentially controls the N-glycosylation of a yeast cell wall glycoprotein. EMBO Rep. 4: 628-632.

El-Sherbeini, M., and Clemas, J.A. 1995. Concomitant nikkomycin Z-supersensitivity of an echinocandin-resistant mutant of *Saccharomyces cerevisiae*. Antimicrob. Agents Chemother. 39: 200-207.

Elorza, M.V., Murgui, A and Sentandreu, R. 1985. Dimorphism in *Candida albicans*: contribution of mannoproteins to the architecture of yeast and mycelial cells. J. Gen. Microbiol. 131: 2209-2216.

Ernst, J.F., and Prill, S.K. 2001. O-glycosylation. Med. Mycol. 39, Suppl. 1: 67-74.

Farkas, I., Hardy, T.A., DePaoli-Roach, A.A., and Roach, P.J. 1990. Isolation of the GSY1 gene encoding yeast glycogen synthase and evidence for the existence of a second gene. J. Biol. Chem. 265: 20879-20886.

Fischer-Parton, S., Parton, R.M., Hickey, P.C., Dijksterhuis, J., Atkinson, H.A., and Read, N.D. 2000. Confocal microscopy of FM4-64 as a tool for analysing endocytosis and vesicle trafficking in living fungal hyphae. J. Microsc. 198: 246-259.

Fonzi, W.A. 1999. *PHR1* and *PHR2* of *Candida albicans* encode putative glycosidases required for proper cross-linking of β-1,3- and β-1,6-glucans. J. Bacteriol. 181: 7070-7079.

Franzusoff, A. 1992. Beauty and the yeast: compartmental organization of the secretory pathway. Semin. Cell Biol. 3: 309-324.

Franzusoff, A., and Schekman, R. 1989. Functional compartments of the yeast Golgi apparatus are defined by the *sec7* mutation. EMBO J. 8: 2695-2702.

Frieman, M.B., McCaffery, J.M., and Cormack, B.P. 2002. Modular domain structure in the *Candida glabrata* adhesin Epa1p, a β-1,6 glucan-cross-linked cell wall protein. Mol. Microbiol. 46: 479-492.

Fujii, T., Shimoi, H., and Iimura, Y. 1999. Structure of the glucan-binding sugar chain of Tip1p, a cell wall protein of *Saccharomyces cerevisiae*. Biochim. Biophys. Acta 1427: 133-144.

Gao, X.D., Katsumoto, T., and Onodera, K. 1995. Purification and characterization of chitin deacetylase from *Absidia coerulea*. J. Biochem. (Tokyo) 117: 257-263.

Garcerá, A., Martinez, A.I., Castillo, L., Elorza, M.V., Sentandreu, R., and Valentín, E. 2003. Identification and study of a *Candida albicans* protein homologous to *Saccharomyces cerevisiae* Ssr1p, an internal cell-wall protein. Microbiology 149: 2243-2250.

Garcia-Rodriguez, L.J., Trilla, J.A., Castro, C., Valdivieso, M.H., Duran, A., and Roncero, C. 2000. Characterization of the chitin biosynthesis process as a compensatory mechanism in the *fks1* mutant of *Saccharomyces cerevisiae*. FEBS Lett. 478: 84-88.

Garcerá-Teruel, A., Xoconostle-Cázares, B., Rosas-Quijano, R., Ortiz, L., León-Ramírez, C., Specht, C.A., Sentandreu, R. and Ruiz-Herrera, J. 2004. Loss of virulence in *Ustilago maydis* mutants affected in the chitin synthase gene Umchs6. Res. Microbiol. In press

Garrett-Engele, P., Moilanen, B., and Cyert, M.S. 1995. Calcineurin, the Ca2+/calmodulin-dependent protein phosphatase, is essential in yeast mutants with cell integrity defects and in mutants that lack a functional vacuolar H(+)-ATPase. Mol. Cell. Biol. 15: 4103-4114.

Gaughran, J.P., Lai, M.H., Kirsch, D.R., and Silverman, S.J. 1994. Nikkomycin Z is a specific inhibitor of *Saccharomyces cerevisiae* chitin synthase isozyme Chs3 *in vitro* and *in vivo*. J. Bacteriol. 176: 5857-5860.

Gaynor, E.C., and Emr, S.D. 1997. COPI-independent anterograde transport: cargo-selective ER to Golgi protein transport in yeast COPI mutants. J. Cell. Biol. 136: 89-802

Gaynor, E.C., Mondesert, G., Grimme, S.J., Reed, S.I., Orlean, P., and Emr, S.D. 1999. *MCD4* encodes a conserved endoplasmic reticulum membrane protein essential for glycosylphosphatidylin ositol anchor synthesis in yeast. Mol. Biol. Cell 10: 627-648.

Gemmill, T.R., and Trimble, R.B. 1999. Overview of N- and O-linked oligosaccharide structures found in various yeast species. Biochim. Biophys. Acta 1426: 227-237.

Gentzsch, M., and Tanner, W. 1996. The *PMT* gene family: protein O-glycosylation in *Saccharomyces cerevisiae* is vital. EMBO J. 15: 5752-5759.

Girbardt, M. 1955. Lebendbeobachtungen on *Polystictus versicolor* (L.). Flora: 142: 540-563.

Girbardt, M. 1957. Der Spitzenkörper von *Polystictus versicolor* (L.). Planta 50: 47-59.

Girbardt, M. 1969. Die Ultrastruktur der Apikal Region von Pilzhyphen. Protoplasma 67: 413-441.

Girrbach, V., and Strahl, S. 2003. Members of the evolutionarily conserved PMT family of protein O-mannosyltransferases form distinct protein complexes among themselves. J. Biol. Chem. 278: 12554-12562.

Glaser, L., and Brown, D.H. 1957. The synthesis of chitin in cell-free extracts of *Neurospora crassa*. J. Biol. Chem. 228: 729-742.

Gloriely, A., and Tabor, M. 2003. Self-similar tip growth in filamentary organisms. Phys. Rev. Lett. 90: 108101-108104.

Gooday, G.W. 1994. Cell membrane. In: The Growing Fungus. N.A.R. Gow and G.M. Gadd, eds. Chapman and Hall, London. p. 63-74.

Gow, N.A.R. 1994. Tip growth and polarity. In: The Growing Fungus. N.A.R. Gow and G.M. Gadd, eds. Chapman and Hall, London. p. 277-299.

Graham, T.R., and Emr, S.D. 1991. Compartmental organization of Golgi-specific protein modification and vacuolar protein sorting events defined in a yeast *sec18* (NSF) mutant. J. Cell Biol. 114: 207-218.

Graham, T.R., Seeger, M., Payne, G.S., MacKay, V.L., and Emr, S.D. 1994. Clathrin-dependent localization of α-1,3-mannosyltransferase to the Golgi complex of *Saccharomyces cerevisiae*. J. Cell Biol. 127: 667-678.

Guo, W., Tamanoi, F., and Novick, P. 2001. Spatial regulation of the exocyst complex by Rho1 GTPase. Nat. Cell. Biol. 3: 353-360.

Hamada, K., Terashima, H., Arisawa, M., Yabuki, N., and Kitada, K. 1999. Amino acid residues in the ω- region participate in cellular localization of yeast glycosylphosphatidylinositol-attached proteins. J. Bacteriol. 181: 3886-3889.

Hamburger, D., Egerton, M., and Riezman, H. 1995. Yeast Gaa1p is required for attachment of a completed GPI anchor onto proteins. J Cell Biol. 129: 629-39.

Harold, F. M. 1997. How hyphae grow: morphogenesis explained? Protoplasma 197: 137-147.

Harris, S.D., Hofmann, A.F., Tedford, H.W., and Lee M.P. 1999. Identification and characterization of genes required for hyphal morphogenesis in the filamentous fungus *Aspergillus nidulans*. Genetics 151: 1015-1025.

Harris, S.L., and Waters, M.G. 1996. Localization of a yeast early Golgi mannosyltransferase, Och1p, involves retrograde transport. J. Cell Biol. 132: 985-998.

Hazan, I., Sepulveda-Becerra, M., and Liu, H. 2002. Hyphal elongation is regulated independently of cell cycle in *Candida albicans*. Mol. Biol. Cell. 13: 134-145.

Hebert, D.N., Foellmer, B., Helenius, A. 1995. Glucose trimming and reglucosylation determine glycoprotein association with calnexin in the endoplasmic reticulum. Cell 81: 425-433.

Herrera-Estrella, L., and Ruiz-Herrera, J. 1983. Light response in *Phycomyces blakesleeanus*: evidence for roles of chitin biosynthesis and breakdown. Exp. Mycol. 7: 362-369.

Herscovics A., and Orlean, P. 1993. Glycoprotein biosynthesis in yeast. FASEB J. 7: 540-550.

Horsch, M., Mayer, C., and Rast, D.M. 1996. Stereochemical requirements of chitin synthase for ligand binding at the allosteric site for *N*-acetylglucosamine. Eur. J. Biochem. 237: 476-482.

Horsch, M., and Sowdhamini, R. 1996. A fold prediction for the catalytic domain of chitin synthases. In: Chitin Enzymology, Vol. 2. RAA Muzzarelli, ed. Atec Edizioni, Grottammare. p. 447-458.

Hutchins, K., and Bussey, H. 1983. Cell wall receptor for yeast killer toxin: involvement of β-1,6-D-glucan. J. Bacteriol. 154: 161-169.

Ichinomiya, M., Motoyama, T., Fujiwara, M., Takagi, M., Horiuchi, H., and Ohta, A. 2002. Repression of chsB expression reveals the functional importance of class IV chitin synthase gene chsD in hyphal growth and conidiation of *Aspergillus nidulans*. Microbiology 148: 1335-1347.

Igual, J.C., Johnson, A.L., and Johnston, L.H. 1996. Coordinated regulation of gene expression by the cell cycle transcription factor SW14 and the protein kinase C MAP kinase pathway for yeast cell integrity. EMBO J. 15: 5001-5013.

Imhof, I., Canivenc-Gansel, E., Meyer, U., and Conzelmann, A. 2000. Phosphatidylethanolamine is the donor of the phosphorylethanolamine linked to the α-1,4-linked mannose of yeast GPI structures. Glycobiology 10: 1271-1275.

Inoue, S., Turgeon, B.G., Yoder, O.C., and Aist, J.R. 1998. Role of fungal dynein in hyphal growth, microtubule organization, spindle pole body motility and nuclear migration. J. Cell Sci.111: 1555-1566.

Ishiguro, J., Saitou, A., Duran, A., and Ribas, J.C. 1997. cps1+, a *Schizosaccharomyces pombe* gene homolog of *Saccharomyces cerevisiae* FKS genes whose mutation confers hypersensitivity to cyclosporin A and papulacandin B. J. Bacteriol. 179: 7653-7662.

Jackson, S.L., and Heath, I.B. 1993. Roles of calcium in hyphal tip growth. Microbiol. Rev. 57: 367-382.

Jentoft, N. 1990. Why are proteins O-glycosylated? Trends Biochem. Sci. 15: 291-294.

Jung, U.S., Sobering, A.K., Romeo, M.J., and Levin, D.E. 2002. Regulation of the yeast Rlm1 transcription factor by the Mpk1 cell wall integrity MAP kinase. Mol. Microbiol. 46: 781-789.

Kamada, Y., Qadota, H., Pitón, C.P., Anraku, Y., Ohya, Y., and Levin, D.E. 1996. Activation of yeast protein kinase C by Rho1 GTPase. J. Biol. Chem. 271: 9193-9196.

Kang, M.S., and Cabib, E.. 1986. Regulation of fungal cell wall growth: a guanine nucleotide-binding, proteinaceous component required for activity of β-1,3-D-glucan synthase. Proc. Natl. Acad. Sci. U.S.A. 83: 5808-5812.

Kanik-Ennulat, C., Montalvo, E., and Neff, N. 1995. Sodium orthovanadate-resistant mutants of *Saccharomyces cerevisiae* show defects in Golgi-mediated protein glycosylation, sporulation and detergent resistance. Genetics 140: 933-943.

Kapteyn, J.C., Hoyer, L.L., Hecht, J.E., Müller, W.H., Andel, A., Verkleij, A.J., Makarow, M., Van Den Ende, H., and Klis, F.M. 2000. The cell wall architecture of *Candida albicans* wild-type cells and cell wall-defective mutants. Mol. Microbiol. 35: 601-611.

Kapteyn, J.C., Ram, A.F., Groos, E.M., Kollar, R., Montijn, R.C., Van Den Ende, H., Llobell, A., Cabib, E and Klis, F.M. 1997. Altered extent of cross-linking of β-1,6-glucosylated mannoproteins to chitin in *Saccharomyces cerevisiae* mutants with reduced cell wall β-1,3-glucan content. J. Bacteriol. 179: 6279-6284.

Kapteyn, J.C., Van Den Ende, H., and Klis, F.M. 1999. The contribution of cell wall proteins to the organization of the yeast cell wall. Biochim. Biophys. Acta 1426: 373-383.

Karaoglu, D., Kelleher, D.J.and Gilmore, R. 1995. Functional characterization of Ost3p. Loss of the 34-kD subunit of the *Saccharomyces cerevisiae* oligosaccharyltransferase results in biased underglycosylation of acceptor substrates. J. Cell. Biol. 130:567-577.

Katayama, S., Hirata, D., Arellano, M., Perez, P., and Toda, T. 1999. Fission yeast α-glucan synthase Mok1 requires the actin cytoskeleton to localize the sites of growth and plays an essential role in cell morphogenesis downstream of protein kinase C function. J.Cell Biol. 144: 1173-86.

Kelleher, D.J., and Gilmore, R. 1994. The *Saccharomyces cerevisiae* oligosaccharyltransferase is a protein complex composed of Wbp1p, Swp1p, and four additional polypeptides. J. Biol. Chem. 269: 12908-12917.

Kelly, R., Register, E., Hsu, M.J., Kurtz, M. and Nielsen, J. 1996. Isolation of a gene involved in 1,3-β-glucan synthesis in *Aspergillus nidulans* and purification of the corresponding protein. J. Bacteriol. 178: 4381-4391.

Ketela, T., Green, R., and Bussey, H. 1999. *Saccharomyces cerevisiae* Mid2p is a potential cell wall stress sensor and upstream activator of the *PKC1-MPK1* cell integrity pathway. J. Bacteriol. 181: 3330-3340.

Klebl, B., Kozian, D., Leberer, E., and Kukuruzinska, M.A. 2001. A comprehensive analysis of gene expression profiles in a yeast N-glycosylation mutant. Biochem. Biophys. Res. Comm. 286: 714-720.

Klis, F.M., Mol, P., Hellingwerf, K., and Brul, S. 2002. Dynamics of cell wall structure in *Saccharomyces cerevisiae*. FEMS Microbiol. Rev. 26: 239-256.

Knauer, R., and Lehle, L. 1999. The oligosaccharyltransferase complex from yeast. Biochim. Biophys. Acta 1426: 259-273

Kollar, R., Petrakova, E., Ashwell, G., Robbins, P.W., and Cabib, E. 1995. Architecture of the yeast-cell wall. The linkage between chitin and β-1,3-glucan. J. Biol. Chem. 270: 1170-1178.

Kollar, R., Reinhold, B.B., Petrakova, E., Yeh, H.J., Ashwell, G., Drgonova, J., Kapteyn, J.C., Klis, F.M., and Cabib, E. 1997. Architecture of the yeast cell wall. β(**1→6**)-glucan interconnects mannoprotein, β(**1→3**)glucan, and chitin. J. Biol. Chem. 272: 17762-17775.

Konomi, M., Fujimoto, K., Toda, T., and Osumi, M. 2003. Characterization and behaviour of α-glucan synthase in *Schizosaccharomyces pombe* as revealed by electron microscopy. Yeast 20: 427-438.

Kopecek, P., and Raclavsky, V. 1999. Comparison of chitin content in the apical and distal parts of fungal hyphae in *Basidiobolus ranarum, Neurospora crassa* and *Coprinus sterquilinus*. Folia Microbiol. (Praha) 44: 397-400.

Kozubowski, L., Panek, H., Rosenthal, A., Bloecher, A., DeMarini, D.J., and Tatchell, K.A. 2003. bni4-glc7 phosphatase complex that recruits chitin synthase to the site of bud emergence. Mol. Biol. Cell 14: 26-39.

Kukuruzinska, M.A., Bergh, M.L., and Jackson, B.J. 1987. Protein glycosylation in yeast. Annu. Rev. Biochem. 56: 915-944.

Kukuruzinska, M.A., and Lennon, K. 1998. Protein N-glycosylation: molecular genetics and functional significance. Crit. Rev. Oral Biol. Med. 9: 415-448.

Larriba, G., Morales, M., and Ruiz-Herrera, J. 1981. Biosynthesis of β-glucan microfibrils by cell free extracts from *Saccharomyces cerevisiae*. J. Gen. Microbiol. 124: 375-383.

Lehle, L., and Bause, E. 1984. Primary structural requirements for N- and O-glycosylation of yeast mannoproteins. Biochem. Biophys. Acta 799: 246-251

Lending, C., Leal-Morales, C.A., Flores-Martinez, A., Bracker, C.E., and Bartnicki-Garcia, S. 1991. Purification and characterization of 16S chitin synthetase particles from cell walls of *Mucor rouxii*. Exp. Mycol. 15: 11-25.

Leon, M, Sentandreu, R., and Zueco, J. 2002. A single *FKS* homologue in *Yarrowia lipolytica* is essential for viability. Yeast 19: 1003-1014.

Leon, M., Jaafar, L., and Zueco, J. 2003. *RHO1* (*YlRHO1*) is a non-essential gene in *Yarrowia lipolytica* and complements *rho1*Δ lethality in *Saccharomyces cerevisiae*. Yeast. 20: 343-350.

Lipke, P.N., and Ovalle, R. 1998. Cell wall architecture in yeast: new structure and new challenges. J. Bacteriol. 180: 3735-3740.

Lopez-Matas, M.A., Eslava, A.P., and Diaz-Minguez, J.M. 2000. *Mcchs1*, a member of a chitin synthase gene family in *Mucor circinelloides*, is differentially expressed during dimorphism. Curr. Microbiol. 40: 169-175.

Lopez-Romero, E, and Ruiz-Herrera, J. 1978. Properties of β-glucan synthetase from *Saccharomyces cerevisiae*. Ant.v. Leeuwenhoek. 44: 329-339.

Lussier, M., Gentzsch, M., Sdicu, A.M., Bussey, H., and Tanner, W. 1995. Protein O-glycosylation in yeast. The PMT2 gene specifies a second protein O-mannosyltransferase that functions in addition to the PMT1-encoded activity. J. Biol. Chem. 270: 2770-2775.

Madrid, M.P., Di Pietro, A., and Roncero, M.I. 2003. Class V chitin synthase determines pathogenesis in the vascular wilt fungus *Fusarium oxysporum* and mediates resistance to plant defence compounds. Mol. Microbiol. 47: 257-266.

Martin, H., Dagkessamanskaia, A., Satchanska, G. Dallies N., and Francois J. 1999. KNR4, a suppressor of *Saccharomyces cerevisiae cwh* mutants, is involved in the transcriptional control of chitin synthase genes. Microbiology 145: 249-258.

Martinez-Cadena, G., and Ruiz-Herrera, J. 1987. Activation of chitin synthetase from *Phycomyces blakesleeanus* by calcium and calmodulin. Arch. Microbiol. 148: 280-285.

Martinou, A., Koutsioulis, D., and Bouriotis, V. 2002. Expression, purification, and characterization of a cobalt-activated chitin deacetylase (Cda2p) from *Saccharomyces cerevisiae*. Protein Exp. Purif. 24: 111-116.

Matlack, K.E., Misselwitz, B., Plath, K., and Rapoport, T.A. 1999. BiP acts as a molecular ratchet during posttranslational transport of prepro-α factor across the ER membrane. Cell 97: 553-564.

Mazur, P., Morin, N., Baginsky, W., el-Sherbeini, M., Clemas, J.A., Nielsen, J.B., and Foor, F. 1995. Differential expression and function of two homologous subunits of yeast β-1,3-D-glucan synthase. Mol. Cell. Biol. 15: 5671-5681.

McGillivray, A.M., and Gow, N.A. R.1986. Applied electrical fields polarize the growth of mycelial fungi. J. Gen. Microbiol. 131: 751-756.

McGillivray, A.M., and Gow, N.A.R. 1987. Cytoplasmic migration and intracellular organelle movements during tip growth of fungal hyphae. Exp. Mycol. 11: 79-100.

Meaden, P., Hill, K., Wagner, J., Slipetz, D., Sommer, S.S., and Bussey, H. 1990. The yeast *KRE5* gene encodes a probable endoplasmic reticulum protein required for β-1,6-D-glucan synthesis and normal cell growth. Mol. Cell. Biol. 10: 3013-3019.

Mellado, E., Dubreucq, G., Mol, P., Sarfati, J., Paris, S., Diaquin, M., Holden, D.W., Rodriguez-Tudela, J.L., and Latge, J.P. 2003. Cell wall biogenesis in a double chitin synthase mutant (*chsG-/chsE-*) of *Aspergillus fumigatus*. Fungal Genet. Biol. 38: 98-109.

Merz, R.A., Horsch, M., Nyhlén, L.E., and Rast, D.M. 1999. Biochemistry of chitin synthase. In: Chitin and Chitinases. P. Jolles and R.A.A. Muzzarelli, eds. Birkhäuser, Basel. p. 9-37.

Meyer, H.A., Grau, H., Kraft, R., Kostka, S., Prehn, S., Kalies, K.U., and Hartmann, E. 2000. Mammalian Sec61 is associated with Sec62 and Sec63. J. Biol. Chem. 275: 14550-14557.

Mio, T., Adachi-Shimizu, M., Tachibana, Y., Tabuchi, H., Inoue, S.B., Yabe, T., Yamada-Okabe, T., Arisawa, M., Watanabe, T., and Yamada-Okabe, H. 1997 Cloning of the *Candida albicans* homolog of *Saccharomyces cerevisiae GSC1/FKS1* and its involvement in β-1,3-glucan synthesis. J. Bacteriol. 179: 4096-4105.

Mizoguchi, J., Saito, T., Mizuno, K., and Hayano, K. 1977. On the mode of action of a new antifungal antibiotic, aculeacin A: inhibition of cell wall synthesis in *Saccharomyces cerevisiae*. J. Antibiot. 30: 308-313.

Momany, M., Westfall, P.J., and Abramowsky, G. 1999. *Aspergillus nidulans swo* mutants show defects in polarity establishment, polarity maintenance and hyphal morphogenesis. Genetics 151: 557-567.

Mondesert, G., and Reed, S.I. 1996. *BED1*, a gene encoding a galactosyltransferase homologue, is required for polarized growth and efficient bud emergence in *Saccharomyces cerevisiae*. J. Cell Biol. 132: 137-151.

Moukadiri, I., Jaafar, L., and Zueco, J. 1999. Identification of two mannoproteins released from cell walls of a *Saccharomyces cerevisiae mnn1 mnn9* double mutant by reducing agents. J. Bacteriol. 181: 4741-4745.

Mouyna, I., Fontaine, T., Vai, M., Monod, M., Fonzi, W.A., Diaquin, M., Popolo, L., Hartland, R.P., and Latge, J.P. 2000. Glycosylpho sphatidylinositol-anchored glucanosyltransferases play an active role in the biosynthesis of the fungal cell wall. J. Biol. Chem. 275: 14882-14889.

Muhlschlegel, F., and Fonzi, W.A. 1997. *PHR2* of *Candida albicans* encodes a functional homolog pf the pH-regulated gene *PHR1* with an inverted pattern of pH-dependent expression. Mol. Cell. Biol. 17: 5960-5967.

Muller, C., Hjort, C.M., Hansen, K., and Nielsen, J. 2002. Altering the expression of two chitin synthase genes differentially affects the growth and morphology of *Aspergillus oryzae*. Microbiology 148: 4025-4033.

Munro, C.A., and Gow, N.A.R. 2001. Chitin synthesis in human pathogenic fungi. Med. Mycol. 39 Suppl. 1: 41-53.

Munro, C.A., Winter, K., Buchan, A., Henry, K., Becker, J.M., Brown, A.J., Bulawa, C.E., and Gow, N.A. 2001. Chs1 of *Candida albicans* is an essential chitin synthase required for synthesis of the septum and for cell integrity. Mol. Microbiol. 39: 1414-1426.

Muñoz, M.D., Hernandez, L.M., Basco, R., Andaluz, E., and Larriba, G. 1994. Glycosylation of yeast exoglucanase sequons in alg mutants deficient in the glucosylation steps of the lipid-linked oligosaccharide. Presence of glucotriose unit in Dol-PP-GlcNAc2Man9Glc3 influences both glycosylation efficiency and selection of N-linked sites. Biochim. Biophys. Acta. 1201: 361-366.

Nagahashi, S., Lussier, M., and Bussey, H. 1998. Isolation of *Candida glabrata* homologs of the *Saccharomyces cerevisiae KRE9* and *KNH1* genes and their involvement in cell wall β-1,6-glucan synthesis. J. Bacteriol. 180: 5020-5029.

Nagahashi, S., Sudoh, M., Ono, N., Sawada, R., Yamaguchi, E., Uchida, Y., Mio, T., Takagi, M., Arisawa, M., and Yamada-Okabe, H. 1995. Characterization of chitin synthase 2 of *Saccharomyces cerevisiae*: implication of two highly conserved domains as possible catalytic sites. J. Biol. Chem. 270: 13961-13967.

Nakanishi-Shindo, Y., Nakayama, K., Tanaka, A., Toda, Y., and Jigami, Y. 1993. Structure of the N-linked oligosaccharides that show the complete loss of α-1,6-polymannose outer chain from *och1, och1 mnn1*, and *och1 mnn1 alg3* mutants of *Saccharomyces cerevisiae*. J. Biol. Chem. 268: 26338-26345.

Nakano, K., Arai, R., and Mabuchi, I. 1997. The small GTP-binding protein Rho1 is a multifunctional protein that regulates actin localization, cell polarity, and septum formation in the fission yeast *Schizosaccharomyces pombe*. Genes Cells. 2: 679-694.

Nakano, K., and Mabuchi, I. 1995. Isolation and sequencing of two cDNA clones encoding Rho proteins from the fission yeast *Schizosaccharomyces pombe*. Gene 155: 119-122.

Nakano, K., Mutoh, T., Arai, R., and Mabuchi, I. 2003. The small GTPase Rho4 is involved in controlling cell morphology and septation in fission yeast. Genes Cells. 8: 357-370.

Northcote, D.H., and Horne, R.W. 1952. The chemical composition and structure of yeast cell wall. Biochem. J. 51: 232-236.

Novick, P., Field, C., and Schekman, R. 1980. Identification of 23 complementation groups required for post translational events in the yeast secretory pathway. Cell 21: 205-215.

Novick, P., and Schekman, R. 1979. Secretion and cell surface growth are blocked in a temperature sensitive mutant of *Saccharomyces cerevisiae*. Proc. Natl. Acad. Sci. U.S.A. 76: 1858-1862.

Oberholzer, U., Marcil, A., Leberer, E., Thomas, D.Y., and Whiteway, M. 2002. Myosin I is required for hypha formation in *Candida albicans*. Eukaryotic Cell 1: 213-228.

Odani, T., Shimma, Y., Wang, X.H., and Jigami, Y. 1997. Mannosylphosphate transfer to cell wall mannan is regulated by the transcriptional level of the *MNN4* gene in *Saccharomyces cerevisiae*. FEBS Lett. 420: 186-190.

Orlean, P. 1997. Biogenesis of yeast wall and surface components. In: The Molecular and Cellular Biology of the Yeast *Saccharomyces*. J. R. Pringle, J. R. Broach and E. W. Jones, eds. Cold Spring Harbor Laboratory Press, Cold Spring Harbor. p. 229-362.

Ortega, J. K. E., Gamow, R. I., and Ahlquist, C. N. 1975. *Phycomyces*: a change in the mechanical properties after a light stimulus. Plant Physiol. 55: 333-337.

Ortega, J.K.E., Zehr, E.G., and Keaninio, R.G. 1989. *In vivo* creep and stress relaxation experiments to determine wall extensibility and yield treshold for the sporangiophore of *Phycomyces*. Biophys. J. 56: 465-475.

Parent, S.A., Nielsen, J.B., Morin, N., Chrebet, G., Ramadan, N., Dahl, A.M., Hsu, M.J., Bostian, K.A., and Foor, F. 1993. Calcineurin-dependent growth of an FK506- and CsA-hypersensitive mutant of *Saccharomyces cerevisiae*. J. Gen. Microbiol. 139: 2973-2984.

Park, B.C., Park, Y.H., and Park, H.M. 2003. Activation of chsC transcription by AbaA during asexual development of *Aspergillus nidulans*. FEMS Microbiol. Lett. 220: 241-246.

Parker, C.G., Fessler, L.I., Nelson, R.E., and Fessler, J.H. 1995. Drosophila UDP-glucose:glycoprotein glucosyltransferase: sequence and characterization of an enzyme that distinguishes between denatured and native proteins. EMBO J. 14: 1294-1303.

Pathak, R., Parker, C.S., and Imperiali, B. 1995. The essential yeast NLT1 gene encodes the 64 kDa glycoprotein subunit of the oligosaccharyl transferase. FEBS Lett. 362: 229-234.

Pereira, M., Felipe, M.S., Brigido, M.M., Soares, C.M., and Azevedo, M.O. 2000. Molecular cloning and characterization of a glucan synthase gene from the human pathogenic fungus *Paracoccidioides brasiliensis*. Yeast 16: 451-462.

Popolo, L., Gilardelli, D., Bonfante, P., and Vai, M. 1997. Increase in chitin as an essential response to defects in assembly of cell wall polymers in the *ggp1Δ* mutant of *Saccharomyces cerevisiae*. J. Bacteriol. 179: 463-469.

Popolo, L., Gualtieri, T., and Ragni, E. 2001. The yeast cell-wall salvage pathway. Med. Mycol. 39, Suppl. 1: 111-121.

Popolo, L., and Vai, M. 1998. Defects in assembly of the extracellular matrix are responsible for altered morphogenesis of a *Candida albicans phr1* mutant. J. Bacteriol. 180: 163-166.

Popolo, L., Vai, M., Gatti, E. Porello, S., Bonfante, P., Bellestrinoi, P., and Alberghina, L. 1993. Physiological analysis of mutants indicates involvement of the *Saccharomyces cerevisiae* GPI-anchored protein gp115 in morphogenesis and cell separation. J. Bacteriol. 175: 1879-1885.

Pruyne, D., and Bretscher, A. 2000a. Polarization of cell growth in yeast. I. Establishment and maintenance of polarity states. J. Cell Sci. 113: 365-375.

Pruyne, D., and Bretscher, A. 2000b. Polarization of cell growth in yeast. II.The role of the cortical and actin cytoskeleton. J. Cell Sci. 113: 571-585.

Qadota, H., Python, C.P., Inoue, S.B., Arisawa, M., Anraku, Y., Zheng, Y., Watanabe, T., Levin, D.E., and Ohya, Y. 1996. Identification of yeast Rho1p GTPase as a regulatory subunit of β-1,3-glucan synthase. Science 272: 279-281.

Ram, A.F., Brekelmans, S.S., Oehlen, L.J., and Klis, F.M. 1995. Identification of two cell cycle regulated genes affecting the β1,3-glucan content of cell walls in *Saccharomyces cerevisiae*. FEBS Lett. 358: 165-170.

Ram, A.F., Kapteyn, J.C., Montijn, R.C., Caro, L.H., Douwes, J.E., Baginsky, W., Mazur, P., van den Ende, H., and Klis, F.M. 1998. Loss of the plasma membrane-bound protein as1p in *Saccharomyces cerevisiae* results in the release of β1,3-glucan into the medium and induces a compensation mechanism to ensure cell wall integrity. J. Bacteriol. 18: 1418-1424.

Ram, A.F., Wolters, A., Ten Hoopen, R., and Klis, F.M. 1994. A new approach for isolating cell wall mutants in *Saccharomyces cerevisiae* by screening for hypersensitivity to calcofluor white. Yeast 10: 1019-1030.

Reddy, V.A., Johnson, R.S., Biemann, K., Williams, R.S., Ziegler, F.D., Trimble, R.B., and Maley, F. 1988. Characterization of the glycosylation sites in yeast external invertase. I. N-linked oligosaccharide content of the individual sequons. J. Biol. Chem. 263: 6978-6985.

Redecker, D., Kodner, R., and Graham, L.E. 2000. Glomelean fungi from the Ordovician. Science 289: 1920-1921.

Reggiori, F., Canivenc-Gansel, E., and Conzelmann, A. 1999. Analysis of the lipid moiety of GPI anchor in the yeast *Saccharomyces cerevisiae*. Meth. Mol. Biol. 116: 91-1105.

Reiss, G., te Heesen, S., Gilmore, R., Zufferey, R., and Aebi, M. 1997. A specific screen for oligosaccharyltransferase mutations identifies the 9 kDa OST5 protein required for optimal activity *in vivo* and *in vitro*. EMBO J. 16: 1164-1172.

Richard, M., De Groot, P., Courtin, O., Poulain, D., Klis, F., and Gaillardin, C. 2002. GPI7 affects cell-wall protein anchorage in *Saccharomyces cerevisiae* and *Candida albicans*. Microbiology 148: 2125-2133.

Riquelme, M., Gierz, G., and Bartnicki-Garcia, S. 2000. Dynein and dynactin deficiencies affect the formation of the Spitzenkörper and distort hyphal morphogenesis of *Neurospora*. Microbiology 146: 1742-1752.

Roberg, K.J., Bickel, S., Rowley, N., and Kaiser, C.A. 1997. Control of amino acid permease sorting in the late secretory pathway of

Saccharomyces cerevisiae by SEC13, LST4, LST7 and LST8. Genetics 147: 1569-1584.

Roelants, F.M., Torrance, P.D., Bezman, N., and Thorner J. 2002. Pkh1 and pkh2 differentially phosphorylate and activate ypk1 and ykr2 and define protein kinase modules required for maintenance of cell wall integrity. Mol. Biol. Cell. 13: 3005-3028.

Roemer, T., and Bussey, H. 1995. Yeast Kre1p is a cell surface O-glycoprotein. Mol. Gen. Genet. 249: 209-216.

Roemer, T., Paravicini, G., Payton, M.A., and Bussey, H. 1994. Characterization of the yeast β-1,3-glucan biosynthetic components, Kre6p and Skn1p, and genetic interactions between the PKC1 pathway and extracellular matrix assembly. J. Cell Biol. 127: 567-579.

Roh, D.H., Bowers, B., Riezman, H., and Cabib, E. 2002. Rho1p mutations specific for regulation of β-(1,3-) glucan synthesis and the order of assembly of the yeast cell wall. Mol. Microbiol. 44: 1167-1183.

Romero, P.A., Sleno, B., and Herscovics, A. 1994. Glycoprotein biosynthesis in *Saccharomyces cerevisiae*. Partial purification of the α-1,6-mannosyltransferase that initiates outer chain synthesis. Glycobiology 4: 135-140.

Roncero, C., Rodriguez-Garcia, L.J., Sanz, M., Castrejon, F., Trilla, J.A., and Duran, A. 2001. The control of chitin synthesis in *Saccharomyces cerevisiae*. In: Chitin Enzymology 2001. R.A.A. Muzzarelli, ed. Atec Edizioni. Grottammare. p. 485-492.

Rothblatt, J.A., and Meyer, D.I. 1986. Secretion in yeast: reconstitution of the translocation and glycosylation of a-factor and invertase in a homologous cell-free system. Cell 44: 619-628.

Ruiz-Herrera, J. 1982. In: Cellulose and Other Natural Polymer Systems. R. M . Brown, ed. Plenum Press, New York. p. 207-223.

Ruiz-Herrera, J. 1984. The role of chitosomes in the apical growth of fungi. In: Microbial Cell Wall Synthesis and Assembly. C. Nombela, ed. Elsevier, Amsterdam. p. 113-120.

Ruiz-Herrera, J. 1991. Biosynthesis of β-glucans in fungi. Ant. v. Leeuwenhoek 60: 72-81.

Ruiz-Herrera, 1992. Fungal Cell Wall: Structure, Synthesis, and Assembly. CRC Press, Boca Raton, FL.

Ruiz-Herrera, J., Gonzalez-Prieto, J.M., and Ruiz-Medrano, R 2001. Significance of the multigenic control of chitin synthase in fungi. In: Chitin Enzymology 2001. R.A.A. Muzzarelli, ed. Grottammare, Atec Edizioni. p. 451-462.

Ruiz-Herrera, J., González-Prieto, J.M., and Ruiz-Medrano, R. 2002. Evolution and phylogenetic relationships of chitin synthases from yeasts and fungi. FEMS Yeast Res. 1: 247-256.

Ruiz-Herrera, J., and Larriba, G. 1995. High molecular weight precursors of glucans in *Saccharomyces cerevisiae*. Ant. v. Leeuwenhoek 68: 231-235.

Ruiz-Herrera, J., and Ruiz-Medrano, R. 2003. Chitin biosynthesis in fungi. In: Handbook of Fungal Biotechnology, 2nd. Ed., P.K. Arora, ed.. Marcel Dekker, New York. In press.

Ruiz-Herrera, J., and San-Blas, G. 2003. Chitin synthesis as a target for antifungal drugs. Current Drug Targets-Infectious Disorders 3: 77-91.

Santos, B., and Snyder, M. 1997. Targeting of chitin synthase 3 to polarized growth sites in yeast requires Chs5p and Myo2p. J. Cell Biol. 136: 95-110.

Santos, B., and Snyder, M. 2000. Sbe2p and sbe22p, two homologous Golgi proteins involved in yeast cell wall formation. Mol. Biol. Cell 11: 435-452.

Saporito-Irwin, S.M., Birse, C.E., Sypherd, P.S., and Fonzi, W.A. 1995. PHR1, a pH-regulated gene of *Candida albicans*, is required for morphogenesis. Mol. Cell. Biol. 15: 601-613.

Saxena, I.M., Brown, M.R.Jr., Fevre, M., Geremia, R.A., and Henrissat, B. 1995. Multidomain architecture of β-glycosyl transferases: implications for mechanisms of action. J. Bacteriol. 177: 1419-1424.

Schimoler-O'Rourke, R., Renault, S., Mo, W., and Selitrennikoff, C.P. 2003. *Neurospora crassa* FKS protein binds to the (1,3-)beta-glucan synthase substrate, UDP-glucose. Curr. Microbiol. 46: 408-412.

Schleip, I., Heiss, E., and Lehle, L. 2001. The yeast *SEC20* gene is required for N- and O-glycosylation in the Golgi. Evidence that impaired glycosylation does not correlate with the secretory defect. J. Biol. Chem. 276: 28751-28758.

Schmidt, A., Schmelzle, T., and Hall, M.N. 2002. The RHO1-GAPs SAC7, BEM2 and BAG7 control distinct Rho1 functions in *Saccharomyces cerevisiae*. Mol. Microbiol. 45: 1433-1441.

Schorr, M., Then, A., Tahirovic, S., Hug, N., and Mayinger, P. 2001. The phosphoinositide phosphatase Sac1p controls trafficking of the yeast Chs3p chitin synthase. Curr. Biol. 11: 1421-1426.

Sekiya-Kawasaki, M., Abe, M., Saka, A., Watanabe, D., Kono, K., Minemura-Asakawa, M., Ishihara, S., Watanabe, T., and Ohya, Y. 2002. Dissection of upstream regulatory components of the Rho1p effector, β1,3-glucan synthase, in *Saccharomyces cerevisiae*. Genetics 162: 663-676.

Sentandreu, R., and Northcote, D.H. 1968. The structure of a glycopeptide isolated from the yeast cell wall. Biochem. J. 109: 419-432.

Shahinian, S. and Bussey, H. 2000. β-1,6-Glucan synthesis in *Saccharomyces cerevisiae*. Mol. Microbiol. 35: 477-489.

Shaw, J.A., Mol, P.C., Bowers, B., Silverman, S.J., Valdivieso, M.H., Duran, A., and Cabib, E. 1991. The function of chitin synthases 2 and 3 in the *Saccharomyces cerevisiae* cell cycle. J. Cell Biol. 114: 111-123.

Shematek, E.M., Braatz, J.A., and Cabib, E. 1980. Biosynthesis of the yeast cell wall. I. Preparation and properties of β-1,3-glucan synthetase. J. Biol. Chem. 255: 888-894.

Shematek, E.M., and Cabib, E. 1980. Biosynthesis of the yeast cell wall. II. Regulation of β-(1, 3)-glucan synthetase by ATP and GTP. J. Biol. Chem. 255: 895-902.

Sietsma, J.H., and Wessels, J.G.H. 1979. Evidence for covalent linkages between chitin and β-glucan in a fungal cell wall. J. Gen. Microbiol. 114: 99-108.

Sietsma, J.H., and Wessels, J.G.H. 1994. Apical wall biogenesis. In: The Mycota, Vol. 1. J G.H. Wessels, and F. Meinhardt, eds. Springer Verlag, Berlin. p. 43-65.

Silberstein, S., and Gilmore, R. 1996. Biochemistry, molecular biology, and genetics of the oligosaccharyltransferase. FASEB J. 10: 849-858.

Skrzypek, M., Lester, R.L., and Dickson, R.C. 1997. Suppressor gene analysis reveals an essential role for sphingolipids in transport of g lycosylphosphatidylinositol-anchored proteins in *Saccharomyces cerevisiae*. J. Bacteriol. 179: 1513-1520.

Smith, A.E., Zhang, Z., Thomas, C.R., Moxham, K.E., and Middelberg, A.P. 2000. The mechanical properties of *Saccharomyces cerevisiae*. Proc. Natl. Acad. Sci. U.S.A. 97: 9871-9874.

Specht, C.A., Liu, Y., Robbins, P., Bulawa, C.E., Iartchouk, N., Winter, K.R., Riggle, P.J., Rhodes, J.C., Dodge, C.L., Culp, D.W., and Borgia, P.T. 1996. The *chsD* and *chsE* genes of *Aspergillus nidulans* and their roles in chitin synthesis. Fungal Genet. Biol. 20: 153-167.

Spellman, P.T., Sherlock, G., Zhang, M.Q., Iyer, V.R., Anders, K., Eisen, M.B., Brown, P.O., Botstein, D., and Futcher, B. 1998. Comprehensive identification of cell cycle-regulated genes of the yeast *Saccharomyces cerevisiae* by microarray hybridization. Mol. Biol. Cell 9: 3273-3297.

Strahl-Bolsinger, S., Gentzsch, M., and Tanner, W. 1999. Protein O-mannosylation. Biochim. Biophys. Acta. 1426: 297-307.

Surarit, R., Gopal, P.K., and Shepherd, M.G. 1988. Evidence for a glycosidic linkage between chitin and glucan in the cell-wall of *Candida albicans*. J. Gen. Microbiol. 134: 1723-1730.

Suresh, K., and Subramanyam, C. 1997. A putative role for calmodulin in the activation of *Neurospora crassa* chitin synthase. FEMS Microbiol. Lett. 150: 95-100.

Suzuki, T., and Lennarz, W.J. 2000. In yeast the export of small glycopeptides from the endoplasmic reticulum into the cytosol is not affected by the structure of their oligosaccharide chains. Glycobiology 10: 51-58.

Tanner, W., and Lehle, L. 1987. Protein glycosylation in yeast. Biochim. Biophys. Acta 906: 81-99.

Tentler, S., Palas, J., Enderlin, C., Campbell, J., Taft, C., Miller, T.K., Wood, R.L., and Selitrennikoff, C.P. 1997. Inhibition of *Neurospora crassa* growth by a glucan synthase-1 antisense construct. Curr. Microbiol. 34: 303-308.

Thompson, J.R., Douglas, C.M., Li, W., Jue, C.K., Pramanik, B., Yuan, X., Rude, T.H., Toffaletti, D.L., Perfect, J.R., and Kurtz, M. 1999 A glucan synthase *FKS1* homolog in *Cryptococcus neoformans* is single copy and encodes an essential function. J. Bacteriol. 181: 444-453.

Torralba, S., and Heath, I.B. (2001). Cytoskeletal and Ca^{2+} regulation of hyphal tip growth and initiation. Curr. Top. Dev. Biol. 51: 135-187.

Torralba, S., Raudaskoski, M., Pedregosa, A.M., and Laborda, F. 1998. Effect of cytochalasin A on apical growth, actin cytoskeleton organization and enzyme secretion in *Aspergillus nidulans*. Microbiology 144: 45-53.

Trilla, J.A., Duran, A., and Roncero, C. 1999. Chs7p, a new protein involved in the control of protein export from the endoplasmic reticulum that is specifically engaged in the regulation of chitin synthesis in in *Saccharomyces cerevisiae*. J. Cell Biol. 145: 1153-1163.

Trimble, R.B., Maley, F., and Chu, F.K. 1983. Glycoprotein biosynthesis in yeast. Protein conformation affects processing of high mannose oligosaccharides on carboxypeptidase Y and invertase. J. Biol. Chem. 258: 2562-2567.

Valdivieso, M.H., Mol, P.C., Shaw, J.H., Cabib, E., and Duran, A. 1991. *CAL1*, a gene required for activity of chitin synthase 3 in *Saccharomyces cerevisiae*. J. Cell Biol. 114: 101-109.

Valentin, E., Herrero, E., Rico, H., Miragall, F., and Sentandreu, R. 1987. Cell wall mannoproteins during the population growth phases in *Saccharomyces cerevisiae*. Arch. Microbiol. 148: 88-94.

Verostek, M.F., Atkinson, P.H., and Trimble, R.B. 1993a. Glycoprotein biosynthesis in the *alg3 Saccharomyces cerevisiae* mutant. I. Role of glucose in the initial glycosylation of invertase in the endoplasmic reticulum. J. Biol. Chem. 268: 12095-12103.

Verostek, M.F., Atkinson, P.H., and Trimble, R.B. 1993b. Glycoprotein biosynthesis in the *alg3 Saccharomyces cerevisiae* mutant. II. Structure of novel $Man_{6-10}GlcNAc_2$ processing intermediates on secreted invertase. J. Biol. Chem. 268: 12104-12115.

Walker, W.F. 1985. 5 S and 5.8 S ribosomal RNA sequences and protist phylogenetics. Biosystems 18: 269-278.

Walter, P., and Blobel G. 1980. Purification of a membrane-associated protein complex required for protein translocation across the endoplasmic reticulum. Proc. Natl. Acad. Sci. U.S.A. 77: 7112-7116.

Wang, Q., Liu, H., and Szaniszlo, P.J. 2002. Compensatory expression of five chitin synthase genes, a response to stress stimuli, in *Wangiella* (*Exophiala*) *dermatitidis*, a melanized fungal pathogen of humans. Microbiology 148: 2811-2817.

Wang, Z., and Szaniszlo, P.J. 2000. *WdCHS3* a gene that encodes a class III chitin synthase in *Wangiella* (*Exophiala*) *dermatitidis*, is expressed differentially under stress conditions. J. Bacteriol. 182: 874-881.

Wang, Z., Zheng, L., Liu, H., Wang, Q., Hauser, M., Kauffman, S., Becker, J.M., and Szaniszlo, P.J. 2001. WdChs2p, a class I chitin synthase, together with WdChs3p (class III) contributes to virulence in *Wangiella* (*Exophiala*) *dermatitidis*. Infect. Immun. 69: 7517-7526.

Wendland, J., and Philippsen, P. 2001. Cell polarity and hyphal morphogenesis are controlled by multiple rho-protein modules in the filamentous ascomycete *Ashbya gossypii*. Genetics 157: 601-610.

Xoconostle-Cázares, B., León-Ramirez, C., and Ruiz-Herrera, J. 1996. Two chitin synthase genes from *Ustilago maydis*. Microbiology 142: 377-387.

Xoconostle-Cazares, B., Specht, C.A., Robbins, P.W., Liu, Y., Leon, C., and Ruiz-Herrera, J. 1997. *Umchs5*, a gene coding for a class IV chitin synthase in *Ustilago maydis*. Fungal Genet. Biol. 22: 199-208.

Zhao, C., Jung, U.S., Garrett-Engele, P., Roe, T., Cyert, M.S., and Levin, D.E.. 1998. Temperature-induced expression of yeast *FKS2* is under the dual control of protein kinase C and calcineurin. Mol. Cell. Biol. 18: 1013-1022.

Ziman, M., Chuang, J.S., Tsung, M., Hamamoto, S., and Scheckman, R. 1998. Chs6p-dependent anterograde transport of Chs3p from the chitosome to the plasma membrane in *Saccharomyces cerevisiae*. Mol. Biol. Cell 9: 1565-1576.

From: Pathogenic Fungi: Structural Biology and Taxonomy
Edited by: G. San-Blas and R.A. Calderone

Chapter 3

Cell Cycle of Fungal Pathogens

Judith Berman and Neil A.R. Gow

Abstract

The cell cycle coordinates morphogenesis so that mitosis and growth are coupled to the doubling of all cellular components. Recent work with certain fungal pathogens has revealed modifications of the basic cell cycle pattern as exhibited in model yeast species that tailors the needs of the pathogen to its growth in the host environment. Many fungal pathogens such as *Candida albicans* are dimorphic or pleomorphic, and so the cell cycle must be regulated to enable cell shape to be modulated while the nuclear cycle is maintained. Filamentous pathogens such as *Aspergillus fumigatus* maintain a nuclear cycle but uncouple cell division from the growth cycle. Therefore fungal pathogens provide interesting examples of how cell cycle events are both conserved and modified to facilitate replication in the context of invasion of human tissues.

1. Introduction

The cell cycle is the central clock against which morphogenetic events are ordered and timed. In the case of fungal pathogens, morphogenetic processes include the switching from growth by budding to filamentous growth forms that include both true hyphae and pseudohyphae, the formation of mating structures, and the production of various asexual and sexual spores. The cell cycle can dictate when such changes in development can take place. In addition the cell cycle can be altered to enable new growth forms to be propagated. For example, cytokinesis, the process of division of the two daughter cells at the end of the cell cycle, may be altered or inhibited to permit filamentous cell development in dimorphic fungal pathogens. This chapter discusses the cell cycle of fungal pathogens and is focused on the dimorphic fungus *Candida albicans*. Clearly the cell cycle is a highly developed branch of modern cell biology and many of the central control mechanisms are highly conserved in all eukaryotes. Indeed some of the key enzymes that regulate the cell cycle are so highly conserved that yeast gene products can functionally complement homologous mutations in human cell lines and *vice versa*. Model fungi, in which many of the basic features of the cell cycle have been studied, include the yeasts *Saccharomyces cerevisiae* and *Schizosaccharomyces pombe* which have well developed genetics but are not primary pathogens of humans. Nonetheless, progress in understanding the cell cycle in pathogenic species is best described in the light of the insights obtained in these model species. However, recent work has begun to show that significant departures from the cell cycle programme of these yeasts can occur in *C. albicans* and other organisms, contributing to our understanding of general cell cycle processes as well as the regulation of fungal pathogenesis.

2. Cell Cycle Progression in Different Morphological Forms of *C. albicans*

2.1. Yeast Cells

In *C. albicans*, the yeast form (also known as the blastospore), is characterized by round to ellipsoid cells that become separated from the mother cell after cytokinesis. *C. albicans* yeast cell cycle progression appears to be very similar to the process in *S. cerevisiae* yeast cells (reviewed in (Lew *et al.*, 1997)). As in *S. cerevisiae*, there are at least 4 major cell cycle-dependent processes or cycles that occur in a discrete order: 1) the budding cycle (directed localization and relocalization of the actin cytoskeleton and secretory pathway components); 2) DNA replication and the chromosome cycle; 3) the spindle pole body (SPB) and microtubule cycle; and 4) the assembly of a septin ring and subsequent septation at that ring. While these major processes have not been studied as extensively in *C. albicans*, they appear to be linked to one another under most growth conditions, such that, as in *S. cerevisiae*, bud emergence, DNA replication initiation, and spindle pole body duplication appear to be coordinated and to occur within the same time frame that is considered to be the G1/S transition (Wain *et al.*, 1976; Hazan *et al.*, 2002).

The *C. albicans* yeast cell cycle (Figure 1) is characterized by growth (cell expansion) occurring primarily in the daughter cell (the bud). As in *S. cerevisiae*, the round to ellipsoid shape of the bud is likely the result of the distribution and reorganization of F-actin and actin-associated proteins that direct surface growth. In an unbudded cell, a septin ring (composed of Cdc3,10,11 and 12 septin proteins) appears at the incipient bud site prior to bud emergence (Figure 1) (Sudbery 2001, Warenda and Konopka, 2002). Actin also becomes highly polarized at the incipient bud site (Anderson and Soll, 1986; Yokoyama *et al.*, 1990; Akashi *et al.*, 1994). The growing bud emerges through the septin ring and actin remains associated with the tips of small budded cells (polarized growth). Spindle pole body duplication and DNA replication initiation occur close to the time of bud emergence, suggesting that the budding, replication and spindle cycles are coordinated at START in *C. albicans* yeast cells as they are in *S. cerevisiae* (Barton and Gull, 1988; Hazan *et al.* 2002).

Spindle dynamics appear to be very similar to those in *S. cerevisiae* as well (Hazan *et al.*, 2002). In medium sized buds, DNA replication is most likely active and spindle pole body separation and the formation of short spindles is observed (Finley and Berman, unpublished data; Hazan *et al.*, 2002). In growing buds, the actin cables remain oriented toward the bud, but cell wall expansion occurs throughout the bud "isotropic growth". Prior to mitosis, the nucleus, and the spindle associated with it, stretch through the neck between the mother and daughter cells. The nucleus oscillates back and forth across the neck until the spindle elongates and disassembles, leaving the separated nuclei in the mother and daughter cells (Finley and Berman, unpublished data).

In large buds there is a transient period when the actin structures become randomly distributed over the entire cell prior to their reorientation toward the mother-bud neck to promote the secretion and cell synthesis necessary for septation and cytokinesis. In *S. cerevisiae*, chitinase is recruited to the bud side of the neck and is important for executing the cell separation that is characteristic

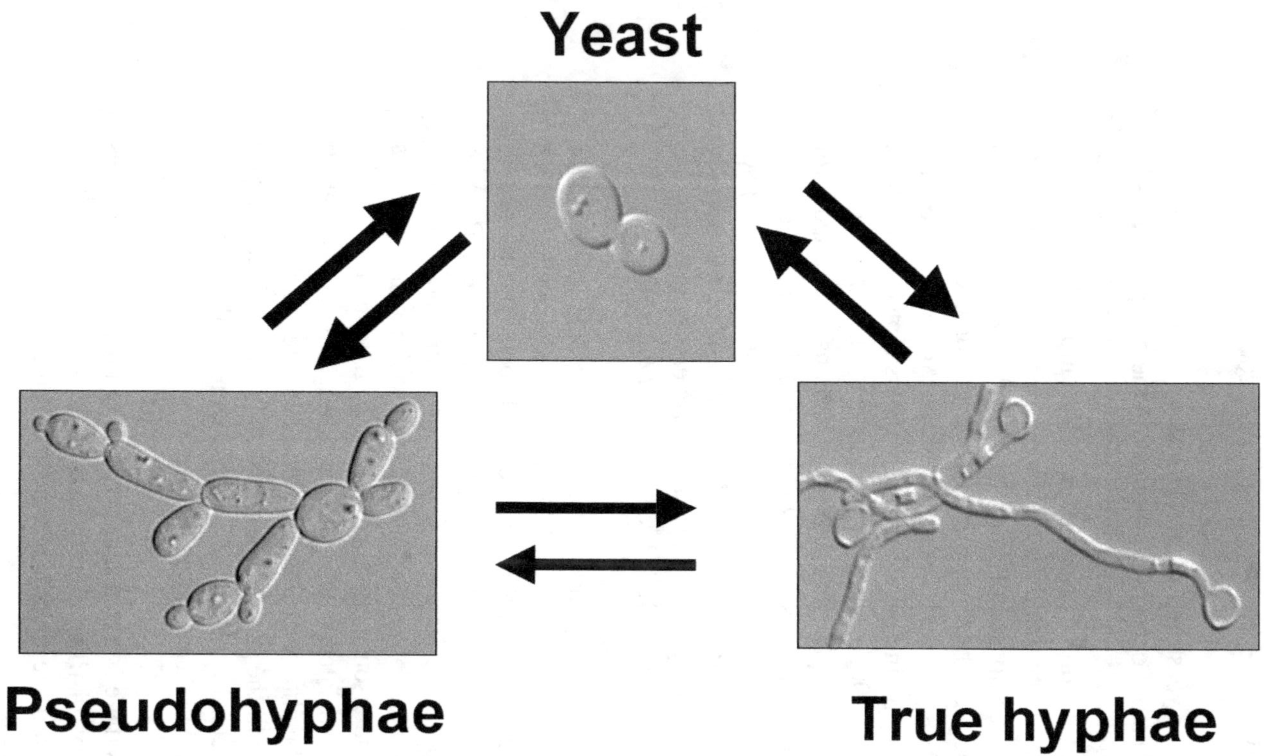

Figure 1. Morphology of the three morphologically distinct vegetative growth forms of *Candida albicans*. Each form is interconvertible and involves modification of cell cycle parameters.

of yeast cells (Kuranda and Robbins, 1991). *C. albicans* has four chitinase genes (McCreath *et al.*, 1995; 1996; Selvaggini *et al.*, 2004) that are presumed to have a similar role. Following cytokinesis, the actin patches remain associated with the neck for a short while and then become randomly distributed or reorganize at the next incipient bud site. The septin rings disassemble after cytokinesis, prior to formation of the septin ring at the next incipient bud site (Gale *et al.*, 2001; Sudbery 2001; Warenda and Konopka, 2002).

As in *S. cerevisiae*, the mother cells enter the next cell cycle before the daughter cells, presumably because mothers are larger than daughters and cells must reach a critical size before entering the next cell cycle (Chaffin, 1984a; Kron and Gow, 1995). Nonetheless, the difference in timing is sufficient to make it very difficult to achieve synchrony of a cell population for more than one cell cycle.

While the *C. albicans* yeast cell cycle shares many features with *S. cerevisiae* cell cycle progression, several differences should be noted. First, *C. albicans* yeast cells have a more rapid division cycle, dividing every ~60 min. in rich medium (*S. cerevisiae* cells generally divide with a ~90 min. doubling time). Second, *S. cerevisiae* yeast cells exhibit budding patterns that are primarily dependent upon genome ploidy (Chant and Pringle, 1995). In haploid *S. cerevisiae*, new buds usually emerge adjacent to the previous bud site (axial budding pattern); in diploid cells the first bud emerges at the opposite end of the cell from the previous bud site (bipolar budding pattern). *C. albicans* strains are generally diploid (for reviews on the status of mating and the sexual cycle of *C. albicans* (Johnson, 2003; Soll *et al.*, 2003)), yet within a population of wild-type cells, both axial and bipolar budding patterns are observed. Interestingly, the budding pattern of *C. albicans* yeast cells is influenced by temperature: at lower temperatures (~25°C), a higher proportion of cells are axial and at higher temperatures more cells in the population are bipolar (Chaffin, 1984b; Herrero *et al.*, 1999; Gale *et al.*, 2001).

2.2. Pseudohyphal Cells

C. albicans pseudohyphal cells have been less well characterized than yeast or hyphal cells. In part, this is because conditions that yield a population of all pseudohyphae are difficult to achieve. Nonetheless, several mutant strains that grow primarily or exclusively as

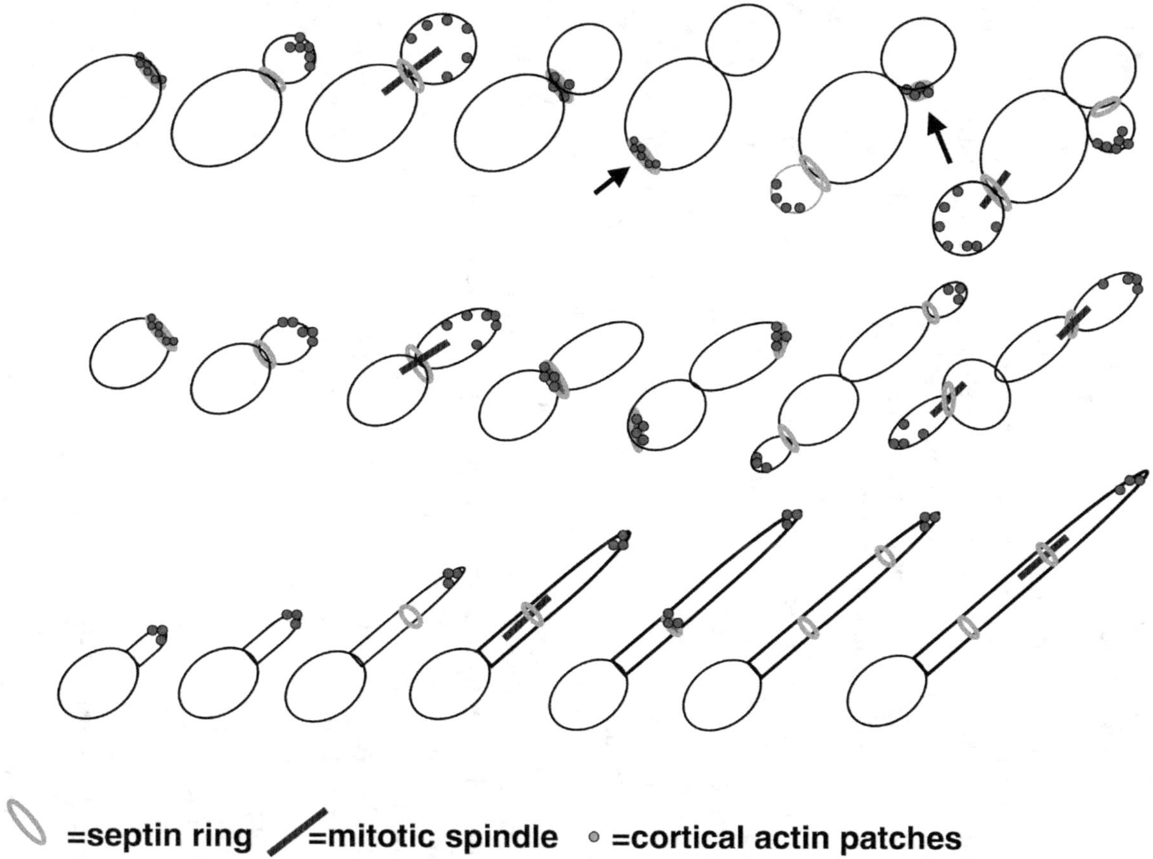

Figure 2. The cell cycle of yeast cells (top panel), pseudohyphae (middle panel) and true hyphae (bottom panel) showing changes in the distribution of actin patches septin rings and position of the mitotic spindle.

pseudohyphae have been described, suggesting that the pseudohyphal state may be a default state of *C. albicans* growth.

Pseudohyphal growth is characterized by elongated budded cells that remain attached to one another in a branching pattern (Figure 1). *C. albicans* pseudohyphae were long thought to be an intermediate state somewhere between yeast and true hyphae (Merson-Davies and Odds, 1989). This idea was reinforced by the fact that pseudohyphal cells can range in length from 5 to 15 microns. Pseudohyphae are generally >2.8 microns wide (Sudbery, 2001). In many cases, long *C. albicans* cells were termed 'filaments' and the distinction between pseudohyphae and true hyphae was not made. However, recent work has demonstrated that pseudohyphae are very different from true hyphae (Sudbery, 2001). There are significant differences between yeast and pseudohyphal cells, as well, although these are less well characterized (Sudbery *et al.*, 2004).

Septin ring placement and actin organization in pseudohyphae resembles that seen in yeast cells. As in yeast cells, septin rings appear prior to bud emergence and septation occurs at the mother bud neck. A constriction at the site of septation is evident in pseudohyphal cells (Figure 2), presumably because as buds emerge through the septin ring the region of growth expands beyond the constricted width proscribed by the septin ring. Actin and secretory vesicle distribution undergoes a series of polarized, followed by isotropic, growth patterns that are similar to those seen in yeast cells, except that the period of polarized actin cortical patch localization to the bud tip lasts for a longer period of time in pseudohyphae relative to yeast cells (Kron *et al.*, 1994). Again, this has not yet been measured carefully. The processes of DNA replication and spindle formation have not been carefully analyzed in pseudohyphal cells.

In *S. cerevisiae*, pseudohyphae form readily in certain strain backgrounds but not in conventional laboratory strains. *S. cerevisiae* pseudohyphal cells appear to spend more time in the G2 phase of the cell cycle, such that daughter cells achieve a size similar to the mother cells prior to cytokinesis (Kron and Gow, 1995). This results in much more synchronous divisions of mother cells relative to daughter cells. *C. albicans* pseudohyphae also have more synchronous division cycles than do yeast-form cells.

Pseudohyphal cells tend to bud in a "unipolar distal" budding pattern, with the first bud formed at the opposite end of the cell from the birth scar and subsequent buds forming at sites adjacent to this first bud. In addition, most pseudohyphal cells, especially if grown on solid medium (rather than subjected to the sheering forces of shaking liquid cultures) remain attached to one another following cytokinesis. This attachment is likely due to the lower levels of chitinase expressed in cells with a pseudohyphal-like morphology (Bensen *et al.*, 2002). A unipolar budding pattern and the attachment of daughter cells gives rise to a characteristic branching pattern in which progeny grow away from the initial mother cell. This feature of pseudohyphal growth is thought to allow the growing culture to 'forage' for nutrients further from the site where the initial cell was placed on the solid medium (Gimeno and Fink, 1992).

In addition to spreading along the surface of solid medium by branching, pseudohyphae can invade the agar medium. The force necessary to invade the solid surface requires surface proteins to adhere to the agar medium and progeny cells to remain attached to one another (Lo and Dranginis, 1998). It is not known whether pseudohyphal cells express surface proteins that are distinct from those expressed in true hyphal cells. However, pseudohyphae do express low levels of Hwp1, a protein expressed at high levels in true hyphae and not expressed in yeast-form cells (J. Berman and P. Sudbery, unpublished data).

2.3. True Hyphal Cells

C. albicans true hyphae are long, narrow tube-like cells (generally ≤ 2 microns wide) with parallel sided walls and no obvious constriction at the site of septation (Figure 2) (Hedden and Buck, 1980). Unlike the yeast and pseudohyphal forms, which appear to have characteristics similar to *S. cerevisiae* yeast and pseudohyphal cells, *C. albicans* hyphal cells have no obvious analogous form in *S. cerevisiae*. Furthermore, because *C. albicans* hyphal cells remain uninucleate while most other filamentous fungi have multinucleate hyphal compartments, there is no model organism with cell cycle progression features highly similar to those of *C. albicans* hyphae.

One of the most obvious features of hyphal cells is that they are not buds. They evaginate from the mother cell to form a "germ tube", the name of the parallel walled structure that is formed prior to formation of the first septum (Odds, 1988). Hyphal cells are narrower than pseudohyphae and lack the constrictions seen at the septa of yeast and hyphal cells. The Morphology Index or Mi is a metric developed to distinguish between true hyphae, pseudohyphae, and yeast (Merson-Davies and Odds, 1989). The length (l), maximum diameter (d) and diameter at the septal junction (s) of each cell is measured and then used in the following formula: $Mi = 2 + 1.78 \log (ls/d^2)$. For hyphal cells, $d \approx s$, because there is no constriction at the septum. In yeast and pseudohyphae, $d > s$. In general, pseudohyphae also have $l > 2d$. Cells that have an Mi of > 3.4 are considered hyphae while those with an Mi value ranging from 2 to 3.4 are considered to be pseudohyphae (Merson-Davies and Odds, 1989) (see Chapter 7 for further discussions on morphology index).

The growth cycle of hyphal cells involves processes that differ significantly from those of yeast and pseudohyphal cells. First, 50% of germ tubes evaginate with a 'random' pattern relative to the previous bud sites. The other 50% evaginate with a bipolar pattern (Chaffin, 1984b; Herrero *et al.*, 1999). This indicates that mechanisms that affect bud site selection have different roles during germ tube evagination. It is interesting to note that genes important for budding pattern site selection, such as *BUD1*, *BUD2*, *BUD5*, and *INT1/BUD4* affect filamentous growth (Gale, C., unpublished observations).

Second, the first true septin ring forms well after evagination of the germ tube and often 10-15 microns from the mother cell. A 'basal septin band' appears at the site where the germ tube begins emerging from the mother cell. This band is composed of three septin proteins (Cdc10p, Cdc11p and Cdc12p), but does not contain all of the components of a true septin ring: it lacks one of the essential septin proteins, Cdc3p (Gale *et al.*, 2001; Sudbery, 2001), and it also lacks the septin ring-associated Gin4 kinase (Wightman *et al.*, 2004) as well as the bud-site selection protein Int1p (C. Gale, unpublished data). The basal septin band appears as striations that run parallel with the mother-daughter axis (Wightman *et al.*, 2004) and that are reminiscent of the septin striations seen in the necks of *S. cerevisiae* mating projections (Longtine *et al.*, 1998). Another similarity between the septin bands in *S. cerevisiae* mating projections and in

C. albicans germ tubes is that the formation of both of these structures is independent of the septin ring-associated Gin4 kinase (Longtine *et al.*, 1998, Wightman *et al.*, 2004). In contrast, the first true septin ring in hyphal cells is found within the germ tube and forms after the tip has grown several microns beyond the site of ring formation. The true septin ring, like septin rings in yeast and pseudohyphae, is composed of all 4 septins (Cdc3, 10, 11 and 12) (Gale *et al.*, 2001; Sudbery, 2001; Warenda and Konopka, 2002), and its formation requires Gin4 kinase (Wightman *et al.*, 2004). As in yeast and pseudohyphal cells, the septin ring in true hyphal cells functions as a scaffold for the actin-myosin ring that eventually contracts to execute cytokinesis. The septin ring itself appears first as a single ring that then splits into two rings flanking the eventual site of septation (Figure 1). An interesting property of hyphal septin rings is that they often persist in the hyphal tube well after septation and through subsequent cell cycles, such that two, three and even four septin rings can be detected in a single hyphal tube (Gale *et al.*, 2001; Sudbery, 2001; Warenda and Konopka, 2002). Furthermore, the hyphal septum contains a micropore that is approximately 25 nm in diameter, which might permit cytosolic continuity but not the movement of organelles such as nuclei and mitochondria between cellular compartments (Gow *et al.*, 1980). However, there is no evidence that the compartments remain continuous, as ultrasonication can release individual compartments which can grow. Furthermore, single compartments respond to electrical fields in a manner that suggests that the hyphal cytoplasm is not connected across septal junctions (Crombie *et al.*, 1990).

After septation, the compartments of true hyphae remain attached to one another and do not readily separate. Unlike pseudohyphal cells, which can be separated from one another by digestion with lyticase (an endoglucanase that digests fungal wall material), true hyphae cannot be easily separated by lyticase treatment (Merson-Davies and Odds, 1989; Sudbery, 2001). Nonetheless, hyphae grown in shaking liquid cultures sometimes break off cell compartments that are then observed to initiate new hyphal or yeast-like growth.

Another striking feature of hyphal cells is that actin patches remain highly polarized at the growing tip throughout the cell cycle: there is no obvious isotropic growth state at all (Anderson and Soll, 1986; Hazan and Liu, 2002; Hazan *et al.*, 2002). While some actin patches appear at the presumptive septum (presumptum) prior to septation, most actin patches remain continuously associated with the growing tip. The myosin light chain protein Mlc1p is found in an organelle with many similarities to the vesicle-rich "Spitzenkörper" structure seen in other filamentous fungi (Crampin, Berman and Sudbery, unpublished data). Furthermore, filipin staining of hyphal tips, like mating projection tips in *S. cerevisiae*, appear to contain lipid raft-like structures in the tip membrane (Bagnat and Simons, 2002; Crampin, Berman and Sudbery, unpublished results). Consistent with the constitutive presence of actin at the growing tip, the polarity establishment proteins Cdc42p and Cdc24p are required for polarized growth and constitutively associate with the tip as well (Hazan and Liu, 2002; Michel *et al.*, 2002; Ushinsky *et al.*, 2002; Bassilana *et al.*, 2003). Furthermore, the concentration of actin and of Cdc42 in hyphal tips appears to be significantly higher than the concentration of these proteins in the tips of growing yeast and pseudohyphal cells (Hazan and Liu, 2002; Bassilana *et al.*, 2003).

After the first cytokinesis in a growing hypha, the tip continues to elongate, with the apical cell entering the next cell cycle, eventually forming a septin ring and dividing the nucleus across the second ring. However, the mother cell does not reenter the cell cycle immediately. It remains quiescent, apparently in the G1 phase of the cell cycle (Barelle *et al.*, 2003). Another feature of cytokinesis is that the apical cell retains a large proportion of the cytoplasm while a large proportion of the mother cell volume is taken up by one or more large vacuoles (Gow *et al.*, 1986; Gow and Gooday, 1987). The mother cell, along with other sub-apical cells, continues to metabolize while remaining in the G1 phase of the cell cycle. After several cycles of apical cell division, a subapical cell may re-enter the cell cycle, by initiating the formation of a hyphal branch, a new germ tube, or a new yeast cell. This appears to occur once the proportion of the cytoplasm occupied by the vacuole has decreased below a threshold level (Barelle *et al.*, 2003). Whether the threshold is measuring vacuolar volume versus cytoplasmic volume or versus nuclear volume is not known. New hyphal branches usually initiate from a site adjacent to an existing hyphal septum (analogous to an axial budding pattern). Because of the delay in reentry into the cell cycle for the non-apical cells, the growth rate of hyphal cultures, especially during the first several cell cycles, approximates a linear, rather than the exponential rate seen in cultures of yeast or pseudohyphal cells (Gow and Gooday, 1982; Gow *et al.*, 1986).

The nuclei in hyphal cells exhibit different dynamics from the nuclei in yeast and pseudohyphal cells. In hyphae, the nucleus must leave the mother cell and divide across the presumptum, which is often 10 or more microns from the mother cell. Following cytokinesis, the mother nucleus returns to the round mother cell, with the rate of mother nucleus reverse movement being much faster than the rate of the forward movement of the daughter nucleus toward the hyphal tip. This observation suggests that the mechanisms or regulation of forward and reverse nuclear migration are different. Microtubule-based motor proteins are likely involved in nuclear migration in one or both directions. In both yeast and hyphal cells, nuclear segregation is defective in cells lacking *DYN1*, the gene encoding the heavy chain of cytoplasmic dynein, a membrane-anchored molecular motor that moves microtubules (Finley and Berman, unpublished results).

The spindle cycle in true hyphae has not been studied extensively. Spindle pole body duplication, which is detected as an intensification of the tubulin signal (via immunofluoresence or using tubulin-GFP fusion proteins), occurs after germ tube evagination (Hazan *et al.*, 2002). Hyphal spindles are much longer than those in yeast cells, and astral microtubules exhibit polarization and depolarization as well as microtubule sliding as they interact with the cell cortex. The microtubule sliding that is observed further suggests the involvement of microtubule motor proteins in nuclear migration (Finley and Berman, unpublished data).

Taken together, the properties of true hyphae, such as the uncoupling of germ tube growth from the G1/S transition, as well as the very different behaviour of the actin cytoskeleton, suggest that the dynamics of cell cycle progression are distinct in hyphae relative to pseudohyphal and yeast form cells. Because of these differences, it is important to distinguish between conditions and mutations that alter this cell cycle aspect of hyphal growth and conditions that alter the degree to which hyphae or pseudohyphae are polarized. For example, there are several mutations, such as deletion of the gene encoding the Fkh2p transcription factor, that result in cells that appear to be constitutively pseudohyphal under all growth conditions. Under some hyphal induction conditions (e.g., pH8, 37°C or Milk Tween medium), *fkh2Δ/fkh2Δ* mutants form pseudohyphae that resemble the *fkh2Δ/fkh2Δ* pseudohyphae that form under yeast growth conditions. Importantly, there are other stimuli (e.g., exposure to 20% serum at 37°C, or *N*-acetyl glucosamine at 37°C) that cause *fkh2Δ/fkh2Δ* mutants to grow as longer pseudohyphal cells. These cells have obvious constrictions at the septa and cell width that are ≥3 microns, and thus are clearly pseudohyphal cells.Yet they respond to signals that stimulate increased polarized growth. Thus, some inducers of hyphal growth (e.g., serum and *N*-acetyl glucosamine) can elicit a polarized growth response that is independent of the cell cycle changes that accompany the shift from yeast or pseudohyphal growth to true hyphal growth.

3. Cyclin Dependent Kinases and Cyclin Proteins

In all eukaryotes, cyclin dependent kinases (CDKs) regulate cell cycle progression and the coordination between cell growth and cell division. In fungi, CDKs also regulate important transitions in morphogenesis and/or developmental state. In *S. cerevisiae*, cell cycle progression is coordinated with cell size at G1/S (START), the point at which DNA replication, spindle pole body duplication, and bud formation initiate in response to G1-cyclin activities. A second, more cryptic control occurs at G2/M. In contrast, in *S. pombe*, the major size control operates in G2 phase and a cryptic size control occurs at G1 that is only revealed in strains such as the *wee1* mutant that has its cell cycle advanced and hence generates small sized cells that have a delayed S phase. In *S. cerevisiae* and *S. pombe,* there is only one major kinase (ScCdc28 = SpCdc2) required for cell cycle progression. In contrast, in *A. nidulans*, two different kinases (*Cdc2=NimX* and *NimA*) are both required for entry into M phase during hyphal growth.

3.1. Roles of Cyclins and Cyclin Dependent Kinases in *S. cerevisiae*

In *S. cerevisiae* the cyclin dependent kinases (CDKs) and their associated cyclin proteins have been studied extensively. The major *S. cerevisiae* CDK is Cdc28, which associates with three G1 cyclins (Clns1-3) and six B-cyclins (Clbs1-6). Another CDK, Pho85, associates with 9 Pcl cyclins (Pcls1-9) as well as with Pho80 to regulate aspects of cell cycle progression as well as phosphate metabolism (Kaffman *et al.*, 1994; Measday *et al.*, 1997; Nishikawa *et al.*, 2002).

Because of partly overlapping capabilities (at least when other cyclins are missing), it has been difficult to determine the exact role of each of the *S. cerevisiae* cyclins. It is clear that different groups of cyclin genes are expressed at different times in the cell cycle and have primary roles in different aspects of cell cycle progression. Beginning with G1, Cln3 regulates the transcription of Clns 1 and 2, which function at START (when they reach a critical concentration) to coordinate cell growth and division by controlling the timing of entry into S-phase relative to cell size (Futcher, 1996; Reed, 1996). Cln3 is unlike the other cyclins in that it is present throughout the cell cycle. This feature allows Cln3/Cdc28 to act as a 'pilot light' to turn on the transcription of Cln1 and Cln2, as well as other transcripts required in late G1 phase (Lew *et al.*, 1997). Cln1 and Cln2 transcript levels are negatively regulated by Clbs1-4, while Cln1 and Cln2 protein levels are thought to be regulated by proteolysis. Targets of the G1 cyclin/CDKs are important for initiating DNA replication, SPB duplication and initiating bud growth.

The B-cyclins can be divided into those that affect S-phase progression (Clb5 and 6) and those that affect mitotic processes such as spindle assembly/disassembly and mitotic exit (Clb1,2,3 and 4). Clb5 and 6 are also required in G1 and early S for the initiation of replication via activation of origins of replication. Clb3 and 4 are activated in S-phase and are required for proper spindle function although this function can be performed by other Clbs if Clb3 and 4 are deleted. Clbs1 and 2 are the only essential pair of B-cyclins: if both are deleted cells do not survive. Clbs1 and 2 are activated last, peaking in M-phase, and are responsible for decreasing the levels of Cln1 and 2 mRNAs by reducing the activity of the SBF transcription factor (Amon *et al.*, 1993). Clb1 and 2 also affect morphogenesis by mediating the switch from polarized to isotropic growth (Lew and Reed, 1995).

3.2. CDK Effects on Morphogenesis in *S. cerevisiae*

The balance of Cln and Clb CDK activities regulates *S. cerevisiae* morphogenesis (Lew *et al.*, 1997). In *S. cerevisiae*, Clns1 and 2 generally promote polarized growth early in the cell cycle while Clb2 down-regulates polarized growth. As cells proceed through S-phase, Cln levels decline and Clb levels rise, causing cells to switch from polarized to isotropic growth. If the switch to isotropic growth is delayed because of excess Clns or low levels of Clbs, cells become more polarized. Consistent with this, overexpression of Clns generates cells with a more polarized morphology and overexpression of Clb2 results in cells with a more isotropic/round morphology. Accordingly, deletion of Clb2, certain mutations of Cdc28, or conditions that inactivate Clb2/Cdc28 (through activation of Swe1) all result in cells that are elongated and resemble pseudohyphal cells (reviewed in Rua *et al.*, 2001). Importantly, cells lacking Cln1, Cln2 and Clb2 had the same elongated phenotype as cells lacking Clb2 alone (Loeb *et al.*, 1999a). In addition, much of the polarized cell phenotype of cells overexpressing Cln1 or Cln2 is dependent upon Swe1, which phosphorylates (and inactivates) Clb2/Cdc28. Thus, the most parsimonious explanation for the role of the cyclins in morphogenesis is that Cln1 and 2 mediate filamentous growth through their negative regulation of the major B-cyclin, Clb2.

3.3. Cyclins and CDKs in *C. albicans*

C. albicans has a homolog of *S. cerevisiae* Cdc28 that can functionally complement mutations in both *S. cerevisiae CDC28* and *S. pombe cdc2* (Sherlock *et al.*, 1994; Damagnez and Cottarel, 1996). *CDC28/CDK1* is an essential gene in *C. albicans* (Roemer *et al.*, 2003). *C. albicans* also has several cyclin dependent kinase-like genes, including *CRK1*, an ortholog of *S. cerevisiae SVG1,* that has an effect on morphogenesis that is independent of the Cph1 and Efg1 transcription factors and independent of Ras1 (Chen *et al.*, 2000). Mutation of *CRK1* also affects virulence and reduces the overall growth rate (Chen *et al.*, 2000). However, it is not clear that Crk1p has a direct role in cell cycle progression. The *C. albicans* sequence (http: //www-sequence.stanford.edu:8080/haploid19.html) also includes a Pho85 homolog and several cyclins that are most similar to the *S. cerevisiae* Pcl cyclins that associate with the *S. cerevisiae* Pho85 CDK. No analysis of the function of *PHO85* in *C. albicans* has been reported.

C. albicans has two G1 cyclins that can provide G1 cyclin function to *S. cerevisiae* (Sherlock *et al.*, 1994). CaCln1 was originally cloned in a screen for genes that interfered with mating-factor induced cell cycle arrest in *S. cerevisiae* cells (Whiteway *et al.*, 1992). The gene, originally named *CCN1* is also equivalent to orf6.4392 and was termed *CaCLN1* by Liu and coworkers (Loeb *et al.*, 1999b). *CaCLN1* mRNA exhibits a cell cycle specific expression pattern like that of *S. cerevisiae CLN1* and *CLN2*, and *Cacln1/Cacln1* mutants exhibit reduced hyphal growth on several solid media (Loeb *et al.*, 1999b). In liquid Lee's medium *Cacln1/Cacln1* mutants initially formed long filaments but after 15 hours the cultures contained primarily pseudohyphal and hyphal cells, unlike the wild-type cultures which continued to produce hyphae (Loeb *et al.*, 1999b). Thus, *C. albicans* Cln1p is not required for filamentous growth, although it appears to promote the maintenance of filamentous growth in wild-type cells. CaCln2p is slightly more similar to *S. cerevisiae* Cln3p than to *S. cerevisiae* Cln1p (4e-18) or Cln2p. Another G1-like cyclin is CaCln21 (orf6.3156), which is most similar to ScCln1 and ScCln2.

There are only two obvious B-cyclins in the *C. albicans* genome sequence: Cyb1 and Cyb4. Cyb1 is most similar to *S. cerevisiae* Clb2 and Cyb4 is most similar to Clb4. *C. albicans CYB1* (orf6.7127) encodes a B-cyclin that can partially complement a mutation in *S. pombe cdc13*, the major G2/M B-cyclin in fission yeast (Damagnez and Cottarel, 1996). Cyb4 was originally identified from the DNA sequence and was given the anonymous name Cyb99 (orf6.7127). The duplication of each cyclin in *S. cerevisiae* is thought to be the result of a duplication of the genome ~100 million years ago (Wolfe and Shields, 1997). This duplication occurred after the *C. albicans* ancestor separated from the *S. cerevisiae* ancestor. Nonetheless, the *C. albicans* sequencing project has not identified any B-type cyclin with obvious similarity to Clbs5 and 6. Thus, DNA replication initiation, and other early events of cell cycle progression that are regulated by Clb5/6 in *S. cerevisiae*, appear to be regulated in a different manner in *C. albicans*. Analysis of the cell cycle roles of the identified cyclins will be needed to elucidate the mechanism(s) used by *C. albicans* to regulate early cell cycle events.

Recent work in the Berman lab indicates that Cyb1 and Cyb4 are negative regulators of polarized growth (Bensen and Berman, unpublished data). Cells depleted of Cyb1 form highly elongated cells that resemble hyphae but that rarely undergo nuclear division or septation. Cells lacking both copies of Cyb4 form primarily pseudohyphae. Cyb1 is an essential gene while Cyb4 is not essential (Bensen and Berman, unpublished data). These results suggest that, while both *C. albicans* B-cyclins are negative regulators of polarized growth, they have different roles in morphogenesis.

3.4. Regulation of Entry into S and M

The cell cycle is under "check-point control" meaning that progression of the cell cycle does not occur until preceding steps are completed successfully. Two major checkpoints exist in eukaryotic cell cycles that regulate entry into the period of DNA replication (S phase) and mitosis (M phase). Entry into S is regulated in G1 at the START event and ensures that cells do not initiate a cell cycle if (a) there is not an adequate supply of nutrients and (b) the bud from the previous cell cycle has not achieved a size equivalent to that of the parent yeast cell. The START event also leads to the evagination of the new bud. Entry into mitosis is regulated at a second checkpoint at the G2/M interface that ensures that mitosis does not occur until DNA replication is complete. A second size-regulated control point exits at the G2/M checkpoint. As discussed above, cellular morphogenesis is also coupled to these two checkpoints (see Lew and Reed, 1993; Sheu and Snyder, 2001). Both checkpoints can therefore be regarded as essential in order to allow critical stages of the cell cycle to be properly executed and to control the shape of the cell.

For budding populations of cells at most growth rates, the buds are born at a size smaller than that of the mother cell. Consequently the cell cycle of the daughter cell is longer than the mother cell since it first grows through an extended G1 phase before it can achieve a threshold size required to initiate START. Therefore, for a batch culture of yeast cells the doubling time of the culture is an average of the two generation times for mother cells (of a shorter generation time) and daughter cells (of a longer generation time). The START event is catalysed by the master switch CDK/Cdc28 along with at least one G1 cyclin. START only occurs when the growth medium contains sufficient nitrogen to permit active growth and when the cells achieve a critical cell size. The Cln3 protein appears to be important in acting as a cell size sensor (Tyers *et al.*, 1993; Rupes, 2002). However, around 500 genes are known to be important in positively and negatively regulating the size of *S. cerevisiae* at START and so there are multiple inputs into the determination of cell size which influences several key aspects of cell cycle control (Jorgensen *et al.*, 2002). The regulation of START by the Cdc28 CDK and associated cyclins is discussed above.

A key event regulated by START is the evagination of the bud. Although not analysed in the same detail as in *S. cerevisiae,* the evagination of buds of *C. albicans* is presumed to be regulated in a similar way. In pseudohyphal cells of *C. albicans*, cell size control over the initiation of filament formation was estimated to be similar to that in the budding phase if the vacuolar volume was omitted from calculations of cell size (Yokoyama and Takeo, 1983). In true hyphae of *C. albicans,* a surprising departure has been described from START-regulated evagination of buds of yeast cells and pseudohyphae. In Liu's laboratory it was shown that germ tubes of true hyphae could emerge from the mother cell before the spindle pole was duplicated, before the chitin ring was formed at the neck, and before DNA synthesis occurred (Hazan *et al.*, 2002). Also, actin polarization leading to hyphal emergence occurred before START, suggesting that the evagination event was uncoupled from the cell cycle. In addition, Cdc28-Tyr19 phosphorylation was also assayed using a phospho-specific antibody as a marker for cell cycle progression and timing in yeast cells and hyphal cells. No marked difference between the timing of cell cycle events of yeast cells (grown at 30°C) and emerging hyphal cells (grown at 37°C) was found, which was interpreted to suggest that cell cycle progression was similar in these morphological forms despite the fact that germ tube evagination occurred prior to START. It was also shown that yeast cells of *C. albicans* that were in phases of the cell cycle other than G1 could also evaginate, again underlining the uncoupling of hyphal growth from the cell cycle. The diameter of germ tubes that were induced early in the cell cycle were narrower than those forming from yeast cells later in the cycle (Hazan *et al.*, 2002; see Chapter 6). Together these data reinforce the view that the regulation of the emergence of buds, that form yeast cells or pseudohyphae are coupled to START, but that the evagination of true hyphae is not. This establishes a surprising paradigm in which hyphal induction and elongation is regulated in parallel with the hyphal cell cycle. The uncoupling of the cell cycle from the evagination event of the *C. albicans* germ tube is more akin to shmoo formation of *S. cerevisiae*.

Another difference in the cell cycle of buds and hyphae of *C. albicans* is evident in the persistence of polarized actin distribution and tip growth throughout the cell cycle in hyphal cells. In buds, the G2/M morphogenetic checkpoint results in the loss of the polarized growth phase and a period of isotropic expansion of the cell (Anderson and Soll, 1986). The same pattern of bud expansion exists in *S. cerevisiae* (Lew and Reed, 1995; Lew *et al.*, 1997). The switching off of the polarized growth phase in the last part of the cell cycle results in a more rounded final shape to the mature bud. In hyphae this G2/M phase morphogenetic checkpoint is by-passed to allow continued apical growth throughout the cell cycle (Staebell and Soll, 1985). Growth by budding in *S. cerevisiae*, and presumably in *C. albicans,* is regulated by the inhibitory activity of Swe1 on the CDK at the G2/M morphogenetic checkpoint that determines when the transition from polarized to isodiametric growth will occur. In *C. albicans*, hyphae form normally in *swe1*Δ/ *swe1*Δ deleted cells (Wightman *et al.*, 2004). Therefore the regulation of hyphal elongation apparently is uncoupled from this Swe1-dependent checkpoint. Pseudohyphal growth of *S. cerevisiae* is also Swe1-independent.

These observations contribute to a growing awareness that there are distinct differences in the cell cycle and other aspects of the cell biology of yeasts, pseudohyphae and true hyphae (Sudbery *et al.*, 2004).

4. Duplication Cycle in Filamentous Fungi

Much less is known about the cell cycle in filamentous fungal pathogens. The cell cycle in molds has been termed "the duplication cycle" by Trinci and others (Fiddy and Trinci, 1976; Trinci *et al.*, 1994; Harris, 1997) in order to distinguish this cycle that occurs in the absence of cell division from that in unicellular microorganisms. The concept of the "cell" is clearly different in filamentous fungi since many hyphae are coenocytic, with many hundreds or thousands of microns of hypha potentially contributing to growth of the apex. This led to the concept of the "peripheral growth zone" of a hypha that is defined as the length of hypha contributing to growth of the hyphal apex (Trinci, 1971). This measurement may or may not approximate the value of the "hyphal growth unit" which is derived from measurement of the total hyphal length of an intact mycelium divided by the total number of hyphal tips. The increase in total hyphal length of a filamentous mold follows an exponential function. The rate of increase in total hyphal length is the same as that for the increase in nucleus number and number of branches (Trinci, 1974). This reflects underlying cell cycle regulation that leads to a constant rate of doubling of all cellular components in a non-nutrient limited medium. Clearly therefore, filamentous fungi have a similar cell cycle to yeasts despite differences in cell shape, the lack of cell separation at the end of the cycle, and the fact that many nuclei may share the same cytoplasm.

Analysis of the cell cycle and polarized growth of filamentous fungi has been focused largely on *Aspergillus nidulans* (see Harris, 1997; Hamer *et al.*, 1999; Turner and Harris, 1999; Momany, 2002 for excellent reviews) and is beyond the scope of this review on pathogenic fungal species. It is likely however that studies in this fungus will closely reflect cell cycle events in *Aspergillus fumigatus*. Filamentous Ascomycetes like *Aspergillus* species have incomplete septa and so functional cells may equate to several contiguous, cytoplasmically connected compartments. However these compartments can be plugged with Woronin bodies meaning that compartments can become isolated as individual functional cells. This generally occurs in older parts of mycelia or as a response to damage that isolates a disrupted region of a mycelium (Gull, 1978; Jedd and Chua, 2000). Therefore the concept of the "cell" in a filamentous mold is subtle and much more complex than of a yeast cell or even a hypha such as those of *C. albicans* that do not have cytoplasmic continuity between adjacent compartments.

In *A. nidulans*, conidia undergo a period of swelling before the germ tube emerges and three rounds of nuclear division occur before the first septation event occurs (Harris *et al.*, 1994). Hydroxyurea and microtubule inhibitors, which block DNA synthesis and nuclear division respectively, prevent septation (Momany and Hamer, 1997). However, temperature sensitive *sepA* mutants can grow and branch normally at the restrictive temperature in the absence of septation. Therefore septation is not required absolutely for the duplication cycle (Sharpless and Harris, 2002). When the *sepA* mutant was shifted to the permissive temperature, septation occurred at normal positions, but only if nuclear division was able to occur. These data suggest that nuclear division marks the sites of septation in hyphae. Key components of the cell cycle regulatory machinery have been identified in *A. nidulans* such as the NimX gene which encodes the homologue of $p34^{Cdc28/Cdc2}$ master kinase of yeast cells have been identified in *A. nidulans* (Hamer *et al.*, 1999). Ongoing genome sequencing projects of *A. fumigatus* and many other filamentous fungal pathogens will surely recognize homologues of CDKs, cyclins and other cell cycle regulatory proteins in fungal pathogens in the coming months and years.

5. Cell Cycle and Fungal Pathogenesis

Regulation of the cell cycle may seem at first to be of indirect relevance to the biology and virulence of fungal pathogens. However, the cell cycle is intimately involved with the regulation of morphogenetic processes that are linked with invasion or correlated with growth *in vivo*. Several fungal pathogens exhibit changes in cell shape from a yeast-like form to a filamentous form, and this dimorphic behaviour has often been considered to be an important virulence factor. However while most studies have examined hypha production in *C. albicans* in the context of its pathogenesis, it is the yeast form of other pathogens that is found *in vivo*. These yeast-form pathogens include *Histoplasma capsulatum*, *Paracoccidioides brasiliensis*, *Cryptococcus neoformans* and *Penicillium marneffei* which all form hyphae in the external environment (Gow *et al.*, 2002; see Chapter 5). In addition, structures such as the spherule of *Coccidioides immitis*, which is produced in human tissues, are critical for dissemination of disease and must involve specific cell cycle regulation to allow endosporulation within a large multinucleated cell. In this context regulation of the cell cycle is a vital component of the cell regulatory machinery that enables fungal pathogens to flourish in the host environment. Further research in this area will therefore help to understand the shape and form of fungal cells as well as the relationship between morphogenesis, the cell cycle and pathogenesis.

6. Acknowledgements

We thank Pete Sudbery, Helen Crampin, Ray Wightman, Eric Bensen and Cheryl Gale for allowing us to cite data prior to publication. Judith Berman is supported by the NIH (DE/AI14666). Neil Gow thanks the Wellcome Trust, BBSRC and MRC for financial support.

References

Akashi, T., Kanbe, T., and Tanaka, K. 1994. The role of the cytoskeleton in the polarized growth of the germ tube in *Candida albicans*. Microbiology 140: 271-280.

Amon, A., Tyers, M., Futcher, B., and Nasmyth, K. 1993. Mechanisms that help the yeast cell cycle clock tick: G2 cyclins transcriptionally activate G2 cyclins and repress G1 cyclins. Cell 74: 993-1007.

Anderson, J.M., and Soll, D.R. 1986. Differences in actin localization during bud and hypha formation in the yeast *Candida albicans*. J. Gen. Microbiol. 132: 2035-2047.

Bagnat, M., and Simons, K. 2002. Cell surface polarization during yeast mating. Proc. Natl. Acad. Sci. USA 99: 14183-14188.

Barelle, C.J., Bohula, E.A., Kron, S.J., Wessels, D., Soll, D.R., Schafer, A., Brown, A.J.P., and Gow, N.A.R. 2003. Asynchronous cell cycle and asymmetric vacuolar inheritance in true hyphae of *Candida albicans*. Eukaryot. Cell 2: 398-410.

Barton, R. and Gull, K. 1988. Variation in cytoplasmic microtubule organization and spindle length in the two forms of the dimorphic fungus *Candida albicans*. J. Cell Sci. 91: 211-220.

Bassilana, M., Blyth, J., and Arkowitz, R.A. 2003. Cdc24, the GDP-GTP exchange factor for Cdc42, is required for invasive hyphal growth of *Candida albicans*. Eukaryot. Cell 2: 9-18.

Bensen, E.S., Filler, S.G., and Berman, J. 2002. A forkhead transcription factor is important for true hyphal as well as yeast morphogenesis in *Candida albicans*. Eukaryot. Cell 1: 77-98.

Chaffin, W.L. 1984a. The relationship between yeast cell size and cell division in *Candida albicans*. Can. J. Microbiol. 30: 192-203.

Chaffin, W.L. 1984b. Site selection for bud and germ tube emergence in *Candida albicans*. J. Gen. Microbiol. 130: 431-440.

Chant, J., and Pringle., J.R. 1995. Patterns of bud-site selection in the yeast *Saccharomyces cerevisiae*. J. Cell Biol. 129: 751-765.

Chen, J., Zhou, S., Wang, Q., Chen, X., Pan, T., and Liu. H. 2000. Crk1, a novel Cdc2-related protein kinase, is required for hyphal development and virulence in *Candida albicans*. Mol. Cell Biol. 20: 8696-8708.

Crombie, T., Gow, N.A.R., and Gooday, G.W. 1990. Influence of applied electrical fields on yeast and hyphal growth of *Candida albicans*. J. Gen. Microbiol. 136: 311-317.

Damagnez, V., and Cottarel, G. 1996. *Candida albicans CDK1* and *CYB1*: cDNA homologues of the cdc2/*CDC28* and cdc13/*CLB1*/*CLB2* cell cycle control genes. Gene 172: 137-141.

Fiddy, C., and Trinci, A.P.J. 1976. Mitosis, septation, branching and the duplication cycle in *Aspergillus nidulans*. J. Gen. Microbiol. 97: 169-184.

Futcher, B. 1996. Cyclins and the wiring of the yeast cell cycle. Yeast 12: 1635-1646.

Gale, C.A., Gerami-Nejad, M., McClellan, M., Vandoninck, S., Longtine, M.S., and Berman, J. 2001. *Candida albicans* Int1p interacts with the septin ring in yeast and hyphal cells. Mol. Biol. Cell. 12: 3538-3549.

Gimeno, C.J., and Fink, G.R. 1992. The logic of cell division in the life cycle of yeast. Science 257: 626_

Gow, N.A.R., and Gooday, G.W. 1982. Growth kinetics and morphology of colonies of the filamentous form of *Candida albicans*. J. Gen. Microbiol. 128: 2187-2194.

Gow, N.A.R., and Gooday, G.W. 1987. Cytological aspects of dimorphism in *Candida albicans*. Crit. Rev. Microbiol. 15: 73-78.

Gow, N.A.R., Henderson, G., and Gooday, G.W. 1986. Cytological interrelationships between the cell cycle and duplication cycle of *Candida albicans*. Microbios 47: 97-105.

Gow, N.A.R., Gooday, G.W., Newsam, R.J., and Gull, K. 1980. Ultrastructure of the septum in *Candida albicans*. Curr. Microbiol. 4: 357-359.

Gow, N.A.R., Brown, A.J.P., and Odds, F.C. 2002. Fungal morphogenesis and host invasion. Curr. Opin. Microbiol. 5: 366-371.

Gull, K. 1978. Form and function of septa in filamentous fungi. In: The Filamentous Fungi, Volume 3. J.E. Smith and D.R. Berry, eds. Arnold, London. p.78-93.

Hamer, J.E., Morrell, J.A., Hamer, L., Wolkow, T., and Momany, M. 1999. Cellularization in *Aspergillus nidulans*. In: The Fungal Colony. N.A.R. Gow, G.D. Robson and G.M. Gadd, eds. Cambridge University Press, Cambridge. p. 201-228.

Harris, S.D. 1997. The duplication cycle in *Aspergillus nidulans*. Fungal Gen. Biol. 22: 1-12.

Harris, S.D., Morrell, J.L., and Hamer, J.E. 1994. Identification and characterisation of *Aspergillus nidulans* mutants defective in cytokinesis. Genetics 136: 517-532.

Hazan, I., and Liu, H. 2002. Hyphal tip-associated localization of Cdc42 is F-actin dependent in *Candida albicans*. Eukaryot. Cell 1: 856-864.

Hazan, I., Sepulveda-Becerra, M., and Liu, H. 2002. Hyphal elongation is regulated independently of cell cycle in *Candida albicans*. Mol. Biol. Cell 13: 134-145.

Hedden, D.M., and Buck, J.D. 1980. A re-emphasis- germ tubes diagnostic for *Candida albicans* have no constrictions. Mycopathologia 70: 95-101.

Herrero, A.B., Lopez, M.C., Fernandez-Lago, L. and Dominguez, A. 1999. *Candida albicans* and *Yarrowia lipolytica* as alternative models for analysing budding patterns and germ tube formation in dimorphic fungi. Microbiology 145: 2727-2737.

Jedd, G., and Chua, N.H. 2000. A new self-assembled peroxisomal vesicle required for efficient resealing of the plasma membrane. Nat. Cell Biol. 2: 226-231.

Johnson, A. 2003. The biology of mating in *Candida albicans*. Nat. Rev. Microbiol. 1: 106-115

Jorgensen, P., Nishikawa, J.L., Breitkreutz, B-J., and Tyers, M. 2002. Systematic identification of pathways that couple cell growth and division in yeast. Science 297: 395-400.

Kaffman, A., Herskowitz, I., Tjian, R., and O'Shea, E.K. 1994. Phosphorylation of the transcription factor PHO4 by a cyclin-CDK complex, PHO80-PHO85. Science 263: 1153-1156.

Kron, S.J., and Gow, N.A.R. 1995. Budding yeast morphogenesis: signaling, cytoskeleton and cell cycle. Curr. Opin. Cell Biol. 7: 845-855.

Kron, S. J., Styles, C.A., and Fink, G.R. 1994. Symmetric cell division in pseudohyphae of the yeast *Saccharomyces cerevisiae*. Mol. Biol. Cell 5: 1003-1022.

Kuranda, M.J., and Robbins, P.W. 1991. Chitinase is required for cell separation during growth of *Saccharomyces cerevisiae*. J. Biol. Chem. 266: 19758-19767.

Lew D.J., and Reed, S.I. 1993. Morphogenesis in the yeast cell cycle: regulation by Cdc28 and cyclins. J. Cell Biol. 120: 1305-1320.

Lew, D.J., and Reed, S.I. 1995. Cell cycle control of morphogenesis in budding yeast. Curr. Opin. Genet. Dev. 5: 17-23.

Lew, D.J., Weinert, T., and Pringle, J.R. 1997. Cell cycle control in *Saccharomyces cerevisiae*.. In: The Molecular and Cellular Biology of the Yeast *Saccharomyces:* Cell Cycle and Cell Biology, J.R. Pringle, J.R. Broach, and E.W. Jones, eds. Cold Spring Harbor Press, Cold Spring Harbor. p 607-696.

Lo, W.S., and Dranginis, A.M. 1998. The cell surface flocculin Flo11 is required for pseudohyphae formation and invasion by *Saccharomyces cerevisiae*. Mol. Biol. Cell 9: 161-171.

Loeb, J.D., Kerentseva, T.A., Pan, T., Sepulveda-Becerra, M., and Liu, H. 1999a. *Saccharomyces cerevisiae* G1 cyclins are differentially involved in invasive and pseudohyphal growth independent of the filamentation mitogen-activated protein kinase pathway. Genetics 153: 1535-1546.

Loeb, J.D., Sepulveda-Becerra, M., Hazan, I., and Liu, H. 1999b. A G1 cyclin is necessary for maintenance of filamentous growth in *Candida albicans*. Mol. Cell Biol. 19: 4019-4027.

Longtine, M.S., Fares, H., and Pringle, J.R. 1998. Role of the yeast gin4p protein kinase in septin assembly and the relationship between septin assembly and septin function. J. Cell Biol. 143: 719-736.

McCreath, K.J., Specht, C.A., Liu, Y., and Robbins, P.W. 1996. Molecular cloning of a third chitinase gene (*CHT1*) from *Candida albicans*. Yeast 12: 501-504.

McCreath, K.J., Specht, C.A., and Robbins, P.W. 1995. Molecular cloning and characterization of chitinase genes from *Candida albicans*. Proc. Natl. Acad. Sci. USA 92: 2544-2548.

Measday, V., Moore, L., Retnakaran, R., Lee, J., Donoviel, M., Neiman, A.M., and Andrews, B. 1997. A family of cyclin-like proteins that interact with the Pho85 cyclin- dependent kinase. Mol. Cell. Biol. 17: 1212-1223.

Merson-Davies, L.A., and Odds, F.C. 1989. A morphology index for characterization of cell shape in *Candida albicans*. J. Gen. Microbiol. 135: 3143-3152.

Michel, S., Ushinsky, S., Klebl, B., Leberer, E., Thomas, D., Whiteway, M., and Morschhauser, J. 2002. Generation of conditional lethal *Candida albicans* mutants by inducible deletion of essential genes. Mol. Microbiol. 46: 269-280.

Momany, M. 2002. Polarity in filamentous fungi: establishment, maintenance and new axes. Curr. Opin. Microbiol. 5: 580-585.

Momany, M., and Hamer, J.E. 1997. The relationship of actin, microtubules and crosswall synthesis during septation in *Aspergillus nidulans*. Cell Mot. Cytoskel. 38: 373-384.

Nishikawa, A., Poster, J.B., Jigami, Y., and Dean, N. 2002. Molecular and phenotypic analysis of CaVRG4, encoding an essential Golgi apparatus GDP-mannose transporter. J. Bacteriol. 184: 29-42.

Odds, F.C. 1988. Candida and Candidosis. Baillière Tindall, London. pp 468

Reed, S.I. 1996. G1/S regulatory mechanisms from yeast to man. Prog. Cell Cycle Res. 2: 15-27.

Roemer, T., Jiang, B., Davison, J., Ketela, T., Veillette, K., Breton, A., Tandia, F., Linteau, A., Sillaots, S., Marta, C., Martel, N.,

Veronneau, S., Lemieux, S., Kauffman, S., Becker, J., Storms, R., Boone, S., and Bussey, H. 2003. Large-scale essential gene identification in *Candida albicans* and applications to antifungal drug discovery. Mol. Biol. 50: 167-182.

Rua, D., Tobe, B.T., and Kron, S.J. 2001. Cell cycle control of yeast filamentous growth. Curr. Opin. Microbiol. 4: 720-727.

Rupeš, I. 2002. Checking cell size in yeast. Trends Genet. 18: 479-485.

Selvaggini, S., Munro, C.A., Pachoud, S., Sanglard, D., and Gow, N.A.R. 2004. Independent regulation of chitin synthase and chitinase activity in *Candida albicans*. Microbiology. In press.

Sharpless, K.E., and Harris, S.D. 2002. Functional characterization and localization of the *Aspergillus nidulans* formin SepA. Mol. Biol.Cell 13: 469-479.

Sherlock, G., Bahman, A.M., Mahal, A., Shieh, J.C., Ferreira, M., and Rosamond. J. 1994. Molecular cloning and analysis of *CDC28* and cyclin homologues from the human fungal pathogen *Candida albicans*. Mol. Gen. Genet. 245: 716-723.

Sheu, Y-J., and Snyder, M. 2001. Control of cell polarity and shape. In: The Mycota VIII. R.J. Howard and N.A.R. Gow, eds. Springer-Verlag, Heidelberg. p. 19-53.

Soll, D.R., Lockhart, S.R. and Zhao, R. 2003. Relationship between switching and mating in *Candida albicans*. Eukaryot. Cell 2: 390-397.

Staebell, M., and Soll, D.R. 1985. Temporal and spatial differences in cell wall expansion during bud and mycelium formation in *Candida albicans*. J. Gen. Microbiol. 131: 1467-1480.

Sudbery, P.E. 2001. The germ tubes of *Candida albicans* hyphae and pseudohyphae show different patterns of septin ring localization. Mol. Microbiol. 4: 19-31.

Sudbery, P., Gow, N.A.R., and Berman, J. 2004. The distinct morphogenic states of *Candida albicans*. Trends Microbiol. In press.

Trinci, A.P.J. 1971. Influence of the width of the peripheral growth zone on the radial growth rate of fungal colonies on solid media. J. Gen. Microbiol. 76: 325-344.

Trinci, A.P.J. 1974. A study of the kinetics of hyphal extension and branch initiation of fungal mycelia. J. Gen. Microbiol. 81: 225-236.

Trinci, A.P.J., Weibe, M.G., and Robson, G.D. 1994. The mycelium as an integrated entity. In: The Mycota I, Growth, Differentiation and Sexuality. J.G.H. Wessels and H. Meinhardt, eds. Springer-Verlag, Berlin. p. 175-193.

Turner, G., and Harris, S.D. 1999. Genetic control of polarized growth and branching in filamentous fungi. In: The Fungal Colony. N.A.R. Gow, G.D. Robson and G.M. Gadd, eds. Cambridge University Press, Cambridge. p. 229-260.

Tyers, M., Tokiwa, G., and Futcher, B. 1993. Comparison of the *Saccharomyces cerevisiae* G1 cyclins: Cln3 may be an upstream activator of Cln1, Cln2 and other cyclins. EMBO J. 12: 1955-1968.

Ushinsky, S.C., Harcus, D., Ash, J., Dignard, D., Marcil, A., Morchhauser, J., Thomas, D.Y., Whiteway, M., and Leberer, E. 2002. *CDC42* is required for polarized growth in human pathogen *Candida albicans*. Eukaryot. Cell 1: 95-104.

Wain, W.H., Price, M.F., Brayton A.R., and Cawson, R.A. 1976. Macromolecular syntheses during the cell cycles of yeast and hyphal phases of *Candida albicans*. J. Gen. Microbiol. 972: 211-217.

Warenda, A.J., and Konopka, J.B. 2002. Septin function in *Candida albicans* morphogenesis. Mol. Biol. Cell 13: 2732-2746.

Whiteway, M., Dignard, D., and Thomas, D.Y. 1992. Dominant negative selection of heterologous genes: isolation of *Candida albicans* genes that interfere with *Saccharomyces cerevisiae* mating factor-induced cell cycle arrest. Proc. Natl. Acad. Sci. USA 89: 9410-9414.

Wightman, R., Bates, S., Amnoraptanapan, P., and Sudbery, P. 2004. In *Candida albicans*, the Nim1 kinases Gin4 and Hsl1 negatively regulate pseudohypha formation and Gin4 also controls septin organization. Journal of Cell Biology. In press.

Wolfe, K.H., and Shields, D.C. 1997. Molecular evidence for an ancient duplication of the entire yeast genome. Nature 387: 708-713.

Yokoyama, K., and Takeo, K. 1983. Differences of asymmetrical division between the pseudomycelial and yeast forms of *Candida albicans* and their effect on multiplication. Arch. Microbiol. 134: 251-253.

Yokoyama, K., Kaji. H., Nishimura, K., and Miyaji, M. 1990. The role of microfilaments and microtubules in apical growth and dimorphism of *Candida albicans*. J. Gen. Microbiol. 136: 1067-1075.

From: Pathogenic Fungi: Structural Biology and Taxonomy
Edited by: G. San-Blas and R.A. Calderone

Chapter 4

Morphogenesis in *Candida albicans*

Tamaki Cho

Abstract

The human pathogen *Candida albicans* displays polymorphic growth. That is, it grows as unicellular yeast but also converts reversibly to filamentous growth (hyphae) by germination of yeast cells, a process that is referred to as morphogenesis and can form an intermediate type of growth or pseudohyphae. This polymorphic growth is regulated by a number of environmental conditions and is considered to be one of virulence factors of the organism. A number of groups have studied the factors that influence morphogenesis in this organism. More recent approaches that emphasize molecular genetics have rapidly appeared in the literature. Functional analysis of genes associated with morphogenesis in *C. albicans* has included the identification of regulatory systems with the objective of constructing protein networks that regulate growth patterns. In this chapter, the literature will be summarized on several aspects of morphogenesis, including, 1) the identification of growth form-specific proteins and morphogenesis signaling pathways, 2) recent studies that examine transcriptional profiling of morphogenesis through microarray approaches, and 3) gene analysis of the early stages of morphological conversion of yeast to hyphae.

1. Morphogenesis in *Candida* and Other Human Pathogens

Fungal vegetative morphotypes are broadly classified in three groups. The monomorphic fungi produce a single type of growth, such as the mycelial (filamentous) fungi and monomorphic yeast. In the case of the mycelial fungi such as *Aspergillus fumigatus* and *Trichophyton rubrum,* these organisms grow vegetatively as mycelia but also reproduce asexually by forming spores called conidia. Monomorphic yeasts include *Cryptococcus neoformans* and *Candida glabrata*, that replicate by budding so that growth is exponential in these organisms. The second group of fungi is referred to as the dimorphic fungi. Many dimorphic pathogenic fungi display a mycelial/hyphal growth in the environment where they exist as saprophytes but grow as yeast when causing disease. Some examples of this group include, *Blastomyces dermatitidis, Histoplasma capsulatum, Paracoccidioides brasiliensis, Penicillium marneffei* and *Sporothrix schenckii* (see Chapter 5 for morphogenesis of these fungi). The type of growth (unicellular or filamentous) is temperature-dependent. Thus, these fungi produce mycelial growth at lower temperatures (20-30°C) but are unicellular at or near body temperature (35-37°C). The yeast monomorphic fungi replicate by budding except for *P. marneffei,* which divides by fission. The third group of fungi is represented by the human pathogens *Candida albicans* and *Candida dubliniensis* that are described as polymorphic.

C. albicans, for example, undergoes a reversible, morphogenetic transition among budding, pseudohyphal, and hyphal growth forms. The morphology index (Mi) of *C. albicans* was determined to express the physical growth characteristics of *C. albicans* (Merson-Davies and Odds, 1989). Mi is calculated using the equation l (maximum length)•s (septal diameter)/d^2 (maximum diameter) and is usually 1 for spherical yeast cells and 4 for true hyphae. Elongated yeast cells, and pseudohyphae are given intermediate values (see Chapter 7 for a detailed account of Mi). Pseudohyphae are chains that join as unidirectional, elongated yeast cells. Germ tubes are elongated growth forms that originate from a parent yeast cell and become extended to form filaments that are true hyphae. Compartments in true hyphae are separated by cross-walls or septa. True hyphae are usually observed as straight filaments. However, spiral hyphae are observed under an anaerobic condition *in vitro* (Kaminishi *et al.*, 1994). Cytological aspects that influence cell shape in *C. albicans* include the dynamic changes in actin localization during bud and hyphae formation in *C. albicans* as shown by Anderson and Soll (1986). Actin granules at early germ tube evagination are clustered in both budding and hyphae-forming cells, but during subsequent bud and hyphal growth, the distinct differences are observed in actin granule localization. During hyphal growth, the actin granules cluster at the hyphal apex, whereas the actin granules that occur in cells that bud are redistributed throughout the cytoplasmic cortex. They suggested differences in cell wall development between bud and hyphae. Hazan *et al.* (2002) also studied the localization of actin and spindle pole bodies in both yeast and hyphal forms of *C. albicans.* They proposed that germ tube with or without constrictions are visible if the hyphal program is on after or before budding. Gow and Gooday (1987) reported that a parent yeast cell with an elongating germ tube is accompanied by a vacuolar space in the cell. Brown (2002a) proposed two types of developmental relationships among yeast, pseudohyphal, and hyphal cells. First, pseudohyphal cells are represented as intermediate forms between budding forms and hyphal forms, or, secondly, pseudohyphal cells have a separate developmental pathway. Ernst (2000a) has discussed the difference between pseudohyphae and hyphae, which are induced by different environmental conditions, and has proposed along with other investigators that yeasts (unicellular) and filamentous forms are the end products of different signal transduction pathways.

Environmental factors that regulate morphogenesis in *C. albicans* include, temperature, pH, oxygen concentration, carbon dioxide levels, inoculum size, and chemicals such as N-acetyl-D-glucosamine, L-proline, glucose, serum (not defined components) (Odds, 1988). For example, a temperature at or near 37°C and an inoculum of $< 10^6$ yeast cells result in the conversion of blastospores (yeasts) to filamentous forms or hyphae. When an initial yeast cells concentration exceeds 10^6 cells/ml, mycelial development of *C. albicans* in the medium is prevented (Odds, 1988). Regulation of germination by the size of the inoculum is apparently due to the production of an extracellular quorum-sensing molecule, farnesol. Prior to these reports in *C. albicans*, most studies on the quorum-sensing phenomenon were reported in bacteria (Fuqua *et al.*, 1994), but Hornby *et al.* (2001) identified an extracellular signal molecule (farnesol: $C_{15}H_{26}O$), which mediates an eukaryotic quorum-sensing system in *C. albicans.* The addition of farnesol to media that normally induce germination suppressed the morphological conversion of yeast cells to hyphae, although the yeast cells continue to grow so that farnesol does not negatively regulate budding.

However, the morphological response of *C. albicans* to these and many other factors is complicated by the use of different strains of the organism to test various types of inducers/inhibitors. Recently, the molecular mechanisms controlling these responses have been demonstrated. *RIM20* (Davis *et al.*, 2000a), *PRR1/RIM8* (Davis *et al.*, 2000a; 2000b), and *PRR2/RIM101* (Ramon *et al.*, 1999) were isolated as genes encoding pH signaling proteins. Prr2/Rim101 is a zinc-finger transcription factor that is an activator of the alkaline-expressed gene *PHR1* (Saporito-Irwin *et al.*, 1995) and a repressor of the acid-expressed gene *PHR2* (Muhlschlegel and Fonzi, 1997). Additional transcription factors have been described. A hyphal growth observed in embedded conditions that might be low oxygen conditions would be contributed to *CZF1* that encodes a zinc finger protein (Whiteway *et al.*, 1992; Brown *et al.*, 1999).

2. Hyphal-Specific Proteins

De Bernardis *et al.* (1994) used a monoclonal antibody to show that cell surface expression of mannoprotein antigens is modulated during morphogenesis of *C. albicans*. Mannoproteins on the cell surface recognize host cells, and function as molecules for colonization and adherence in addition to other functions (Calderone, 1993). The adherence mannoproteins are discussed below. Since some of these proteins are hyphal-specific, this fact may account for the increased adherence of hyphae to host cells in comparison to yeast cells. Hence, changing of the cell surface antigens of *C. albicans* during morphogenesis could be one of the ways in which the conversion process contributes to the virulence of *C. albicans*.

Morphological-specific proteins in *C. albicans* include cell surface and cell wall proteins, invasive enzymes (see Schaller *et al.*, 2004), and biosynthetic enzymes. *CaCHS2*, *HYR1*, *HWP1*, *ECE1*, and *ALS3* were isolated as genes encoding hyphal-specific proteins. Bowen *et al.* (1992) designed degenerate primers to perform polymerase chain reaction (PCR) from conserved sequences of three genes: *Saccharomyces cerevisiae CHS1* (chitin synthase gene) and *CHS2*, and *CaCHS1* (*C. albicans CHS1*). PCR of genomic DNA from 14 fungal species resulted in the amplification of fragments that were homologous to the chitin synthases, among which *CaCHS2* was identified. A hydrophobicity plot showed that *CaCHS2* has a hydrophilic tail at the amino terminus that is not homologous to any of the other synthases (Chen-Wu *et al.*, 1992). *CaCHS2* was preferentially expressed in a medium which stimulates hyphal growth (a synthetic starch and vitamin medium, 40°C) using a low inoculum density (4x10^6 cells/ml) of yeast cells grown to stationary phase in a nutrient broth at 37°C for 24 h. Under similar growth conditions, expression of *CaCHS1* was minimal indicating that *CaCHS2* is probably hyphal-specific while *CaCHS1* is not (Chen-Wu *et al.*, 1992). However, *CaCHS2* is not essential for growth as yeast or hyphae since a *cachs2* mutant of *C. albicans* grew normally (Gow *et al.*, 1994).

Bailey *et al.* (1996) isolated *HYR1* (hyphally regulated gene) as a false positive in an immunological screen for genes encoding the β-subunit of *C. albicans* G proteins. They performed a database analysis of the predicted amino acid sequence and suggested the presence of a hydrophobic C-terminus that was consistent with the attachment of a glycosylphosphotidylinositol anchor. Moreover, the mRNA of *HYR1* was detected only in hyphae induced in two hyphal growth conditions. Those conditions were, yeast cells grown to late exponential phase in YPD (yeast extract, peptone, dextrose) medium at 25°C and then induced to form hyphae in YPD with serum at 37°C (2x10^7 cells/ml), or yeast cells grown to late exponential phase in Soll's medium (Soll, 1986) at 25°C, pH 4.5, were induced to form hyphae in the same medium but at 37°C and at pH 6.5 (2x10^7 cells/ml). *HYR1* was disrupted by the Urablaster method (Fonzi and Irwin, 1993) and the mutants tested for growth rates and morphological transition in two hyphal-inducing growth conditions. However, the growth rates and morphological transition in the *hyr1* null mutant were unaffected. Hence, *HYR1* encodes a nonessential component of the hyphal cells.

HWP1, ECE1 and *ALS* genes were isolated using morphologic-specific library and probes from *C. albicans* (Table 1). The isolation of *HWP1* and *ECE1* was performed using morphologic-specific cDNA library from stationary-phase yeast, but the isolation of *ALS1* was performed using a cDNA library constructed from mid exponential-phase yeast cells. *HWP1* (hyphal wall protein) was isolated by a screening of hyphal-specific cDNA library using hyphal-specific antiserum which was obtained from rabbits immunized with germ tubes of *C. albicans* (Sundstrom and Aliaga, 1992). Hyphal growth conditions for construction of the latter library and probes to isolate *HWP1* were as follows. Yeast cells were grown to stationary phase in YPD medium at 30°C for 48 h, then incubated in M-199 medium without serum (pH7) at 37° C for 60 min. *HWP1* encodes an outer cell

Table 1. Gene isolation using morphology-specific libraries and probes

	Library		Probe	
	Yeast-specific	Hyphal-specific	Yeast-specific	Hyphal-specific
HWP1 (Sundstrom and Aliaga, 1992)		YPD at 30 °C for 48 h (stationary phase) ▼ M-199 (pH7) at 37 °C for 60 min ↓ mRNA → cDNA		YPD at 30 °C for 48 h (stationary phase) ▼ M-199 (pH7) at 37 °C for 60 min  → antiserum
ECE1 (Birse *et al*., 1993)	YPD at 30 °C for 48 h (stationary phase) ↓ total RNA → cDNA	YPD at 30 °C (stationary phase) ▼ M-199 (pH7) at 37 °C for 60 min ↓ total RNA → cDNA	YPD at 30 ℃ for 48 h (stationary phase) ▼ Lee (pH4.5) at 25 ℃ for 90 min ↓ total RNA → cDNA	YPD at 30 °C (stationary phase) ▼ Lee (pH6.5) at 37 °C for 90 min ↓ total RNA → cDNA
ALS1 (Hoyer *et al*., 1995)		YPD at 25 °C (mid-log phase) ▼ RPMI at 37 °C for 20 min ↓ total RNA → cDNA	YPD at 25 ℃ (mid-log phase) ↓ total RNA → cDNA	YPD at 25 °C (mid-log phase) ▼ RPMI at 37 °C for16 h ↓ total RNA → cDNA

wall mannoprotein with similarities to mammalian small proline-rich proteins, which resemble substrates of epithelial cell transglutaminase (TGase) (Staab *et al.*, 1996). Hwp1p appeared only on hyphal surfaces of *C. albicans*. Thus, the small proline-rich region of Hwp1 was believed to serve as a substrate for the mammalian TGase (Staab *et al.*, 1999). A disruptant of *HWP1* did not express the protein on its germ tubes and was unable to form stable, covalent adherence to human buccal epithelial cells *in vitro*. Hence, Hwp1 contributes to adherence through the formation of covalent linkages between germ tubes of *C. albicans* and mammalian cells (Staab *et al.*, 1999). However, the null mutant of *HWP1* grew at the same growth rate and germinated equally well as wild type cells (Sharkey *et al.*, 1999).

ECE1 (extent of cell elongation) was isolated by differential hybridization screening of a *C. albicans* cDNA library (Birse *et al.*, 1993). Morphological specific-cDNA probes were prepared from the RNA of cells grown as yeasts (25°C, pH 4.5 in Lee's medium, Lee *et al.*, 1975), or as hyphae (37°C, pH 6.5 in Lee's medium). cDNA libraries were prepared from RNA of stationary phase cells in YPD medium at 30°C, or from stationary-phase cells that were induced to form germ tubes in M-199 medium at 37°C for 60 min. The *ECE1* gene expression was hyphal-specific and was observed within 30 min after cells had been induced to form hyphae. Alterations in morphogenesis in an *ECE1* null mutant were not apparent, indicating that the gene is not essential for cell elongation or hyphal formation. Moreover, the similarities of the putative *ECE1* protein are not found in the GenBank data base. Hence *ECE1* is a *C. albicans*–specific gene.

Hoyer *et al.* (1995) first characterized a gene that was similar in sequence to the sex α-agglutinin of *S. cerevisiae*. The *ALS1* (agglutinin-like sequence gene) from *C. albicans* is expressed in response to culture conditions. Since that initial report, other *ALS*-like sequences have been identified so that it is more appropriate to refer to an *ALS*-family of proteins that now total 9 proteins. The first *ALS* gene was identified as follows. A hyphal-specific cDNA library was constructed from mid-log phase cells incubated for 20 min in RPMI medium at 37°C and 5% CO_2. A morphological-specific cDNA probe was obtained by using either total RNA from yeast cells grown to mid-log phase in YPD medium at 25°C or total RNA from long, branching hyphae cultured in RPMI medium at 37°C and 5% CO_2 for 16 h. As stated above, *ALS1* encodes a protein similar to the α-agglutinin, a cell-surface adhesion glycoprotein that mediates mating of haploid cells of *S. cerevisiae* (Hauser and Tanner, 1989; Lipke *et al.*, 1989) and contains a central domain of a tandemly repeated sequence not present in *AGα1*-encoded α-agglutinin in *S. cerevisiae* (Hoyer *et al.*, 1995). Based on the sequence of the central tandem repeat domain, the *ALS* family is divided into subfamilies that include at least nine genes as stated above (*ALS1-ALS9*) (Hoyer, 2001). A function of Als proteins is adhesion, but the *ALS* genes are differentially regulated in *C. albicans* by physiologically relevant conditions such as growth media, morphological forms, and stage of growth (Hoyer, 2001). The *efg1/efg1* mutant failed to express *ALS1* (Fu *et al.*, 2002). Therefore, *ALS1* is a downstream effector of the *EFG1* filamentation pathway. The *als1/als1* mutant was reduced in both adherence and filamentation *in vitro,* and the mutant also had reduced virulence in the early stages of infection *in vivo* (Fu *et al.*, 2002). *ALS3* was isolated from a fosmid library by PCR screening using primers based on the consensus tandem-repeat sequence of *ALS1* (Hoyer *et al.*, 1995). *ALS3* was expressed in hyphal inducing conditions, such as imidazole buffer containing N-acetylglucosamine, serum, L-proline, or glucose plus glutamine (Hoyer *et al.*, 1998). *ALS8* is highly similar to *ALS3* (Hoyer, 2001).

3. Signaling Pathways for Stimulation of Hyphal Development

Morphogenesis of *C. albicans* is regulated by many kinds of environmental signals, including temperature, pH, nutrient starvation, concentration of oxygen, or host proteins (Ernst, 2000b; Liu, 2001; Calera and Calderone, 2002; Brown, 2002a; 2002b). These signals interact with membrane receptors to activate gene transcription through signal transduction pathways. There are four signal transduction pathways required for the induction of hyphal development in *C. albicans*, including the Efg1-mediated cyclic adenosine 3',5'-monophosphate (cAMP)-dependent protein kinase A (PKA) pathway, Cph1-mediated mitogen-activated protein kinase (MAPK) pathway, a Prr2-mediated pH pathway, and a Czf1-mediated microaerophilic response pathway (Liu *et al.*, 1994; Stoldt *et al.*, 1997; Brown Jr *et al.*, 1999; Ramon *et al.*, 1999).

One of the key transcription factors that was studied in relationship to morphogenesis was *EFG1* (enhanced filamentous growth) (Stoldt *et al.*, 1997). *EFG1* was originally identified as a gene that accelerated pseudohyphae formation in diploid strains of *S. cerevisiae* that were transformed with a *C. albicans* genomic library constructed using a high-copy number *S. cerevisiae* vector. Moreover, they first reported the conserved sequence among orthologs such as Efg1 in *C. albicans*, StuA in *Aspergillus nidulans* (Miller *et al.*, 1992), and Phd1 and Sok2 in *S. cerevisiae* (Gimeno and Fink, 1994; Ward *et al.*, 1996). These genes are homologous to the basic helix-loop-helix (bHLH) motif of eukaryotic transcription factors (Ferre-D'Amare *et al.*, 1993). StuA regulates morphogenesis of *A. nidulans* (Miller *et al.*, 1992), Phd1 enhances pseudohyphal formation in *S. cerevisiae* (Gimeno and Fink, 1994), but Sok2 supresses this process (Ward *et al.*, 1996). On the other hand, the promoter regions of most hyphae-specific genes, such as *HWP1*, *HYR1*, *ALS3*, and *ALS8*, contain E-boxes (5′-CANNTG-3′, N = A,C,G,T) that are targets of bHLH proteins (Leng *et al.*, 2001). This observations was obtained using the computational analysis of yeast regulatory sequences (Van Helden *et al.*, 2000). However, some bHLH proteins can regulate genes that do not contain an E-box (Calera and Calderone, 2002). When logarithmically growing yeast cells of *C. albicans* were inoculated in hypha-inducing media, significant *EFG1* promoter activity was observed prior to hyphal formation, whereas the activity in yeast cells remained constant (Stoldt *et al.*, 1997). Null mutants of *efg1* were constructed following disruption of one allele with the *ADE2* gene, while removing the *EFG1* promoter of the second allele and replacing it with the regulatable *PCK1* promoter (Stoldt *et al.*, 1997). The disruptant formed elongated cells and pseudohyphal filaments but did not form true hyphae in YPD or YP media containing 5% horse serum at 37°C. The same results were obtained on a solid medium containing 4% bovine calf serum at 37°C (Lo *et al.*, 1997). *C. albicans* transformants overexpressing *EFG1* formed true hyphae in these media (Lo *et al.*, 1997; Stoldt *et al.*, 1997). Efg1p plays a major role in promoting filamentous growth of *C. albicans* at 37°C. However, Efg1p repressed filamentous growth at 25°C (Giusani *et al.*, 2002; Sohn *et al.*, 2003) (see Table 2).

Other regulatory genes that control morphogenesis include the TPK genes. There are three catalytic isoforms of the PKA encoded by three genes (*TPK1, TPK2,* and *TPK3*) in *S. cerevisiae* (Toda *et al.*, 1987). These isoforms are not essential for viability in this organism (Robertson and Fink, 1998). However, *TPK2* is essential for pseudohyphal development, whereas *TPK1* and *TPK3* repress this same process (Robertson and Fink, 1998; Pan and Heitman, 1999). On the other hand, in *C. albicans* there are two catalytic isoforms of the

Table 2. The functions of temperature regulated genes

Temperature	True hyphal growth	Yeast growth	No growth
42 °C		WT (YPD) *nrg1Δ* (YPD)	*tup1Δ* (YPD)
37 °C	WT (YPD+serum, Soll-pH6.5, or Lee-pH6.5) *EFG1*-ON *RAS1*-ON *CPH2*-ON *tup1Δ* (YPD+serum)	WT (YPD) *efg1Δ tup1Δ* (YPD+serum) *efg1Δ* (YPD+serum)	
30 °C	*tup1Δ* (YPD or SD)	WT (YPD or SD) *cph1Δ efg1Δ* (Spider) *efg1Δ tup1Δ* (YPD)	
29 °C	Multiple copies of *PRR2* (199)		
25 °C	*efg1Δcph1Δ* (YPS, embedded) *CZF1*-ON	WT (YPD, Soll-pH4.5, or Lee-pH4.5) *EFG1*-ON WT (YPS, surface)	

Shift up

Lo *et al.*, 1997, Feng *et al.*, 1999, Barkani *et al.*, 2000, Braun and Johnson, 2000, Murad *et al.*, 2001b, Sohn *et al.*, 2003.

PKA encoded by two genes, *TPK1* and *TPK2* (Bockmuhl *et al.*, 2001), that are not essential for growth (Sonneborn *et al.*, 2000; Bockmuhl *et al.*, 2001). Deletion of both alleles of *TPK2* caused a significant delay in hyphal development in liquid Spider medium at 37°C; however, the defect in morphological transition caused by deletion of *TPK2* is not complete and depends upon temperature and the type of medium (liquid or solid) (Sonneborn *et al.*, 2000). In contrast, mutants lacking both *TPK1* alleles showed defective hyphal development on solid inducing medium, whereas in liquid medium, hyphal development was slightly affected (Bockmuhl *et al.,* 2001). Therefore, *C. albicans* *TPK1* and *TPK2* share growth functions under different inducing environments. As for the relationship of *TPK2* and *EFG1*, suppression of the defective *tpk2* phenotype (true hyphal formation is defective), occurred by overexpression of *EFG1*, whereas *TPK2* overexpression does not reconstitute the *efg1* phenotype (Sonneborn *et al.*, 2000). Hence, Efg1p lies downstream of the PKA pathway involving Tpk2. A putative phosphorylation site for PKA, threonine-206, is conserved within an Efg1 domain, and a mutation of that residue in Efg1 seems essential as a putative phosphorylation target to induce hyphal growth (Bockmuhl and Ernst, 2001).

In *S. cerevisiae*, there are two *RAS* genes, *RAS1* and *RAS2* (Kataoka *et al.*, 1984). *RAS2* acts upstream of both the MAP kinase and the PKA pathways (Whistler and Rina, 1997; Robertson and Fink, 1998; Rupp *et al.*, 1999). In *C. albicans*, the Ras1 protein is homologous to the *S. cerevisiae* Ras1 and Ras2 (Feng *et al.*, 1999). The *Candida ras1/ras1* mutant fails to form true hyphae; hence, the gene is essential for hyphal differentiation in response to serum, whereas the gene is not essential for yeast cell growth (Feng *et al.*, 1999).

CPH1 and *CPH2* (*Candida* pseudohyphal regulator) were isolated as genes involved in suppression of pseudohyphal formation of a diploid *S. cerevisiae* that was transformed with a *C. albicans* genomic library on a high-copy number *S. cerevisiae* vector (Liu *et al.*, 1994). *CPH1* is a homolog of *S. cerevisiae STE12*. *STE12* encodes a transcription factor that is required for the expression of mating-specific genes. In the two haploid mating types of *S. cerevisiae* (a- and α-cells) under an appropriate environment, a signal is initiated by two different peptide pheromones (a- and α-factors) mediated by pheromone receptors. The signal is transmitted through a receptor-coupled GTP binding protein and MAP kinase pathway that diverts into two pathways, one resulting in a sterile phenotype in which cells are arrested in the G1 phase of cell cycle, and a second in which mating occurs via the Ste12p transcription factor. In *S. cerevisiae* diploids (a/α-cells) under conditions of nitrogen starvation, external signals bypass the pheromone receptors and GTP binding proteins and, instead, induce pseudohyphae formation through the MAP kinase pathway (Gimeno *et al.*, 1992; Liu *et al.*, 1993). Cph1p lies downstream of the MAP kinase pathway that includes the upstream proteins, Cst20 (p21-activated kinase), Hst7 (MAP kinase kinase), Cek1 (MAP kinase). *CPH1* complements the mating defect of *ste12* haploid strains of *S. cerevisiae* and also complements the defect in pseudohyphal formation of *ste12/ste12* diploid strains (Liu *et al.*, 1994). The *cph1/cph1* mutant of *C. albicans* showed yeast growth on Spider agar, but still formed hyphae in Lee's liquid medium and in a serum-containing medium.

CPH2 is a basic helix-loop-helix transcription factor of the Myc subfamily (Lane *et al.*, 2001b). Experiments with the *cph2/cph2* mutant of *C. albicans* showed that Cph2 regulates hyphal development in *C. albicans*. However, many hypha-specific genes do not have potential Cph2 binding sites in their upstream regions. Moreover, the function of Cph2 was independent of the Cph1- and Efg1-mediated pathways (Lane *et al.*, 2001b).

Changes in the pH of the environment affect the morphological development in *C. albicans*. Ramon *et al.* (1999) identified and cloned *PRR2* (pH response regulator) as a homolog of *pacC* that encodes a zinc finger-containing transcription factor, which mediates pH-dependent gene expression in *A. nidulans* (Tilburn *et al.*, 1995). Their study was based on the hypothesis that the pH response pathway is conserved between *A. nidulans* and *C. albicans*. A homologous sequence encoding the zinc finger region of *pacC* was obtained from a BLASTN search of the *C. albicans* genome sequence database (http://www-sequence.stanford.edu/group/candida). That sequence was amplified by PCR and used as a probe for hybridization screening of a *C. albicans* genomic library. *PRR2* (Ramon *et al.*, 1999) in *C. albicans* induced alkaline-expressed genes such as *PHR1* (Saporito-Irwin *et al.*, 1995) and repressed acid-expressed genes such as *PHR2* (Muhlschlegel and Fonzi, 1997). *PRR2* was not essential for growth but was expressed in a pH-conditional manner (Ramon *et al.*, 1999). Expression of *PRR1* was maximal at acid pH but reduced at alkaline pH (Porta *et al.*, 1999). A *prr1* null mutant did not form hyphae on inducible media including M199, Spider, and 10% serum, whereas the wild type cells produced hyphal growth (Porta *et al.*, 1999). *PRR2* could act as a multicopy suppressor of the temperature effect on filamentation, allowing extensive filamentation at 29°C (Barkani *et al.*, 2000). For filamentation in *C. albicans*, temperature is an important factor in addition to pH. Barkani *et al.* (2000) showed that multiple copies of *PRR2* suppressed temperature restriction on filamentation at 29°C which is below the temperature that induces hyphal growth. Moreover, pH-regulated dimorphism but not pH-dependent gene expression is required for the activity of the transcription factor Efg1p (Barkani *et al.*, 2000).

CZF1 (*C. albicans* zinc finger) encodes a zinc-finger containing protein that has no homolog in the *S. cerevisiae* genome and the product blocks α-factor-mediated cell cycle arrest in *S. cerevisiae* (Whiteway *et al.*, 1992). A *efg1 cph1* double mutant in *C. albicans*, which is nonfilamentous at 37°C under standard laboratory conditions, formed filaments within the tongue of immunosuppressed, gnotobiotic piglets and also when the mutant was embedded in agar at 25°C (Riggle *et al.*, 1999). Brown Jr *et al.* (1999) constructed *C. albicans* genomic DNA library to identify genes involved in the regulation of filamentous growth at 25°C when cells were embedded in agar. Those genes were then used to transform *C. albicans*. The transformants formed filamentous colonies at 25°C after embedding in YPS (yeast extract, peptone, and sucrose) agar. *CZF1* was found as an ectopically expressed gene when *C. albicans* cells were grown in the hyphal form within a matrix of YPS agar at 25°C but was not expressed when the cells were on the surface of the same medium at the same temperature. Moreover, *CZF1* was not needed to promote germ tube formation at 37°C in liquid media such as 10% serum, N-acetyl-D-glucosamine, or proline (Brown Jr *et al.*, 1999).

4. Repressors of Morphogenesis

Morphological transition in *C. albicans* includes both inductive (mentioned above) and repressive regulatory genes. Regulatory genes that repress hyphal development have been identified and found to be *TUP1*, *NRG1*, and *RFG1* in *C. albicans*.

S. cerevisiae TUP1 (dTMP deoxythymidine-5'- monophosphate uptake) has pleiotropic activities and represses the transcription of many different genes including those involved in glucose and oxygen repressions, those that are induced upon DNA damage, a-specific mating, haploid-specific genes, and flocculation genes (Wickner, 1974; Keleher *et al.*, 1992; Trumbly, 1992; Stratford, 1992; Herschbach *et al.*, 1994). Braun and Johnson (1997) isolated *TUP1* in *C. albicans* as a homolog of the *S. cerevisiae TUP1*. They used homologous sequences of the *S. cerevisiae TUP1* and *Kluyveromyces lactis TUP1* for the design of degenerate PCR primers. The PCR products generated from *C. albicans* genomic DNA were used as a probe to isolate the full-length gene from a *C. albicans* genomic library. *TUP1* serves as a transcriptional repressor in both *C. albicans* and *S. cerevisiae* (Braun and Johnson, 1997). However, the phenotypes produced by a *TUP1* deletion in *C. albicans* were different from *tup1* mutants in *S. cerevisiae*. Growth of the *tup1* mutant of *C. albicans* was normal at 37°C but was enhanced in glycerol, whereas the *S. cerevisiae tup1* showed a growth sensitivity at 37°C and poor growth on glycerol. Moreover, growth at 42°C of the *C. albicans tup1* mutant was arrested, whereas wild-type cells were normal (see Table 2). Disruption of *TUP1* in *C. albicans* resulted in filamentous growth (mostly elongated and straight hyphae) in YPD medium at 30°C, whereas wild type cells grew as a yeast form in the same growth conditions. Moreover, no detectable change in filamentous morphology was observed in both wild type and mutant cells in media that induce germ tubes and true hyphae, such as Lee's medium and a serum-containing medium (Braun and Johnson, 1997). Cell morphologies and expression patterns of mutants lacking each combination of *TUP1, CPH1*, and *EFG1* genes, which as stated above are critical in controlling filamentous growth, suggested that these three genes make independent contributions to the filamentous growth of *C. albicans* (Braun and Johnson, 2000). In their experiment, a *tup1* mutant always grew as filaments on non-inducing agar media (SD minimal medium) at 30°C, but the triple mutant (*tupl, cphl,* and *efgl*) produced elongated yeast cells on the same condition. Braun and Johnson (2000) proposed two models, termed central control and network control, of the regulatory circuits leading to filamentous growth in *C. albicans*, based on the analysis of filament-specific gene expression using strains lacking each combination of *TUP1*, *EFG1*, and *CPH1*. Moreover, in these regulatory mutants, the expression of structural genes associated with cell-surface proteins, including *HYR1*, *ECE1, HWP1*, and *ALS1,* were strongly dependent on *EFG1*. At present, *TUP1* is thought to regulate a complex network of signaling pathways for filamentous growth.

NRG1 (negative regulation for glucose repression) from *C. albicans* was isolated by two research groups. Braun *et al.* (2001) identified nucleotide-sequences like *MIG1* (Keleher *et al.*, 1992; Nehlin and Ronne, 1990), *SKO1* (Nehlin *et al.*, 1992; Proft and Serrano, 1999) and *NRG1* (Park *et al.*, 1999) genes, which encoded *TUP1*-dependent DNA-binding repressors in *S. cerevisiae*, using the *C. albicans* genome sequence data (http://www-sequence.stanford.edu/group/candida). They then deleted the *C. albicans* orthologs of the three *S. cerevisiae* genes in *C. albicans* and assessed their phenotypes. The *nrg1* disruptant of *C. albicans* grew as filaments on YPD (for 3 days) or SD agar (for 4 days) at 30°C. However, the growth form of the *nrg1* disruptant still made normal blastospores on the same media, whereas the proportion of filamentous cells in the *tup1* disruptant was nearly 100% (Braun *et al.*, 2001). On the other hand, *MIG1* or *SKO1* relatives were not involved in filamentous growth in *C. albicans*. Based on the fact that a stress promotes yeast-hypha transition in *C. albicans* (Odds, 1988; Brown and Gow, 1999), Murad *et al.* (2001b) identified genes with the C_4T sequence, which corresponds to the stress response element in *S. cerevisiae* (Marchler *et al.*, 1993). The sequence C_4T is contained in promoters of *HYR1* (Bailey *et al.*, 1996) and *ALS8* (Leng *et al.*, 2001) that are activated during serum-induced hyphal growth in *C. albicans* (Murad *et al.*, 2001b). Further, they screened a *C. albicans* cDNA library, which was constructed with mRNA from hyphal cells of *C. albicans* induced by the addition of serum at 37°C (Swoboda *et al.*, 1993), using an oligonucleotide containing C_4T. Murad *et al.* (2001b) isolated *NRG1* from a *C. albicans* genomic library using the colony hybridization method (Sambrook *et al.*, 1989). A Nrg1 response element (NRE) is found in the promoters of hyphae-specific genes and genes encoded by other virulence factors (Murad *et al.*, 2001b). The NRE sequence in *C. albicans* is (A/C)(A/C/G)C_3T (Murad *et al.*, 2001b), whereas the stress response element-sequence in *S. cerevisiae* is C_4T or C_3TC (Park *et al.*, 1999). The *nrg1* disruptant formed chains of pseudohyphal cells in YPD medium at 25°C (Murad *et al.*, 2001b). Moreover, Murad *et al.* (2001b) analyzed the *nrg1/nrg1/MET3-NRG1* strain to confirm that the pseudohyphal growth correlated with the deletion of *NRG1*. *MET3* is repressed by methionine and cysteine (Care *et al.*, 1999). The strain grew in the yeast form in SC medium (Kaiser *et al.*, 1994) lacking methionine and cysteine but formed pseudohyphae in SC medium containing these amino acids.

RFG1 (reduced filamentous growth) in *C. albicans* was isolated by two groups. Kadosh and Johnson (2001) isolated *RFG1* as a gene encoding a potential negative regulator of filamentous growth in *C. albicans*. They constructed a library of *C. albicans* genomic

DNA fragments in an *S. cerevisiae* high-copy-number vector. The *S. cerevisiae* transformants were grown under nitrogen-poor conditions (Gimeno *et al.*, 1992) in order to induce the pseudohyphae. A single clone in approximately 9,400 transformants showed a reduced filamentous growth phenotype. That clone was found to have a 3.4-kb fragment that contained a 1.8-kb open reading frame (*RFG1*). Rfg1 has a high-mobility group (HMG) DNA-binding domain that is 52% identical to that encoded by the *S. cerevisiae ROX1* gene (Balasubramanian *et al.*, 1993), a repressor of hypoxic genes (Zitomer and Lowry, 1992). However, Rfg1 in *C. albicans* does not appear to regulate hypoxic gene expression but instead appears to be involved in the control of filamentous growth and virulence (Kadosh and Johnson, 2001). *RFG1* and *ROX1* genes, when overexpressed, repress filamentous growth in *S. cerevisiae*, but overexpression of *RFG1* resulted in a dramatic reduction in filament formation compared to *ROX1* (Kadosh and Johnson, 2001). *RFG1* regulated expression of various filament-specific genes such as *HWP1, RBT1, RBT4, RBT5, HYR1, ECE1,* and *ALS1* (Kadosh and Johnson, 2001). *RFG1* hetero- and homozygous deletion *C. albicans* strains were severely defective for filament formation on Spider medium, serum, Lee's mannitol (pH 4.5), and Lee's mannitol (pH 6.8) plates (Kadosh and Johnson, 2001). Hence Rfg1 functions as a conditional-specific transcriptional regulator of hyphae-specific genes.

Zitomer and Lowry (1992) have studied the regulation of hypoxic genes including *ROX1* in *S. cerevisiae*. Khala and Zitomer (2001) isolated *RFG1* as the *ROX1* homologue from *C. albicans,* which can grow under hypoxic condition s.*ROX1* repression requires the Tup1/Ssn6 general repression complex in *S. cerevisiae* (Deckert *et al.*, 1998). *ROX1* encodes a hypoxic repressor that is capable of binding to a *S. cerevisiae* hypoxic operator DNA sequence. *RFG1* complemented the hypoxic repression function in a *S. cerevisiae rox1* deletion strain. However, the disruptant of *RFG1* in *C. albicans* grew as hyphae. The presence of *C. albicans ROX1* (*RFG1*) resulted in less than one-third ability to repress the ANB1-lacZ fusion, a reporter hypoxic gene, in a *rox1Δtup1Δ S. cerevisiae* strain compared to the *TUP1* wild-type strain (Khala and Zitomer, 2001). Hence repression by *RFG1* in *S. cerevisiae* is Tup1 dependent.

5. DNA Microarray Studies

DNA microarrays are a rapid, quantitative tool used to detect transcript levels of an entire genome. cDNA fragments of genes are fixed on nylon-membrane, plastic, or glass slides. cDNA probes are made from total RNA or polyA$^+$ RNA that is extracted from cells or infected tissues and then used in hybridization with the arrays. It is possible to obtain gene-expression profiling from the analysis of hybridization-patterns. Transcript profiling in *C. albicans* undergoing the morphological transition by DNA microarrays has been reported in a few papers (Murad *et al.*, 2001b; Lane *et al.*, 2001a; Nantel *et al.*, 2002; Sohn *et al.*, 2003).

Gene targets of the transcriptional repressors Tup1, Mig1 and Nrg1, which act as negative regulators of yeast-hypha transition in *C. albicans*, were sought using nylon-membrane arrays containing 2002 genes (Murad *et al.*, 2001a). These disruptants and wild-type cells were grown at 30°C to mid-exponential phase in YPD medium, and then RNA was extracted and used as probes of a DNA microarray. Wild type and *mig1/mig1* cells grew in the yeast form, but the *tup1/tup1* and *nrg1/nrg1* cells grew in filamentous form. An analysis of the functions of the subset-genes regulated by Tup1, Mig1and Nrg1 in *C. albicans* indicated that Tup1, Nrg1, and Mig1 play differential roles in the regulation of metabolism, cellular morphogenesis and stress responses in *C. albicans*. These results were consistent with the roles of their homologues in *S. cerevisiae* (Treitel and Carlson, 1995; Gancedo, 1998; Marquez *et al.*, 1998). Thus, Mig1 and Nrg1 contribute to the control of metabolic functions. Mig1 and Tup1 play roles in carbon catabolite repression, and Nrg1 and Tup1 contribute to some stress responses. The most important data of these DNA microarrays were that Nrg1 and Tup1 strongly regulated the hyphal-specific genes *ALS3*, *ALS8*, *ECE1* and *HWP1*. Therefore, Nrg1 and Tup1 play important roles in the regulation of yeast-hypha morphogenesis in *C. albicans*.

Lane *et al.* (2001a) analyzed how the signaling pathways that include Cph1, Cph2, and Efg1 regulate differential gene expression during filamentation in *C. albicans* using nylon-membrane arrays containing about 700 genes. To extract RNA for probes, wild-type cells and the signaling mutants *cph1/cph1*, *cph2/cph2*, and *efg1/efg1* were grown in YPD or Lee's medium at 25°C for yeast growth and in YPD +10% serum, Lee's medium with 1% mannitol, SS (YNB with ammonium sulfate and succinic acid, pH6.5), or SSA (SS with amino acids) at 37°C for hyphal induction. Clustering analysis (Eisen *et al.*, 1998) was used in the analysis of gene expressions. By using their original medium for the *cph1/cph1* mutant, they showed that Cph1p regulates the same differentially regulated genes that are under the control of Efg1p. Moreover, they found that the transcription of *TEC1* is regulated by *EFG1* and *CPH2*. *TEC1* (Ty1 enhancement control) is a member of the TEA/ATTS family of transcription factors, which bind to the conserved sequences, CATTCC and CATTCT, and is predominantly expressed in the hyphal form of *C. albicans* (Schweizer *et al.*, 2000).

Nantel *et al.* (2002) used glass microarray transcription profiling of over 6000 genes in *C. albicans* cells undergoing the yeast-to-hyphal transition in response to temperature and the culture medium. To extract RNA from yeast or hyphae, wild-type cells and the *efg1*Δ or *efg1*Δ/*cph1*Δ deletions were grown to the stationary phase in YPD at 30°C. Then the cultures were diluted and incubated in YPD at 30° C or YPD + 10% serum at 37°C. Those strains were also grown to the stationary phase in Lee's medium with 10% glucose at 25°C, and then the cultures were diluted and incubated in Lee's medium with 10% glucose at 25°C for yeast growth or at 37°C for hyphal growth. Hyphal induction by Lee's medium, followed by a switch from 25°C to 37°C, generated a slightly different pattern of gene expression than that found with the serum plus 37°C treatments. An explanation of this could be that the growth of hyphae in Lee's medium was late (24 h). To induce significant expression of genes, two environmental changes, temperature shift and the addition of serum, were necessary. These investigators followed the timing of the change in gene expression associated with initiation of hyphal growth. Germ tube formation in serum at 37°C was observed at 60 min; also most genes were induced by 60 min. They showed that Efg1p and Cph1p are needed for the activation of the transcriptional program that is induced by the presence of serum.

Cell wall dynamics in *C. albicans* during the yeast-to-hyphal transition was analyzed using DNA glass microarrays (Sohn *et al.*, 2003). A DNA microarray comprising 117 probes from genes implicated in cell wall biogenesis was established. Wild-type cells and *efg1*Δ, *cph1*Δ, or *efg1*Δ/*cph1*Δ deletions were grown to stationary phase in YPD at 30°C. The cultures were inoculated and incubated in YPD at 30°C for 8 h or in α-MEM + 2% glucose at 25°C for 8 h as yeast growth conditions, and in YPD +10% serum at 37°C for 4 h, or in α-MEM + 2% glucose at 37°C for 8 h as hyphae-inducing conditions. Then, RNA was extracted and reverse transcribed to use as probes. The results of the DNA microarray were performed using cluster analysis (Eisen *et al.*, 1998). They identified two genes, *YWP1* and *HWP2*, that encoded yeast and hyphal cell wall proteins, respectively. Their data showed that *EFG1* plays a major role in the transcriptional regulation of cell wall proteins in contrast to *CPH1*. Moreover, it is interesting that *EFG1* is essential for the transcription of both hyphae-specific genes (*HWP1, HWP2*) and the yeast form-specific gene (*YWP1*). Therefore, it seems that *EFG1* regulates the transcription of the appropriate genes under the given conditions for growth of each morphological form.

6. Growth Arrest and Morphological Differentiation

6.1. Intracellular Conditions Prior to Germination

Exponentially grown yeast cells of *C. albicans* do not readily form germ tubes at 37°C compared with stationary phase cells (Chaffin and Sogin, 1976), but Soll and Herman (1983) found that starvation for a minimum of 10 to 20 min was required for germination of exponentially grown cells. They assumed that a transient cessation of growth results in an increased capacity of cells to resume growth in either the budding or the mycelial form. Moreover, a transient cessation of growth in cells at all points in the cell cycle lead to pluripotency (in this case, the capacity to resume growth in either the budding or the mycelial form). While the concept of pluripotency is a very interesting explanation of the mechanism of yeast-hyphal transition in *C. albicans*, its underlying molecular mechanism has not been analyzed.

Hazan *et al.* (2002) reported on the relationship of germination to the cell cycle in *C. albicans* using unbudded G1 cells that were synchronized in a rich medium including raffinose at 30°C. Germ tubes at 37°C can evaginate before spindle pore body duplication, chitin ring formation, and DNA replication, and then tip-associated actin polarization in each hyphal element occurs before the events of the G1/S transition. Hence, hyphal elongation occurred throughout the cell cycle.

6.2. Germination of Mid-Log Yeast Cells

Serum, N-acetyl-D-glucosamine (Mattia and Cassone, 1979), or starvation (Soll and Herman, 1983; Cho *et al.*, 1992) cause log-phase yeast cells of *C. albicans* to form germ tubes at 37° C. Germination studies in a glucose (Glc) medium (Glc plus ammonium chloride and basal salts medium: L-thiazolidine-4-carboxylic acid, L-proline, sodium bicarbonate, sodium phosphate monobasic, and sodium acetate) and a N-acetyl-D-glucosamine (GlcNAc) medium (GlcNAc and base medium) at 37°C with mid-log yeast cells, showed two sharp features (Cho *et al.*, 1992). First, around 90% of starved cells formed germ tubes, whereas about 80% of cells in the Glc medium did not form germ tubes unless they were first starved. Therefore, there was a difference between starved and non-starved cells, in the percentage of cells that formed germ tubes. However, over 80% of starved or non-starved cells in the GlcNAc medium formed germ tubes. Second, changes in intracellular cyclic adenosine monophosphate (cAMP) content varied in the two media. The cAMP content of both starved and non-starved cells increased for a period of 60 min after inoculation in Glc medium but then decreased for 120 min. In GlcNAc medium, the cAMP content of both starved and non-starved cells showed a tendency to increase. The difference in the changes in cAMP content of cells in Glc medium and GlcNAc medium may be accounted for by the data of Sullivan *et al.* (1983), who showed that the changes in cAMP concentration could be due to a requirement for cellular metabolism to adjust to the different carbon and nitrogen sources at 37°C. Mid-log phase cells in which growth was initially inhibited by hydroxyurea (DNA synthesis inhibitor), sodium malonate (respiratory inhibitor), and 2-deoxy-D-glucose (glucose metabolic inhibitor) were induced to form germ tubes in Glc medium at 37°C (Cho *et al.*, 1994). It is postulated that starvation or metabolic inhibitors result in a cessation of the log growth as yeast, which precedes the induction of germination in *C. albicans*. Similarly, differentiation in mammalian cells may occur through the induction of cell cycle arrest (DeLuca *et al.*, 1996). Growth cessation and differentiation in certain mammalian cells are associated with the hypophosphorylation of proteins of the cell cycle-regulated retinoblastoma tumor suppressor (Rb) gene family (Chen *et al.*, 1989; Goodrich and Lee, 1993). Exponentially grown cells pretreated with protein phosphorylation inhibitors including forskolin (cAMP-mediated dephosphorylation), okadaic acid (a protein phosphatase 1 and 2A inhibitor), or D-erythro-sphingosine

(Rb product phosphorylation inhibitor) readily germinated (Cho *et al.*, 2000). Therefore, a similar type of mechanism may be operative in *C. albicans*.

6.3. CGR1 Isolated Using RNA Fingerprinting of Arbitrarily Primed PCR

Ralph *et al.* (1993) indicated that cell cycle arrest would likely be accompanied by the differential regulation of many genes. One method to study differential gene regulation is RNA fingerprinting by arbitrarily primed PCR (RAP) method, which is a reliable way to identify many of the genes that change in steady-state expression during the cell cycle. To elucidate the relationship between growth arrest of yeast cells of *C. albicans* and its morphological differentiation, the isolation of a gene expressed during growth cessation of exponential phase cells from *C. albicans* was attempted using RAP with minor modifications (Cho *et al.*, 2001). P130 is a member of Rb gene family (Chao *et al.*, 1993) and has been reported as a gene expressed in the G0/G1 phase or in the early G1 phase of mammalian cells (Chen *et al.*, 1996). Primers in RAP were derived from p130 DNA sequences, homologous to subdomains of other members of Rb gene family (Cho *et al.*, 2001). *CGR1* (*Candida* growth regulation) is a single, continuous open reading frame of 1020 bp that encodes a predicted protein of 340 amino acids with a molecular mass of 38.4 kDa (Cho *et al.*, 2001). A 26% identity and 47% similarity between Cgr1p and the senescence marker protein 30 (SMP30) from human, were observed. *CGR1* has been placed in the SMP30/CGR1 family of the NCBI database. SMP30 was purified as a novel aging-characteristic protein with a molecular mass of 30 kDa from rat liver (Fujita *et al.*, 1992). The biological function of SMP30 is thought to be the regulation of Ca^{2+} homeostasis by enhancing the Ca^{2+}-pumping activity in the plasma membrane (Fujita *et al.*, 1998). Cgr1p and the hypothetical 40.3 kDa protein encoded by *YBR053c* in *S. cerevisiae* share 27% identity and 44% similarity. The predicted protein of *YBR053c* contains a putative cell cycle-dependent (M/G1) CDC28 motif; however, the function of *YBR053c* is still unknown. *CGR1* in yeast-mycelial transition of exponentially grown *C. albicans* was expressed during starvation prior to induction of germination and in yeast cells prior to outgrowth from exponentially grown yeast cells cultured in GlcNAc medium without starvation (Cho *et al.*, 2001). Therefore, *CGR1* may be associated with growth cessation of yeast cells in *C. albicans*.

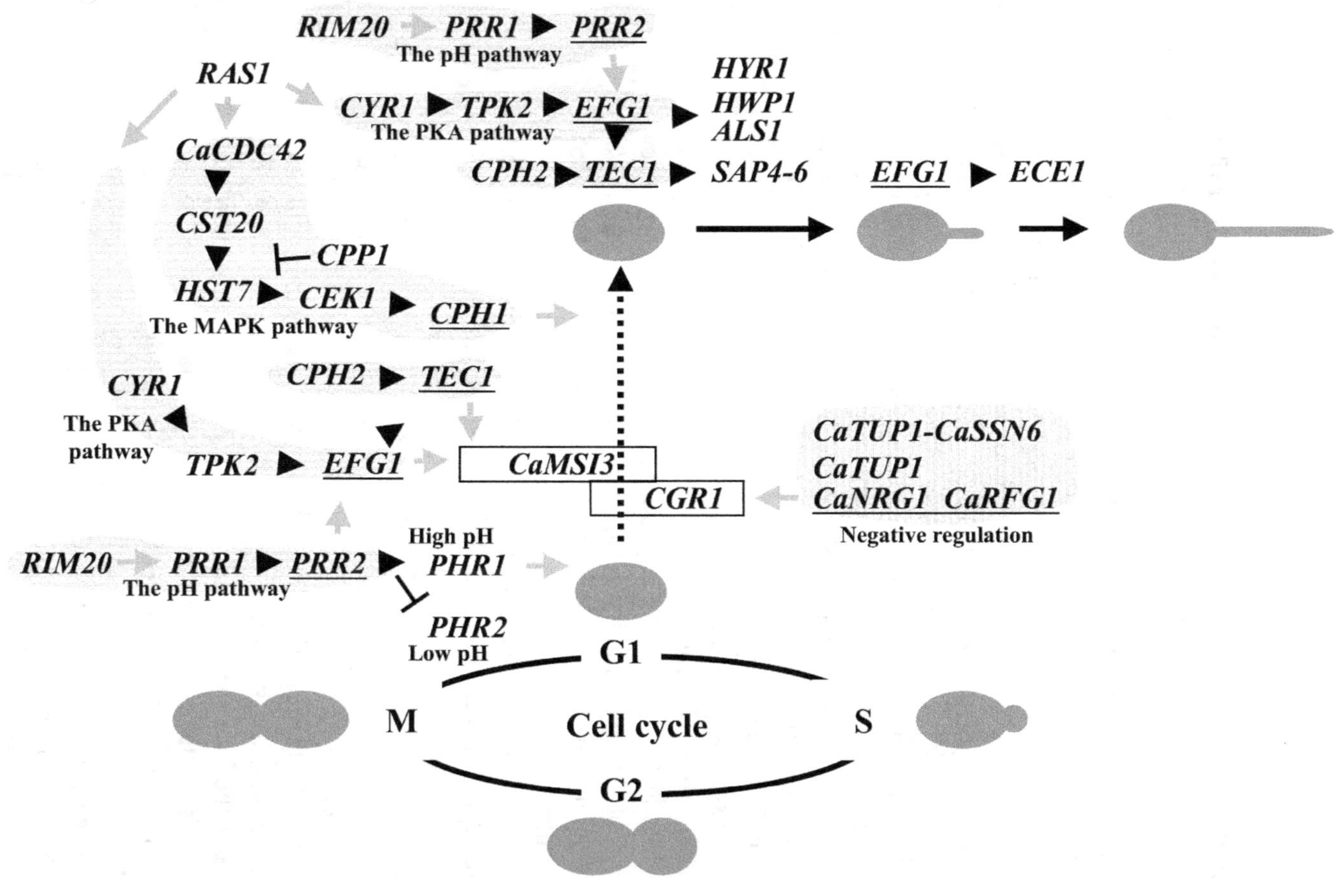

Figure 1. The putative early stages of the yeast-mycelial transition in *C. albicans*. Potential signal pathways, *CGR1*, and *CaMSI3* are integrated in this scheme. Morphological transition from yeast growth in mid-exponential phase, which would be in a tightly regulated cel cycle, to hyphal growth is shown by a dotted line. Black arrows show proven regulation between genes, whereas gray arrows show putative relationship

6.4. *CaMSI3* Isolated Using the Yeast Two-Hybrid System

CaMSI3 (*C. albicans MSI3*) was isolated using the yeast two-hybrid system to identify proteins that interact with Cgr1p (Cho *et al*., 2003). To construct a cDNA library for a prey plasmid, total RNA was isolated from *C. albicans* yeast cells that were grown to mid-exponential phase in YPD medium at 37°C and then incubated in distilled water for 20 min at 37°C, since the *CGR1* gene expression increases when cells are starved for 20 min (Cho *et al*., 2001). The DNA sequence of *CaMSI3* encodes a predicted protein of 702 amino acids with a molecular mass of 78.6 kDa (Cho *et al*., 2003). No introns were detected. CaMsi3p contains an HSP70 motif (IEVIGGSSRIPAVKT) and a potential heat shock element (NGAANNTTCN, N = A/C/G/T) present in the upstream region of the initiation codon of *CaMSI3* (Cho *et al*., 2003). CaMsi3p is 63% identical to Msi3p (Shirayama *et al*., 1993)/Sse1p (Mukai *et al*., 1993) of *S. cerevisiae* that is a new member of the heat shock protein 70 (HSP70) family. CaMsi3p is 28% identical to Ssa1p and Ssb1p of *C. albicans*. But the homology of the Ssa1p and Ssb1p of *C. albicans* is 58%. Hence *CaMSI3* encodes a new HSP70 of *C. albicans* (Cho *et al*., 2003). The *S. cerevisiae* msi3$^-$ mutant has a temperature-sensitive (ts) phenotype at 37°C (Shirayama *et al*., 1993). *CaMSI3* complemented the ts phenotype of the msi3$^-$ mutant of *S. cerevisiae* (Cho *et al*., 2003), but it has not been determined whether *CaMSI3* provides functions related to a heat shock function in *C. albicans* at 37°C.

6.5. *CGR1, CaMSI3*, and Signaling Pathway

Figure 1 depicts the putative early stages of the yeast-mycelial transition in *C. albicans*; additionally, several signal pathways, *CGR1*, and *CaMSI3* are integrated in this scheme. Morphological transition from yeast growth in mid-exponential phase (which would be in a tightly regulated cell cycle) to hyphal growth, is shown by a dotted line. *CGR1* was isolated at this stage. Cgr1 was considered to interact with CaMsi3 based upon the results with the yeast two-hybrid method (Cho *et al*., 2003). Upstream of *CGR1* is the ACCCCT sequence, which corresponds to an Nrg1 response element: (A/C)(A/C/G)C_3T (Murad *et al*, 2001a). Upstream of *CaMSI3* is the *EFG1* binding site, the E box (CANNTG) (Massari and Murre, 2000; Robinson and Lopes, 2000), and the *TEC1* binding site (CATTCC) (Kohler *et al*, 2002). In Figure 1, these putative promoters and their relationship

Table 3. The role of morphogenesis in the virulence of *Candida albicans*

	Genome project	Genomic drug discovery
	Polymorphic genotype	
	Phenotypic variability	W-O transition: switching model
Background		
Drug	Change of antifungal susceptibility	
	Drug resistance	Change of target molecules · *ERG*: ergosterol biosynthetic pathway Drug efflux transport system · *CDR*: transporter genes
Host defense System	Adhesion	· *INT1*: on yeast and hyphae · *ALS, HWP*: on hyphae
	Tissue invasion Tissue damage	Morphological transition Transcription factors · *CPH* , *EFG1*: hyphal genes-on · *TUP1*: hyphal genes-off Secreted enzyme: hydrolytic enzyme · *SAP* · *PL, Lip*

to the signaling pathways are illustrated so that *CGR1* and *CaMSI3* are connected. Black arrows show proven regulation between genes, whereas gray arrows show putative relationship. Our real time RT-PCR analysis showed that expression of *CGR1* increased during early stages of morphological transition but was relatively unchanged in cells that had germinated (Cho *et al.*, unpublished).

7. Role of Morphogenesis in Virulence

In Table 3 the molecular experiments on virulence of *C. albicans* are shown. The *C. albicans* genome project has potential to reveal new virulence factors (see Schaller *et al.*, 2004) and also contribute to the discovery of new drugs (Li and Calderone, 2004). Virulence factors of *C. albicans* are arranged in two categories. The first category relates to the ability of *C. albicans* to acquire drug resistance, while the second indicates the virulence factors that are expressed as a consequence of the interactions of *C. albicans* with the host defences. Morphological transition is included in the second group.

With regards to the avirulence of hypha-deficient mutants in genes that are included in signal transduction systems and transcription factors, the relationship between *EFG1* and the *SAP* genes (secreted aspartic proteinase) was investigated (Felk *et al.*, 2002). *SAP4*, *SAP5*, and *SAP6* are hyphae-specific genes (Hube e*t al.*, 1994; White and Agabian, 1995). Sap4-6 antigens were found on hyphal cells of *C. albicans*, whereas Sap1-3 antigens were found on both yeast and hyphal cells (Felk *et al.*, 2002). Efg1 regulates the *TEC1* transcription factor whose motif is contained in the promoters of *SAP4,5* and *6* genes (Schweizer *et al.*, 2000). Mutants lacking *EFG1* and *in vivo* reduced hyphal-growth and transcription levels of *SAP4*, *SAP5,* and *SAP6* had reduced virulence (Felk *et al.*, 2002). Mutants lacking *SAP4* to *SAP6* had reduced invasiveness but produced hyphae (Felk *et al.*, 2002). Therefore, these data show that hyphae-associated proteinases contribute to tissue damage by *C. albicans* (Felk *et al.*, 2002). *C. albicans* septin proteins localize to a tight ring at the bud and pseudohyphae necks and in emerging germ tubes of hyphae (Warenda and Konopka, 2002). Mutants lacking *CDC10* and *CDC11*, which encode septin proteins of *C. albicans*, showed hyphal growth but without invasion and obvious symptoms of disease (Warenda *et al.*, 2003). Therefore, these data may show that tissue penetration and damage by hyphae require not only this morphological form, but also certain proteins and enzymes in/on the hyphae. Transcription factors associated with hyphal formation would also control genes encoding these proteins in hyphae.

As morphogenesis of *C. albicans* depends on environmental factors, the elucidation of gene function related to this process can be different *in vitro* versus *in vivo*. In this regard, Nakayama *et al.* (2000) developed a system to control gene expression in *C. albicans* by the use of tetracycline-regulatable promoters. This system can provide data on *in vivo* expressed genes by feeding doxycycline to animals in order to repress promoter activity of a gene that is being studied. Therefore, a gene-function analysis by the system would seem to be a reality.

References

Anderson, J.M., and Soll, D.R. 1986. Differences in actin localization during bud and hypha formation in the yeast *Candida albicans*. J. Gen. Microbiol. 132: 2035-2047.

Bailey, D.A., Feldmann, P.J.F., Bovey, M., Gow, N.A.R., and Brown, A.J. P. 1996. The *Candida albicans HYR1* gene, which is activated in response to hyphal development, belongs to a gene family encoding yeast cell wall proteins. J. Bacteriol. 178: 5353-5369.

Balasubramanian, B., Lowry, C.V., and Zitomer, R.S. 1993. The Rox1 repressor of the *Saccharomyces cerevisiae* hypoxic genes is a specific DNA-binding protein with a high-mobility-group motif. Mol. Cell. Biol. 13: 6071-6078.

Barkani, A.E., Kurzai, O., Fonzi, W.A., Ramon, A., Porta, A., Frosch, M., and Muhlschlegel, F.A. 2000. Dominant active alleles of *RIM101* (*PRR2*) bypass the pH restriction on filamentation of *Candida albicans*. Mol. Cell. Biol. 20: 4635-4647.

Birse, C.E., Irwin, M.Y., Fonzi, W.A., and Sypherd, P.S. 1993. Cloning and characterization of *ECE1*, a gene expressed in association with cell elongation of the dimorphic pathogen *Candida albicans*. Infect. Immun. 61: 3648-3655.

Bockmuhl, D.P., and Ernst, J.F. 2001. A potential phosphorylation site for an A-type kinase in the Efg1 regulator protein contributes to hyphal morphogenesis of *Candida albicans*. Genetics 157: 1523-1530.

Bockmuhl, D.P., Krishnamurthy, S., Gerads, M., Sonneborn, A., and Ernst, J.F. 2001. Distinct and redundant roles of the two protein kinase A isoforms Tpk1p and Tpk2p in morphogenesis and growth of *Candida albicans*. Mol. Microbiol. 42: 1243-1257.

Bowen, A. R., Chen-Wu, J.L., Momany, M., Young, R., Szaniszlo, P.J., and Robbins, P.W. 1992. Classification of fungal chitin synthases. Proc. Natl. Acad. Sci. USA 89: 519-523.

Braun, B.R., and Johnson, A.D. 1997. Control of filament formation in *Candida albicans* by the transcriptional repressor *TUP1*. Science 277: 105-109.

Braun, B.R., and Johnson, A.D. 2000. *TUP1, CPH1* and *EFG1* make independent contributions to filamentation in *Candida albicans*. Genetics 155: 57-67.

Braun, B.R., Kadosh, D., and Johnson, A.D. 2001. *NRG1*, a repressor of filamentous growth in *C. albicans*, is down-regulated during filament induction. EMBO J. 20: 4753-4761.

Brown, A.J.P. 2002a. Expression of growth form-specific factors during morphogenesis in *Candida albicans*. In: *Candida* and Candidiasis. R.A. Calderone, ed. ASM Press, Washington, D.C. p. 87-93.

Brown, A.J.P. 2002b. Morphogenetic signaling pathways in *Candida albicans*. In: *Candida* and Candidiasis. R.A. Calderone, ed. ASM Press, Washington, D.C. p. 95-106.

Brown, A.J.P., and Gow, N.A. 1999. Regulatory networks controlling *Candida albicans* morphogenesis. Trends Microbiol. 7: 333-338.

Brown, D.H., Jr, Giusani, A.D., Chen, X., and Kumamoto, C.A. 1999. Filamentous growth of *Candida albicans* in response to physical environmental cues and its regulation by the unique *CZF1* gene. Mol. Microbiol. 34: 651-662.

Calderone, R.A. 1993. Recognition between *Candida albicans* and host cells. Trends Microbiol. 1: 55-58.

Calera, J.A., and Calderone, R.A. 2002. Signaling and the biology of human fungal pathogens. In: Fungal Pathogenesis: Principles and Clinical Applications. R.A. Calderone, and R.L. Cihlar, eds. Marcel Dekker, Inc., New York, Basel. p.115-137.

Care, R.S., Trevethick, J., Binley, K.M., and Sudbery, P.E. 1999. The MET3 promoter: a new tool for *Candida albicans* molecular genetics. Mol. Microbiol. 34: 792-798.

Chaffin, W.L., and Sogin, S.L. 1976. Germ tube formation from zonal rotor fractions of *Candida albicans*. J. Bacteriol. 126: 771-776.

Chao, R., Khan, W., and Hannum, Y.A. 1993. Retinoblastoma protein dephosphorylation induced by d-erythro-sphingosine. J. Biol. Chem. 267: 23459-23462.

Chen, P-L., Scully, P., Shew, J-Y., Wang, J.Y.J., and Lee, W-H. 1989. Phosphorylation of the retinoblastoma gene product is modulated during the cell cycle and cellular differentiation. Cell 58: 1193-1198.

Chen, P.L., Riley, D.J., Chen, Y., and Lee, W.H. 1996. Retinoblastoma protein positively regulates terminal adipocyte differentiation through direct interaction with C/EBPs. Genes Dev. 10: 2794-2804.

Chen-Wu, J.L., Zwicker, J., Bowen, A.R., and Robbins, P.W. 1992. Expression of chitin synthase genes during yeast and hyphal growth phases of *Candida albicans*. Mol. Microbiol. 6: 497-502.

Cho, T., Hagihara, Y., Hamatake, H., and Watanabe, K. 1994. The relationship between the glucose uptake system and growth cessation in *Candida albicans*. J. Med. Vet. Mycol. 32: 461-466.

Cho, T., Hamatake, H., Hagihara, Y., and Kaminishi, H. 2000. Inhibitors of protein phosphorylation including the retinoblastoma protein induce germination of *Candida albicans*. Med. Mycol. 38: 41-45.

Cho, T., Hamatake, H., Kaminishi, H., Hagihara, Y., and Watanabe, K. 1992. The relationship between cyclic adenosine 3',5'-monophosphate and morphology in exponential phase *Candida albicans*. J. Med. Vet. Mycol. 30: 35-42.

Cho, T., Sudoh, M., Tanaka, T., Nakashima, Y., Chibana, H., and Kaminishi, H. 2001. Isolation and expression of a gene (*CGR1*) regulated during the yeast-hyphal transition in *Candida albicans*. Biochim. Biophys. Acta 1517: 288-292.

Cho, T., Toyoda, M., Sudoh, M., Nakashima, Y., Calderone, R.A., and Kaminishi, H. 2003. Isolation and sequencing of the *Candida albicans MSI3*, a putative novel member of the HSP70 family. Yeast 20: 149-156.

Davis, D., Wilson, B.R., and Mitchell, A.P. 2000a. RIM101-dependent and –independent pathways govern pH responses in *Candida albicans*. Mol. Cell. Biol. 20: 971-978.

Davis, D., Edwards Jr, J.E., Michell, A.P., and Ibrahim, A.S. 2000b. *Candida albicans RIM101* pH response pathway is required for host-pathogen interactions. Infect. Immun. 68: 5953-5959.

De Bernardis, F., Molinari, A., Boccanera, M., Stringaro, A., Robert, R., Senet, J-M., Arancia, G., and Cassone, A. 1994. Modulation of cell surface-associated mannoprotein antigen expression in experimental candidal vaginitis. Infect. Immun. 62: 509-519.

Deckert, J., Torres, A.M.R., Hwang, S.M., Kastaniotis, A.J., and Zitomer, R.S. 1998. The anatomy of a hypoxic operator in *Saccharomyces cerevisiae*. Genetics 150: 1429-1441.

DeLuca, A., Esposito, V., Baldi, A., and Giordano, A. 1996. The retinoblastoma gene family and its role in proliferation, differentiation and development. Histol. Histopathol. 11: 1029-1034.

Eisen, M.B., Spellman, P.T., Brown, P.O., and Botstein, D. 1998. Cluster analysis and display of genome-wide expression patterns. Proc. Natl. Acad. Sci. USA 95: 14863-14868.

Ernst, J.F. 2000a. Regulation of dimorphism in *Candida albicans*. In: Dimorphism in Human Pathogenic and Apathogenic Yeasts, vol. 5. J.F. Ernst and A. Schmidt, eds. Contributions to Microbiology, Karger, Basel. p. 98-111.

Ernst, J.F. 2000b. Transcription factors in *Candida albicans* – environmental control of morphogenesis. Microbiology 146: 1763-1774.

Felk, A., Kretschmar, M., Albrecht, A., Schaller, M., Beinhauer, S., Nichterlein, T., Sanglard, D., Korting, H.C., Schafer, W., and Hube, B. 2002. *Candida albicans* hyphal formation and the expression of the Efg1-regulated proteinases Sap4 to Sap6 are required for the invasion of parenchymal organs. Infect. Immun. 70: 3689-3700.

Feng, Q., Summers, E., Guo, B and Fink, G. 1999. Ras signaling is required for serum-induced hyphal differentiation in *Candida albicans*. J. Bacteriol. 181: 6339-6346.

Ferre-D'Amare, A., Prendergast, G.C., Ziff, E.B., and Burley, S.K. 1993. Recognition by Max of its cognate DNA through a dimeric b/HLH/Z domain. Nature 363: 38-45.

Fonzi, W.A., and Irwin, M.Y. 1993. Isogenic strain construction and gene mapping in *Candida albicans*. Genetics 134: 717-728.

Fu, Y., Ibrahim, A.S., Sheppard, D.C., Chen, Y.C., French, S.W., Cutler, J.E., Filler S.G., and Edwards, J.E.Jr. 2002. *Candida albicans* Als1p: an adhesion that is a downstream effector of the *EFG1* filamentation pathway. Mol. Microbiol. 44: 61-72.

Fujita, T., Inoue, H., Kitamura, T., Sato, N., Shimosawa, T., and Maruyama, N. 1998. Senescence marker protein-30 (SMP30) rescues cell death by enhancing plasma membrane Ca^{2+} -pumping activity in Hep G2 cells. Biochem. Biophys. Res. Commun. 250: 374-380.

Fujita, T., Uchida, K., and Mayuyama, N. 1992. Purification of senescence marker protein-30 (SMP30) and its androgen-independent decrease with age in the rat liver. Biochim. Biophys. Acta 1116: 122-128.

Fuqua, W.C., Winans, S.C., and Greenberg, E.P. 1994. Quorum sensing in bacteria: the LuxR-LuxI family of cell density-responsive transcriptional regulators. J. Bacteriol. 176: 269-275.

Gancedo, J.M. 1998. Yeast carbon catabolite repression. Microbiol. Mol. Biol. Rev. 62: 334-361.

Gimeno, C.J., and Fink, G.R. 1994. Induction of pseudohyphal growth by overexpression of *PHD1*, a *Saccharomyces cerevisiae* gene related to transcriptional regulators of fungal development. Mol. Biol. Cell. 14: 2100-2112.

Gimeno, C.J., Ljungdahl, P.O., Styles, C.A., and Fink, G.R. 1992. Unipolar cell divisions in the yeast *Saccharomyces cerevisiae* lead to filamentous growth: regulation by starvation and RAS. Cell 68: 1077-1090.

Giusani, A.D., Vinces, M., and Kumamoto, C.A. 2002. Invasive filamentous growth of *Candida albicans* is promoted by Czf1p-dependent relief of Efg1p-mediated repression. Genetics 160: 1749-1753.

Goodrich, D.W., and Lee, W-H. 1993. Molecular characterization of the retinoblastoma susceptibility gene. Biochim. Biophys. Acta 1155: 43-61.

Gow, N.A., and Gooday, G.W. 1987. Cytological aspects of dimorphism in *Candida albicans*. Crit. Rev. Microbiol. 15: 73-78.

Gow, N.A.R., Robbins, P.W., Lester, J.W., Brown, A.J.P., Fonzi, W.A., and Chapman, T. 1994. A hyphal-specific chitin synthase gene (*CHS2*) is not essential for growth, dimorphism, or virulence of *Candida albicans*. Proc. Natl. Acad. Sci. USA 91: 6216-6220.

Hauser, K., and Tanner, W. 1989. Purification of the inducible α–agglutinin of *Saccharomyces cerevisiae* and molecular cloning of the gene. FEBS Lett. 225: 290-294.

Hazan, I., Sepulveda-Becerra, M., and Liu, H. 2002. Hyphal elongation is regulated independently of cell cycle in *Candida albicans*. Mol. Biol. Cell 13: 134-145.

Herschbach, B.M., Arnaud, M.B., and Johnson, A.D. 1994. Transcriptional repression directed by the yeast alpha 2 protein *in vitro*. Nature 370: 309-311.

Hornby, J.M., Jensen, E.C., Lisec, A.D., Tasto, J.J., Jahnke, B., Shoemaker, R., Dussault, P., and Nickerson, K.W. 2001. Quorum sensing in the dimorphic fungus *Candida albicans* is mediated by farnesol. Appl. Environ. Microbiol. 67: 2982-2992.

Hoyer, L.L. 2001. The *ALS* gene family of *Candida albicans*. Trends Microbiol. 9: 176-180.

Hoyer, L.L., Payne, T.L., Bell, M., Myers, A.M., and Scherer, S. 1998. *Candida albicans ALS3* and insights into the nature of the *ALS* gene family. Curr. Genet. 33: 451-459.

Hoyer, L.L., Scherer, S., Shatzman, A.R., and Livi, G.P. 1995. *Candida albicans ALS1*: domains related to a *Saccharomyces cerevisiae* sexual agglutinin separated by a repeating motif. Mol. Microbiol. 15: 39-54.

Hube, B., Monod, M., Schofield, D.A., Brown, A.J.P., and Gow, N.A.R. 1994. Expression of seven members of the gene family encoding secretory aspartyl proteinases in *Candida albicans*. Mol. Microbiol. 14: 87-99.

Kadosh, D, and Johnson, A.D. 2001. Rfg1, a protein related to the *Saccharomyces cerevisiae* hypoxic regulator Rox1, controls filamentous growth and virulence in *Candida albicans*. Mol. Cell. Biol. 21: 2496-2505.

Kaiser, C., Michaelis, S., and Mitchell, A. 1994. Methods in Yeast Genetics. Cold Spring Harbor Laboratory Press, Cold Spring Harbor, NY.

Kaminishi, H., Iwata, A., Tamaki, T., Cho, T., and Hagihara, Y. 1994. Spiral hyphae of *Candida albicans* formed in anaerobic culture. Mycoses 37: 349-352

Kataoka, T., Powers, S., McGill C., Fasano, O., Strathern, J., Broach, J., and Wigler, M. 1984. Genetic analysis of yeast *RAS1* and *RAS2* genes. Cell 37: 437-445.

Keleher, C.A., Redd, M.J., Schultz, J., Carlson, M., and Johnson, A.D. 1992. Ssn6-Tup1 is a general repressor of transcription in yeast. Cell 68: 709-719.

Khala, R.A., and Zitomer, R.S. 2001. The DNA binding protein Rfg1 is a repressor of filamentation in *Candida albicans*. Genetics 157: 1503-1512.

Kohler, T., Wesche, S., Taheri, N., Braus, G.H., and Mosch, H-U. 2002. Dual role of the *Saccharomyces cerevisiae* TEA/ATTS family transcription factor Tec1p in regulation of gene expression and cellular development. Eukaryotic Cell 1: 673-686.

Lane, S., Birse, C., Zhou, S., Matson, R., and Liu, H. 2001a. DNA array studies demonstrate convergent regulation of virulence factors by Cph1, Cph2, and Efg1 in *Candida albicans*. J. Biol. Chem. 276: 48988-48996.

Lane, S., Zhou, S., Pan, T., Dai, Q., and Liu, H. 2001b. The basic helix-loop-helix transcription factor Cph2 regulates hyphal development in *Candida albicans* partly via Tec1. Mol. Cell. Biol. 21: 6418-6428.

Lee, K.L, Buckley, H.R., and Campbell, C.C. 1975. An amino acid liquid synthetic medium for the development of mycelial and yeast forms of *Candida albicans*. Sabouraudia 13: 148-153.

Leng, P., Lee, P.R., Wu, H., and Brown, A.J.P. 2001. Efg1, a morphogenetic regulator in *Candida albicans*, is a sequence-specific DNA binding protein. J. Bacteriol. 183: 4090-4093.

Li and Calderone. 2004. Antifungal drugs, targets and target discovery. In: Pathogenic Fungi: Host Interactions and Emerging Strategies for Control. G. San-Blas, and R.A. Calderone, eds. Caister Academic Press, Wymondham, Norfolk, UK. p. 335-365.

Lipke, P.N., Wojciechowicz, D., and Kurjan, J. 1989. *AG*α*1* is the structural gene for the *Saccharomyces cerevisiae*, α–agglutinin, a cell surface glycoprotein involved in cell-cell interactions during mating. Mol. Cell. Biol. 9: 3155-3165.

Liu, H. 2001. Transcriptional control of dimorphism in *Candida albicans*. Curr. Opin. Microbiol. 4: 728-735.

Liu, H., Kohler, J., and Fink, G.R. 1994. Suppression of hyphal formation in *Candida albicans* by mutation of a *STE12* homolog. Science 266: 1723-1726.

Liu, H., Styles, C.A., and Fink, G.R. 1993. Elements of the yeast pheromone response pathway required for filamentous growth of diploids. Science 262: 1741-1744.

Lo, H-J., Kohler, J.R., DiDomenico, B., Loebenberg, D., Cacciapuoti, A., and Fink, G.R. 1997. Nonfilamentous *C. albicans* mutants are avirulent. Cell 90: 939-949.

Marchler, G., Schuller, C., Adam, G., and Ruis, H. 1993. A *Saccharomyces cerevisiae* UAS element controlled by protein kinase A activates transcription in response to a variety of stress conditions. EMBO J. 12: 1997-2003.

Marquez, J.A., Pascual-Ahuir, A., Proft, M., and Serrano, R. 1998. The Ssn6-Tup1 repressor complex of *Saccharomyces cerevisiae* is involved in the osmotic induction of HOG-dependent and -independent genes. EMBO J. 17: 2543-2553.

Massari, M.E., and Murre, C. 2000. Helix-loop-helix proteins: regulators of transcription in eukaryotic organisms. Mol. Cell. Biol. 20: 429-440.

Mattia, E., and Cassone, A. 1979. Inducibility of germ-tube formation in *Candida albicans* at different phases of yeast growth. J. Gen. Microbiol. 113: 439-442.

Merson-Davies, L.A., and Odds, F.C. 1989. A morphology index for characterization of cell shape in *Candida albicans*. J. Gen. Microbiol. 135: 3143-3152.

Miller, K.Y., Wu, J., and Miller, B.L. 1992. *StuA* is required for cell pattern formation in *Aspergillus*. Genes Dev. 6: 1770-1782.

Muhlschlegel, F.A., and Fonzi, W.A. 1997. *PHR2* of *Candida albicans* encodes a functional homolog of the pH-regulated gene *PHR1* with an inverted pattern of pH-dependent expression. Mol. Cell. Biol. 17: 5960-5967.

Mukai, H., Kuno, T., Tanaka, H., Hirata, D., Miyakawa, T., and Tanaka, C. 1993. Isolation and characterization of *SSE1* and *SSE2*, new members of the yeast *HSP70* multigene family. Gene 132: 57-66.

Murad, A.M., d'Enfert, C., Gaillardin, C., Tournu, H., Tekaia, F., Talibi, D., Marechal, D., Marchais, V., Cottin, J., and Brown, A.J.P. 2001a. Transcript profiling in *Candida albicans* reveals new cellular functions for the transcriptional repressors CaTup1, CaMig1 and CaNrg1. Mol. Microbiol. 42: 981-993.

Murad, A.M.A., Leng, P., Straffon, M., Wishart, J., Macaskill, S., MacCallum, D., Schnell, N., Talibi, D., Marechal, D., Tekaia, F., d'Enfert, C., Gaillardin, C., Odds, F.C., and Brown, A.J.P. 2001b. *NRG1* represses yeast-hypha morphogenesis and hypha-specific gene expression in *Candida albicans*. EMBO J. 20: 4742-4752.

Nakayama, H., Mio, Y., Nagahashi, S., Kokado, M., Arisawa, M., and Aoki, Y. 2000. Tetracycline-regulatable system to tightly control gene expression in the pathogenic fungus *Candida albicans*. Infect. Immun. 68: 6712-6719.

Nantel, A., Dignard, D., Bachewich, C., Harcus, D., Marcil, A., Bouin, A-P., Sensen, C.W., Hogues, H., van het Hoog, M., Gordon, P., Rigby, T., Benoit, F., Tessier, D.C., Thomas, D.Y., and Whiteway, M. 2002. Transcription profiling of *Candida albicans* cells undergoing the yeast-to-hyphal transition. Mol. Biol. Cell 13: 3452-3465.

Nehlin, J.O., Carlberg, M., and Ronne, H. 1992. Yeast *SKO1* gene encodes a bZIP protein that binds to the CRE motif and acts as a repressor of transcription. Nucleic Acids Res. 20: 5271-5278.

Nehlin, J.O., and Ronne, H. 1990. Yeast *MIG1* repressor is related to the mammalian early growth response and Wilms' tumour finger proteins. EMBO J. 9: 2891-2898.

Odds, F.C. 1988. *Candida* and Candidosis, 2nd ed. Bailliere Tindall, London.

Pan, X., and Heitman, J. 1999. Cyclic AMP-dependent protein kinase regulates pseudohyphal differentiation in *Saccharomyces cerevisiae*. Mol. Cell. Biol. 19: 4874-4887.

Park, S.H., Koh, S.S., Chun, J.H., Hwang, H.J., and Kang, H.S. 1999. Nrg1 is a transcriptional repressor for glucose repression of *STA1* gene expression in *Saccharomyces cerevisiae*. Mol. Cell. Biol. 19: 2044-2050.

Porta, A., Ramon, A.M., and Fonzi, W.A. 1999. *PRR1*, a homolog of *Aspergillus nidulans palF*, controls pH-dependent gene expression and filamentation in *Candida albicans*. J. Bacteriol. 181: 7516-7523.

Proft, M., and Serrano, R. 1999. Repressors and upstream repressing sequences of the stress-regulated *ENA1* gene in *Saccharomyces cerevisiae*: bZIP protein Sko1p confers HOG-dependent osmotic regulation. Mol. Cell. Biol. 19: 537-546.

Ralph, D., McClelland, M., and Welsh, J. 1993. RNA fingerprinting using arbitrarily primed PCR identifies differentially regulated RNAs in mink lung (Mv1Lu) cells growth arrested by transforming growth factor B1. Proc. Natl. Acad. Sci. USA 90: 10710-10714.

Ramon, A.M., Porta, A., and Fonzi, W.A. 1999. Effect of environmental pH on morphological development of *Candida albicans* is mediated via the PacC-related transcription factor encoded by *PRR2*. J. Bacteriol. 181: 7524-7530.

Riggle, P.J., Andrutis, K.A., Chen, X., Tzipori, S.R., and Kumamoto, C.A. 1999. Invasive lesions containing filamentous forms produced by a *Candida albicans* mutant that is defective in filamentous growth in culture. Infect. Immun. 67: 3649-3652.

Robertson, L.S., and Fink, G.R. 1998. The three yeast A kinases have specific signaling functions in pseudohyphal growth. Proc. Natl. Acad. Sci. USA 95: 13783-13787.

Robinson, K.A., and Lopes, J.M. 2000. *Saccharomyces cerevisiae* basic helix-loop-helix proteins regulate diverse biological processes. Nucleic Acids Res. 28: 1499-1505.

Rupp, S., Summers, E., Lo, H-J., Madhani, H., and Fink, G. 1999. MAP kinase and cAMP filamentation signaling pathways converge on the unusually large promoter of the yeast *FLO11* gene. EMBO J. 18: 1257-1269.

Sambrook, J., Fritsch, E.F., and Maniatis, T. 1989. Molecular Cloning: A Laboratory Manual, 2nd ed. Cold Spring Harbor Laboratory Press, Cold Spring Harbor, NY.

Saporito-Irwin, S.M., Birse, C.E., Sypherd, P.S., and Fonzi, W.A. 1995. *PHR1*, a pH-regulated gene of *Candida albicans*, is required for morphogenesis. Mol. Cell. Biol. 15: 601-613.

Schaller, M., Korting, H.C., and Hube, B. 2004. Virulence factors that promote invasion of *Candida albicans*. In: Pathogenic Fungi: Host Interactions and Emerging Strategies for Control. G. San-Blas, and R.A. Calderone, eds. Caister Academic Press, Wymondham, Norfolk, UK. p.97-128.

Schweizer, A., Rupp, S., Taylor, B.N., Rollinghoff, M., and Schroppel, K. 2000. The TEA/ATTS transcription factor CaTec1p regulates hyphal development and virulence in *Candida albicans*. Mol. Microbiol. 38: 435-445.

Sharkey, L.L., McNemar, M.D., Saporito-Irwin, S.M., Sypherd, P.S., and Fonzi, W.A. 1999. *HWP1* functions in the morphological development of *Candida albicans* downstream of *EFG1*, *TUP1*, and *RBF1*. J. Bacteriol. 181: 5273-5279.

Shirayama, M., Kawakami, K., Matsui, Y., Tanaka, K., and Toh-e, A. 1993. MSI3, a multicopy suppressor of mutants hyperactivated in the RAS-cAMP pathway, encodes a novel HSP70 protein of *Saccharomyces cerevisiae*. Mol. Gen. Genet. 240: 323-332.

Soll, D.R. 1986. The regulation of cellular differentiation in the dimorphic yeast *Candida albicans*. Bioessays 5: 5-11.

Soll, D.R., and Herman, M.A. 1983. Growth and the inducibility of mycelium formation in *Candida albicans*:a single-cell analysis using a perfusion chamber. J. Gen. Microbiol. 129: 2809-2824.

Sohn, K., Urban, C., Brunner, H., and Rupp, S. 2003. *EFG1* is a major regulator of cell wall dynamics in *Candida albicans* as revealed by DNA microarrays. Mol. Microbiol. 47: 89-102.

Sonneborn, A., Bockmuhl, D.P., Gerads, M., Kurpanek, K., and Sanglard, S. 2000. Protein kinase A encoded by TPK2 regulates dimorphism of *Candida albicans*. Mol. Microbiol. 35: 386-396.

Staab, J.F., Bradway, S.D., Fidel, P.L., and Sundstrom, P. 1999. Adhesive and mammalian transglutaminase substrate properties of *Candida albicans* Hwp1. Science 283: 1535-1538.

Staab, J.F., Ferrer, C.A., and Sundstrom, P. 1996. Developmental expression of a tandemly repeated, proline- and glutamine-rich amino acid motif on hyphal surfaces of *Candida albicans*. J. Biol. Chem. 271: 6298-6305.

Stoldt, V.R., Sonneborn, A., Leuker C.E., and Ernst, J.F. 1997. Efg1p, an essential regulator of morphogenesis of the human pathogen *Candida albicans*, is a member of a conserved class of bHLH proteins regulating morphogenetic processes in fungi. EMBO J. 16: 1982-1991.

Stratford, M. 1992. Yeast flocculation: reconciliation of physiological and genetic viewpoints. Yeast 8: 25-38.

Sullivan, P.A., Chiew, Y.Y., Molloy, C., and Templeton, M.D. 1983. An analysis of the metabolism and cell wall composition of *Candida albicans* during germ-tube formation. Can. J. Microbiol. 29: 1514-1525.

Sundstrom, P., and Aliaga, G.R. 1992. Molecular cloning of cDNA and analysis of protein secondary structure of *Candida albicans* enolase, an abundant, immunodominant glycolytic enzyme. J. Bacteriol. 174: 6789-6799.

Swoboda, R.K., Bertram, G.., Hollander, H., Greenspan, D., Greenspan, J.S., Gow, N.A., Gooday, G.W., and Brown, A.J. 1993. Glycolytic enzymes of *Candida albicans* are nonubiquitous immunogens during candidiasis. Infect. Immun. 61: 4263-4271.

Tilburn, J., Sarkar, S., Widdick, D.A., Espeso, E.A., Orejas, M., Mungroo, J., Penalva, M.A., and Arst, H.N.Jr. 1995. The *Aspergillus PacC* zinc finger transcription factor mediates regulation of both acid- and alkaline-expressed genes by ambient pH. EMBO J. 14: 779-790.

Toda, T., Cameron, S., Sass, P., Zoller, M., and Wigler, M. 1987. Three different genes in *Saccharomyces cerevisiae* encode the catalytic subunits of the cAMP-dependent protein kinase. Cell 50: 277-287.

Treitel, M.A., and Carlson, M. 1995. Repression by SSN6-TUP1 is directed by MIG1, a repressor/activator protein. Proc. Natl. Acad. Sci. USA 92: 3132-3136.

Trumbly, R.J. 1992. Glucose repression in the yeast *Saccharomyces cerevisiae*. Mol. Microbiol. 6: 15-21.

Van Helden, J., Andre, B., and Collado-Vides, J. 2000. A web site for the computational analysis of yeast regulatory sequences. Yeast 16: 177-187.

Ward, M.P., Gimeno, C.J., Fink, G.R., and Garrett, S. 1996. *SOK2* may regulate cyclic AMP-dependent protein kinase-stimulated growth and psuedohyphal development by repressing transcription. Mol. Cell. Biol. 15: 6854-6863.

Warenda, A.J., Kauffman, S., Sherrill, T.P., Becker, J.M., and Konopka, J.B. 2003. *Candida albicans* septin mutants are defective for invasive growth and virulence. Infect. Immun. 71: 4045-4051.

Warenda, A.J., and Konopka, J.B. 2002. Septin function in *Candida albicans* morphogenesis. Mol. Biol. Cell 13: 2732-2746.

Whistler, J.L., and Rina, J. 1997. Ras2 and Ras1 protein phosphorylation in *Saccharomyces cerevisiae*. J. Biol. Chem. 272: 18790-18800.

White, T.C., and Agabian, N. 1995. *Candida albicans* secreted aspartyl proteinases: isoenzyme pattern is determinated by cell type, and levels are determined by environmental factors. J. Bacteriol. 177: 5215-5221.

Whiteway, M., Dignard, D., and Thomas, D.Y. 1992. Dominant negative selection of heterologous genes: isolation of *Candid albicans* genes that interfere with *Saccharomyces cerevisiae* mating factor-induced cell cycle arrest. Proc. Natl. Acad. Sci. USA 89: 9410-9414.

Wickner, R.B. 1974. Mutants of *Saccharomyces cerevisiae* that incorporate deoxythymidine-5'-monophosphate into deoxyribonucleic acid *in vivo*. J. Bacteriol. 117: 252-260.

Zitomer, R.S., and Lowry, C.V. 1992. Regulation of gene expression by oxygen in *Saccharomyces cerevisiae*. Microbiol. Rev. 56: 1-11.

From: Pathogenic Fungi: Structural Biology and Taxonomy
Edited by: G. San-Blas and R.A. Calderone

Chapter 5

Morphogenesis in Other Agents of Systemic Mycoses

Gioconda San-Blas and Gustavo Niño-Vega

Abstract

Important dimorphic pathogenic fungi like *Histoplasma capsulatum*, *Blastomyces dermatitidis*, *Sporothrix schenkii*, *Paracoccidioides brasiliensis*, *Penicillium marneffei*, *Coccidioides immitis*, and *Wangiella dermatitidis* have not been studied as systematically as *Candida albicans* or *Aspergillus* species from the point of view of molecular mechanisms controlling the dimorphic process. There is only fragmentary information that is available on assorted mechanisms probably involved in the dimorphic transition. The recent focus of study on dimorphism has been of genes associated with cell wall and membrane lipid synthesis, structure, cytoskeleton, and metabolic pathways. Research is needed in both signal transduction pathways and on the molecular level of the genome and the supramolecular level of the cell in order to really understand the morphogenetic processes carried out during the life cycles of many of these fungi.

1. Introduction

Although there are an estimated 100,000 fungal species, only a select group of fungi cause the human systemic mycoses. Many of these pathogens depend upon morphogenetic transitions to express their virulence. Notable among them are important pathogens like *Candida albicans* or several species of *Aspergillus* whose morphogenetic pathways have been subjected to intensive research, reviewed in other chapters of this book.

Fungal pathogens face a sudden and drastic environmental temperature change at the time of infection. They must adapt rapidly to dramatic shifts in temperatures and a variety of environmental stress conditions to survive and invade mammalian tissues (Chauhan and Calderone, 2004). Morphogenesis in pathogenic fungi has some unique features: it is reversible, is not an essential component of the life cycle of the organism, but rather represents an opportunistic adaptation to environmental conditions, and conversion to a different form (generally yeast) is required for progressive infection (Ignatov and Keath, 2002b). The ability of pathogens to initiate infection, invade host tissues and survive in mammalian hosts is critically linked to the induction of specific gene products. In dimorphic fungi, developmentally regulated gene expression is particularly important. Since morphogenetic changes in many fungi are interwoven with the adaptive survival strategies that make them successful pathogens, many studies comprising biochemical, molecular, developmental and immunological approaches have been undertaken to understand the relationships between dimorphism and virulence.

All organisms, including bacteria, fungi, and mammalian cells, respond to cues from the extracellular environment. These cues are then transduced from the cell surface to the interior of the cell, resulting in patterns of altered gene expression and protein activity, which, in turn, cause important cellular responses to the external environment. In eukaryotic cells, the cAMP and MAPK cascade modules are key elements in mediating the transduction of many signals generated at the cell surface to the nucleus (Banuett, 1998). These and other cascades are the subject of intense research in *Saccharomyces cerevisiae*, as well as in *C. albicans* and a few other dimorphic pathogenic fungi, as reflected elsewhere in this book. As a consequence of such signalling events, morphogenetic transition is linked to the synthesis of different compounds that may influence the final shape of the cell. Among the structures that are important to the cell shape is the cell wall, a unique structure that has been thoroughly studied not only for its relevance to morphogenesis but also because its biosynthesis may be the target of highly specific drugs, with potential use as antibiotics. In fact, the recent introduction of caspofungin acetate (Cancidas®, Merck) has marked a breakthrough in antifungal therapy, as a drug whose mechanism of action is based on its capacity to inhibit β-1,3-glucan synthase activity, therefore blocking the synthesis of this unique structural component of the cell wall of several important pathogenic fungi (Denning, 1997). Besides signal transduction and cell wall formation, morphogenesis, including the actin and microtubules of the cytoskeleton, constitute a third area of study of the pathogenic fungi (Bartnicki-García, 2002).

The molecular mechanisms controlling the dimorphic process have not been studied as extensively with the important dimorphic pathogenic fungi like *Blastomyces dermatitidis, Coccidioides immitis*, *Histoplasma capsulatum*, *Paracoccidioides brasiliensis*, *Penicillium marneffei*, *Sporothrix schenkii*, and *Wangiella dermatitidis* as with other human pathogens, like *C. albicans*. Thus, only fragmentary mechanistic information is available on the regulation of the dimorphic transition in the large group of organisms mentioned above. For example, information on some phase-specific genes is available. While providing partial insight on the maintenance and pathogenesis of a given species, this type of data may be unrevealing about the dimorphic process itself. The aim of this chapter is to organize this diverse information into common features of the morphogenetic process in fungal pathogens, with the assumption that although many different signals can induce morphogenetic changes, the strategies for connecting the external signals to the change in cell differentiation are broadly conserved among fungi (Sánchez-Martínez and Pérez-Martín, 2001).

2. *Histoplasma capsulatum*

The dimorphic fungus *H. capsulatum* is the etiological agent of histoplasmosis, a respiratory infection that is worldwide in distribution (Ignatov and Keath, 2002b). As a saprophyte in soil, it is mycelial (M), but it becomes a pathogenic budding yeast (Y) in susceptible hosts. These morphological phases can be reproduced reversibly *in vitro* by shifting the temperature from 25°C, at which it is mycelial, to 37°C, which induces growth as a budding yeast. Inhaled microconidia convert into the pathogenic Y phase by replicating in the permissive intracellular environment, eventually lysing the alveolar macrophage that acts as the first line of defence by the human host. Once the process of morphogenetic conversion starts, the fungal cell wall changes dramatically in its structure and biochemical content by producing α-1,3-glucan that is a relevant feature of the Y phase of virulent *H. capsulatum* strains (Klimpel and Goldman, 1988). The temperature induced conversion also results in a replacement of the β-1,3-glucan present in the cell wall of its M phase.

Changes in cell shape are likely to require alterations in cytoskeletal components; the arrangement of microtubules at the cell surface may direct subsequent morphological events while actin may also play a role. *H. capsulatum* possesses two tubulin genes, *TUB1* and *TUB2*, that codify for α- and β-tubulin, respectively (Harris *et al.*, 1989a), that are expressed constitutively (Harris *et al.*, 1989b). Single α- and β-tubulins are observed in the Y phase and additional α_2- and β_2-tubulin isoforms are seen in the M phase, as probable posttranslationally modified products of the α- and β-tubulin genes. Their expression is developmentally regulated during the dimorphic phase transition; their mRNA increases six to eightfold during the Y to M conversion and decreases two to threefold during the reverse process (Harris *et al.*, 1989b). The other important component of the cytoskeleton, actin, is the product of a single gene in *H. capsulatum* (El-Rady and Shearer, 1997). Its predicted protein has 375 amino acids with a high degree of homology to other actin proteins. Actin mRNA levels were relatively constant throughout the examined stages of the M to Y transition, except for a four-fold decrease in M cultures after a 4 h shift to 37°C.

Many initial studies on *H. capsulatum* focused on nutritional requirements for each morphotype. All strains appear to be auxotrophic for cysteine in the Y phase, while the M phase is independent of this amino acid requirement (Maresca and Kobayashi, 1989). M to Y transition can be permanently arrested by the sulfhydryl-blocking agent *p*-chloromercuriphenylsulfonate (Medoff *et al.*, 1986). The difference in cysteine metabolism may be the result of differential expression of sulfite reductase, which is expressed in hyphae but not in yeast cells of the organism, or by differential expression of cysteine oxidase which is abundant in yeast but absent in the vegetative form. As a result, cysteine affects respiration which in *H. capsulatum* proceeds through a branched electron transport system consisting of mitochondrial and alternative oxidase "shunt" respiratory pathways. This process is tied to heat shock proteins 70 and 83 that are expressed following the uncoupling of oxidative phosphorylation and expression of yeast-specific cysteine reductase (Caruso *et al.*, 1987). Recently, the prediction of an alternative oxidase pathway was fulfilled. Johnson *et al.* (2003) isolated an alternative oxidase gene, *AOX1*, whose ORF predicts a 356 amino acid polypeptide product, with a mass of 40,814 Da. The alternative oxidase enzyme is typically a mitochondrial enzyme. As such, the *H. capsulatum AOX1* translation product has a predicted cleavable mitochondrial targeting sequence consisting of the first 54 N-terminal residues. Therefore, the fully processed enzyme should have a mass of approximately 36 kDa. As expected, treatment of *H. capsulatum* cells with hydrogen peroxide, antimycin A or cyanide led to an elevation of *AOX1* transcript levels, as a result of an oxidative stress caused by inhibition of mitochondrial electron transport with antimycin or cyanide, or by the effect of the antioxidant peroxide. Oxidative stress is also encountered later in the infective process due to the production of reactive oxygen species during the host inflammatory response. In this situation, catalase, which converts hydrogen peroxide to water and oxygen, is a crucial component of oxidative stress tolerance (Demple, 1999). In *H. capsulatum*, three catalase genes are present: *CATA*, *CATB* (encoding the exocellular M antigen), and *CATP*, each one is present as a single copy in the fungal genome (Johnson *et al.*, 2002). All of these are expressed during the mycelial phase of growth under standard conditions, i.e., in the absence of any deliberate oxidative challenge. During yeast phase growth, the *CATB* and *CATP* genes are constitutively expressed, but *CATA* RNA is barely detectable unless a recent oxidative challenge with hydrogen peroxide has occurred (Johnson *et al.*, 2002). Another mitochondrial enzyme that belongs to the respiratory pathway, NAD(+)-isocitrate dehydrogenase, was detected by a cDNA that codes for a subunit of this enzyme (Johnson and McEwen, 1999). It also contained a cleavable targeting sequence of 35 residues that resembled the mitochondrial matrix targeting sequence of proteins in other organisms. Its relation to the dimorphic process was not proven.

Heat shock proteins are expressed at low levels at physiological temperatures, and their transcription is increased under stressful conditions and at high temperatures. In fungi, these are induced very early during infection and play a critical role during adaptation to the new environment. Since morphological transition leads to the pathogenic form, the regulation and biological significance of the heat-shock gene family are important for understanding morphogenesis, adaptation and host-parasite relationships in pathogenic fungi. Transcriptional analysis of cloned *hsp70* and *hsp82* (Maresca *et al.*, 1994) and *hsp60* (Deepe *et al.*, 1996) have shown that in *H. capsulatum*, the temperature at which the maximal heat-shock response occurs depends on the virulence of the strain, and varies between 34°C (for nonvirulent Downs strain) and 37°C (for virulent G222B and G217B strains). Heat shock affects RNA metabolism; under conditions of severe heat shock, pre-mRNAs derived from intron-containing genes are not spliced and accumulate in the cytoplasm, by stabilizing the pre-mRNA-spliceosome complex (Maresca *et al.*, 1994). HSP70 may be one of the proteins involved in protecting splicing events, and it is interesting to note that mycelial cells of *H. capsulatum* have constitutively high levels of HSP70 (Shearer *et al.*, 1987). This protein also has a role as dominant antigen, inducing cellular immune responses in mice and mediating protective immunity against *H. capsulatum* (Gómez *et al.*, 1992). Heat shock hampers the capacity of cells to synthesize ATP, and in *S. cerevisiae* it reduces its capacity to couple oxidative phosphorylation to electron transport when grown under respiratory conditions, an effect that is proportional to the severity and length of the heat shock treatment (Patriarca and Maresca, 1990). In *H. capsulatum*, heat shock proteins induced by mild heat shock pre-treatment or by starvation, preserve the capacity of the cells to retain coupled oxidative phosphorylation when the cells are subsequently exposed to higher temperatures. In addition to protection of mitochondrial ATPase activity by the induction of a thermotolerant state, a strict correlation exists between temperature sensitivity and decreased transcription of the major heat shock genes in strains Downs and G222B, results that may reflect a cellular mechanism for maintaining functional conformation of mitochondrial proteins during and following heat stress (Patriarca *et al.*, 1992).

Heat shock proteins have also been linked to membrane lipids in *H. capsulatum*. The physical properties of lipids and the dynamic nature of membranes have important effects on the functions of many membrane-associated proteins. Homeoviscous adaptation, a common feature of all organisms that adapt to different temperatures, is generally accomplished by incorporation of different types of fatty acids into the membrane during phospholipid synthesis or by regulation of the degree of unsaturation of the fatty acid itself (Gargano *et al.*, 1995). By modifying the transition temperature of membrane phospholipids, homeoviscous adaptation of the membrane maintains a given fluid environment for enzyme function at any physiological temperature. In *H. capsulatum*, the capacity to undergo a stress response is determined by the fatty acid composition of its membrane (Maresca *et al.*, 1994). Strains with altered saturated fatty acid : unsaturated fatty acid (SFA: UFA) ratios have a different level of heat shock gene expression, and a different pattern of adaptation during the M to Y conversion. Addition of SFAs shifts the optimal temperature for heat shock gene transcription from 34°C to 37°C in the temperature-sensitive Downs strains, while addition of UFAs drastically reduces heat shock gene expression at 37° C in the temperature-tolerant G217B strain (Maresca *et al.*, 1994). Exogenous addition of SFA induced a strong increase in heat shock mRNA transcription when cells were heat-shocked at 37°C, whereas treatment with UFA reduced or eliminated the level of heat shock gene transcription at this temperature (Carratù *et al.*, 1996). These results led these authors to propose that a primary target in the biological perception of temperature is the membrane, in which a heat shock sensor is present and whose activation, following heat shock, represents the first step of the signal transduction pathway that leads to the phosphorylation of a heat shock factor, in a similar way as extracellular signals affect the activity of MAPK cascades in yeast or in *C. albicans* (Banuett, 1998).

In animal and fungal cells, UFAs are formed by fatty acid desaturases which are lipophilic microsomal enzymes, whose main function is to maintain membrane fluidity throughout the cell cycle. The Δ^9-desaturase gene product is the major enzyme for converting SFA into UFA and is responsible for the introduction of the first double bond between carbons 9 and 10 of palmitoyl (16:0) or stearoyl (18:0)-CoA to form palmitoleic (16:1) or oleic (18:1) acid, respectively. This enzyme is encoded by *Ole1*, its deduced product has a molecular weight of 53.7 kDa, and a 72% similarity to *S. cerevisiae* Δ^9-desaturase. In contrast with the *S. cerevisiae* gene, the *H. capsulatum Ole1* gene contains one intron. Both proteins contain two regions of 50 amino acids with long stretches of hydrophobic residues, forming loops that are proposed to anchor the protein in the endoplasmic reticulum (Gargano *et al.*, 1995). Transcription of the *Ole1* in the thermotolerant G217B and thermosensitive Downs strains is similar both in yeast phase cells and during the temperature shift down. Nevertheless, the *Ole1* is transcriptionally inactive in mycelia of G217B at 25°C, while it is actively transcribed in the Downs strains at that temperature. An analysis of the promoter of *H. capsulatum Ole1* (Tosco *et al.*, 1997) identified the presence of a canonical AP1 binding site (TGACTAA), located at –705 of 5′ upstream from the 5′ of ATG, in the fragment of the Downs *Ole1* gene, that was absent in the corresponding G217B gene. AP1 is a *cis*-acting transcriptional factor that binds nuclear proteins, contributing to the oxidative stress and is probably involved in the control of expression of the Δ^9-desaturase gene. AP1 was later detected with the use of electrophoretic mobility shift assays (EMSAs) (Ignatov and Keath, 2002a).

In *H. capsulatum*, the temperature shift that initiates the dimorphic transition is accompanied by changes in the intracellular levels of cAMP that is about five times higher in mycelia than in the yeast phase (Maresca and Kobayashi, 1989). Although the mechanism of action is unknown in this fungus, it is suggested that cAMP is involved in the activation of signal transduction mechanisms like the cAMP-dependent protein kinases reported in other dimorphic fungi (D'Souza and Heitman, 2001; Rocha *et al.*, 2001). One possible mechanism is the transcriptional regulation of the Δ^9-desaturase gene by cAMP (Storlazzi *et al.*, 1999). By deletion analysis of the promoter region of Δ^9-desaturase, these authors identified a sequence of eleven nucleotides (Desaturase Regulatory Element or DRE; CAAATCATCGG) that was necessary but not sufficient for the expression of the *H. capsulatum* Δ^9-desaturase gene. They also identified stress expression [STRE] boxes (CCCCT), located at –166 in the Δ^9-desaturase promoters of strain G217B of *H. capsulatum*. In *S. cerevisiae*, STRE boxes are known to be the only *cis*-regulatory sequences involved in cAMP response (Schmitt and McEntee, 1996). Addition of a high concentration of cAMP to yeast phase cells lowers the amount of *H. capsulatum Ole1* mRNA, suggesting that cAMP is involved in the transcriptional regulation of Δ^9-desaturase in this fungus, presumably through STRE box-mediated response. Additional evidence for the existence of transcriptional factors linked to cAMP signalling comes from gel shift assays of nuclear extracts from *H. capsulatum*, in which two such transcriptional factors, AP-2 and CREB, were detected (Ignatov and Keith, 2002a).

Fungal glycosphingolipids, particularly glycosylinositol phosphorylceramides (GIPCs) and monohexosyl-cerebrosides (cerebrosides or CMHs) have also been related to the yeast-mycelium transition in dimorphic fungi. In *H. capsulatum*, studies by means of 1- and 2-dimensional nuclear magnetic resonance (NMR) spectroscopy, electrospray ionisation mass spectrometry (ESI-MS) and low energy tandem collision-induced dissociation mass spectrometry (ESI-MS/CID-MS) (Toledo *et al.*, 2001c) led to the identification of cerebrosides. In both the mycelial and yeast phases, they were characterized as β-glucopyranosyl ceramides (GlcCers), containing (4*E*,8*E*)-9-methyl-4,8-sphingadienine as the long chain base, attached to 18-carbon 2-hydroxy fatty *N*-acyl components. However, while the fatty acid of the yeast GlcCer was virtually all *N*-2'-hydroxyoctadecanoate, the GlcCer from the mycelial phase was characterized as *N*-2'-hydroxy-(*E*)-Δ^3-octadecenoate. This result suggests that the Y to M transition, but not the reverse, is accompanied by an up regulation of an as yet uncharacterised ceramide or cerebroside 2-hydroxy fatty *N*-acyl -(*E*)-Δ^3-desaturase activity. At the same time, a role as a specific messenger is proposed for GlcCer, with activation or deactivation of the desaturase being one step in a signalling cascade directing the morphogenetic transition (Toledo *et al.*, 2001c).

Several techniques have been used to identify genes that are differentially expressed or up-regulated during infection and dimorphic transition. However, a direct relationship of these genes with the morphogenetic process is not always straightforward. Instead, some of these phase-specific genes are easily related to virulence. Colonna-Romano *et al.* (1998) used PCR-based differential display of reverse transcription products (ddRT-PCR) to identify several gene products that are strongly and preferentially expressed by yeast cells during infection of primary cultures. One of these genes, *p100*, has high homology to a family of transcriptional activators, and may function as a regulatory switch to influence gene expression by the parasite in the host environment (Porta *et al.*, 1999). The *cdc2* gene, involved in the control of entry of a cell into the M phase, was characterized in *H. capsulatum* (Di Lallo *et al.*, 1994). It encodes a protein of 324 amino acids with a predicted mass of 36.9 kDa, and is transcriptionally regulated during the morphologic transition, being more actively transcribed in the yeast than in the mycelial phase. The *yps-3* gene, specific for the Y phase (Keath and Abidi, 1994; Keath *et al.*, 1989; Weaver *et al.*, 1996), was studied in the low-virulent Downs strain, and in the intermediate- and highly-virulent strains G186B and G217B. In the Y phase of highly virulent G217B strain, but not in its M phase, three distinct mRNAs of 1.3, 1.05, and 0.95 kb were detected. The Downs strain, instead, failed to express the *yps-3* gene at any stage while the intermediate-virulent strain G186B expressed only a single 1.25 kb mRNA at low levels in the Y phase (Keath *et al.*, 1989). Failure of the Downs *yps3* gene to be expressed was due to its harbouring a 287 bp insertion sequence that disrupted a long ORF defined by the *yps3* G217B cDNA (Keath and Abidi, 1994). A 30 kDa nuclear protein (p30M) in the mycelial phase of strain G217B was detected, which binds to the 5′ regulatory regions of *yps-3* and *yps 21:E-9* (another yeast-specific gene) (Abidi *et al.*, 1998). A decanucleotide motif, TCCTTTTTTT, was identified in the upstream regulatory regions of these *yps* genes in a phase-specific manner, suggesting that p30M, may govern aspects of gene transcription in *H. capsulatum*, acting as a transcriptional repressor. The *yps3* gene is expressed as early as one day after switching mycelial cultures to the yeast phase, and remains high throughout the transition (Keath *et al*., 1989). The promoter of the *yps3* gene shares regions of homology with the *BAD1* (*WI1*) gene of *B. dermatitidis* (Rooney *et al*., 2001), which encodes a 120 kDa polypeptide containing a repetitive adhesin domain and region with homology to the epidermal growth factor family of polypeptides and the YPS3 protein. Transformation of *Histoplasma* with *BAD1* resulted in conservation of yeast phase-specific expression of *BAD1* under conditions where the *BAD1* gene was transcriptionally silent in mycelia. It may be possible that *yps3*, like *BAD1*, is involved in signalling events that prompt the conversion program (Rooney *et al*., 2001). By means of a monoclonal antibody, the *yps3* gene product was localized as a cell wall protein in the highly virulent strain G217B (Weaver *et al*., 1996). Several features of the predicted amino acid sequence of the yps3-polypeptide suggest that it is cotranslationally inserted into the rough endoplasmic reticulum (ER) and posttranslationally modified by transfer to glycolipid, after which it may become covalently attached to the outer face of the cell membrane.

Mycelium-specific genes are also under scrutiny in *H. capsulatum*. A screening of randomly selected clones identified cDNA fragments representing 16 different mycelial up-regulated genes (Tian and Shearer, 2001) which, by probability analysis was extended to 27 up-regulated genes. Of these, the *MS8* transcript was very strongly expressed in mycelia but not detected on Northern blots with yeast RNA. The putative MS8 protein had a predicted size of 21.3 kDa, and was extremely rich in glycine and glutamine (over 40% of the protein). *ms8* knockout mutants formed apparently normal yeast at 37°C but gave rise to aberrant mycelia at 25 °C (Tian and Shearer, 2002).

H. capsulatum calcium-binding protein (CBP; Mw 7.8 kDa), encoded by the *cbp1* gene, is abundantly expressed by yeast in liquid culture (Batanghari and Goldman, 1997; Magrini and Goldman, 2001) as well as in the hostile conditions of the phagolysosome (Kügler *et al*., 2000) where it may scavenge Ca^{2+}. Mycelia, on the contrary, do not secrete CBP and require calcium for growth. The gene *CBP1* has been cloned and sequenced, and a potential calcium binding site is predicted from the secondary structure of CBP (Batanghari *et al.*, 1998). *CBP1* is transcriptionally regulated and is virtually undetected in the mycelial phase (Patel *et al.*, 1998). It contains a 102 bp region, between –786 and –887 upstream of the start codon, which significantly increases promoter activity by an undefined mechanism. This gene, however, was not simply related to a temperature shift but rather developmentally regulated, as it was only fully active after complete morphological conversion to the yeast form (Klüger *et al.*, 2000). The relation between calcium limitation and intracellular parasitism was probed by a strategy to disrupt *CBP1* in *H. capsulatum* using a telomeric linear plasmid and a two-step genetic selection (Sebghati *et al.*, 2000). The resulting *cbp1* yeasts no longer grew when deprived of calcium, and were also unable to destroy macrophages *in vitro* or proliferate in a mouse model of pulmonary infection, suggesting that calcium acquisition is an important strategy for microbial survival into the host.

Calmodulin (CaM), a small heat-resistant acidic protein, is the major CBP in non-muscle cells, and is thought to be ubiquitous in eukaryotic organisms where it is highly conserved. *H. capsulatum CaM* was cloned and sequenced. Its deduced amino acid sequence was almost identical to those of *Aspergillus nidulans* and *Neurospora crassa* (El-Rady and Shearer, G., 1996). CaM mRNA levels were approximately two-fold greater in the yeast than in the mycelial phase of the fungus, suggesting a possible role in dimorphism, perhaps through an increase in the activity of a CaM-dependent cyclic nucleotide phosphodiesterase in response to the higher level of CaM. The increased phosphodiesterase activity would in turn cause a reduction in the levels of intracellular cAMP, a secondary messenger proposed as a critical intracellular signal for dimorphic transition (Medoff *et al.*, 1986; Maresca and Kobayashi, 1989).

By means of a 10,000-element genomic shotgun microarray, approximately 500 clones were identified whose expression was differentially induced in either the yeast or the mycelial forms, including several potential regulatory genes (Hwang *et al.*, 2003). Their data revealed several categories of genes showing significantly higher expression in mycelia than in yeast. Several orthologs of genes involved in conidiation, a process confined to mycelia, were observed such as *FLU1* whose ortholog in *A. nidulans* (*fluG*) regulates conidial differentiation and secondary metabolite production. This gene functions as an upstream member of a conidiation pathway that includes the transcription factor *flbC* and the regulator *wetA* whose *H. capsulatum* orthologs (*FBC1* and *WET1*) were among the set of genes expressed in mycelia. Also an ortholog of phenylalanine-tRNA synthase (*TRF1*) required for polar hyphal growth in *A. nidulans* was found. Some mycelia-specific genes were related to physiology, such as genes that encode for nitrite reductase and phosphatases, reflecting perhaps the different growth requirements of each phase. Genes with higher expression in the yeast phase were also identified, some of which were involved in sulphur metabolism such as cysteine dioxygenase, choline sulphatase, ATP sulphurylase, methionine permease. Other yeast-specific genes were related to other aspects of metabolism and nutrient availability such as lysine permease, oxoprolinase, dehydrogenases, and several ATP-binding cassette and major facilitator superfamily efflux transporters. Northern analysis revealed that, in addition to their differential expression, six mycelial-regulated genes (*FLU1*, *NIR1*, *WET1*, *FDH1*, *DPP1* and *OXO1*) exhibited differences in transcript size in yeast and mycelia. These were due to differences in the length of the 5′ untranslated region that reflected differences in the start site of transcription. These observations suggest that *H. capsulatum* uses both differential expression and differential usage of transcript initiation sites to regulate the transcriptional profile of the two morphological forms (Hwang *et al.*, 2003).

Some additional information is available on the genome of *H. capsulatum*. Studies with ribosomal DNA polymorphism (Spitzer *et al.*, 1989), *yps-3* RFLP analysis (Keath *et al.*, 1992), and arbitrary primer PCR (Kersulyte *et al.*, 1992; Woods, 2002), demonstrated that strains of *H. capsulatum* are actually quite diverse. Analysis of chromosome number also showed major differences between strains, Downs strains having at least seven chromosomes, G186B at least four, and G217B at least three (Steele *et al.*, 1989). By DNA renaturation kinetics, genomic reconstruction with three putative single-copy genes (calmodulin, α-tubulin, and β-tubulin) and flow cytometry, nuclear DNA from strains G186AS (a low virulent strain derived from G186B) and Downs were studied (Carr and Shearer, 1998). G186AS was found to have a genome of approximately 2.3 x 10^7 bp with 0.5% repetitive sequences. The Downs strain, however, was 40% larger, with more than 16 times more repetitive DNA. Both strains were haploid. However, data obtained by genomic reconstruction with the α- and β-tubulin probes indicate that these genes are present in two copies in Downs and in one single copy in G186AS. This suggests that some partial duplication of chromosomal DNA could account for the extra 9 Mb genomic DNA in Downs. Currently, the *Histoplasma* genome project is ongoing (http://genome.wustl.edu/projects/hcapsulatum). It is expected that this project will contribute to probe the relationship between dimorphism and its impact on fungal pathogenesis.

3. *Paracoccidioides brasiliensis*

This pathogenic dimorphic fungus is the causative agent of paracoccidioidomycosis, a human systemic mycosis geographically confined to Latin America, where it constitutes one of the most prevalent deep mycoses (for a review, see San-Blas *et al.*, 2002). The frequency of mucocutaneous lesions suggests that inhalation of air-borne propagules is the main way in which infection is contracted. After penetrating the host, the fungus must convert to its yeast form, a fundamental step for the successful establishment of the infection, a shift that is solely temperature-dependent.

Studies on the biochemical events leading to *P. brasiliensis* morphogenesis started with cell wall studies (San-Blas and San-Blas, 1994). As in *H. capsulatum* (Klimpel and Goldman, 1988) and *B. dermatitidis* (Hogan and Klein, 1994), the cell wall structure of *P. brasiliensis* changes dramatically when the dimorphic transition takes place. In fact, the possible role of α-1,3-glucan as a dimorphic determinant (Kanetsuna *et al.*, 1972) and virulence factor (San-Blas *et al.*, 1977) was first reported in this fungus. This polysaccharide is present in the *P. brasiliensis* yeastlike cell wall but disappears when the fungus changes to its mycelial phase, as it is substituted by a β-1,3-glucan which is almost exclusively present in this morphotype (San-Blas and San-Blas, 1994). In *P. brasiliensis*, the synthesis *in vitro* of β-glucan by β-1,3-glucan synthase requires UDP-glucose as the preferred nucleotide precursor (San-Blas, 1979), and this particulate enzyme is housed in the plasma membrane (Sorais-Landáez and San-Blas, 1993). In common with some ascomycetes, the reaction is inhibited by GTP and other nucleotides (San-Blas and San-Blas, 1986), in sharp contrast to the general role played by these compounds as stimulators of fungal cell wall synthesis (Szaniszlo *et al.*, 1985). So far, only one gene (*FKSPb1*), homologous to the β-glucan synthase genes *FKS1* and *FKS2* from *Saccharomyces cerevisiae* and *FKSa* from *A. nidulans*, has been cloned and sequenced from *P. brasiliensis* (Pereira *et al.*, 2000). This gene has an open reading frame of 5,942 bp; its complete sequence is interrupted by two putative introns. It also contains a promoter region with consensus sequences such as canonical TATA (-126) and CAAT (-244) boxes. At position –809 the sequence GCCAAG, which mediates pH-dependent gene expression in *A. nidulans*, was found. This was interpreted as being important in dimorphic transition occurring during the establishment of host infection, while possibly participating in a complex mechanism that controls the expression of *P. brasiliensis* genes related to cell wall assembly. The deduced sequence of 1,926 amino acids (predicted MW 212 kDa) shows 85% similarity to FksAp from *A. nidulans*, and 71% to Fks1p and Fks2p from *S. cerevisiae*. Six potential N-glycosylation sites were found that may be the sites of post-translational modifications, which confer stability on the resulting protein. Also, a PTS-Hpr (phosphotransferase system-phosphoryl carrier protein) element was found that may be involved in translocation across the cell membrane, to help in the vectorial synthesis and transport of cell wall components through the plasma membrane in a way that supports the orderly assembly of the wall. Pereira *et al.* (2000) also found two other DNA fragments which may correspond to distinct glucan synthase genes. This is in line with research on other fungal systems in which genes encoding glucan synthases appear to comprise multigene families (Katayama *et al.*, 1999). The occurrence of more than one *FKS*-like sequence in *P. brasiliensis* may suggest differential gene regulation in the *FKS* family of this pathogen.

In all pathogenic fungi, chitin represents a major structural component of the cell wall with functions in fungal morphogenesis, wall integrity and conidiophore development (Ruiz-Herrera and San-Blas, 2003). Chitin synthesis in fungi is a rather complex process, that is regulated by multigene families encoding chitin synthase isoenzymes, (some of which are redundant), and whose activities may be spatially ordered and otherwise strictly regulated to bring about the fulfillment of the several roles ascribed to them. Based on differences in regions of high sequence conservation, chitin synthases have been organized into seven classes within two domains, whose functional implications are not yet clear in all cases (Figure 1) (Ruiz-Herrera *et al.*, 2002; Niño-Vega *et al.*, 2004a).

Five chitin synthase genes, representing different classes of enzyme (*PbrCHS1* in class I , *PbrCHS2* in class II, *PbrCHS3* in class IV, *PbrCHS4* in class VII and *PbrCHS5* in class V) are active in *P. brasiliensis* (Niño-Vega *et al.*, 1998; 2000; 2004a); they help in the

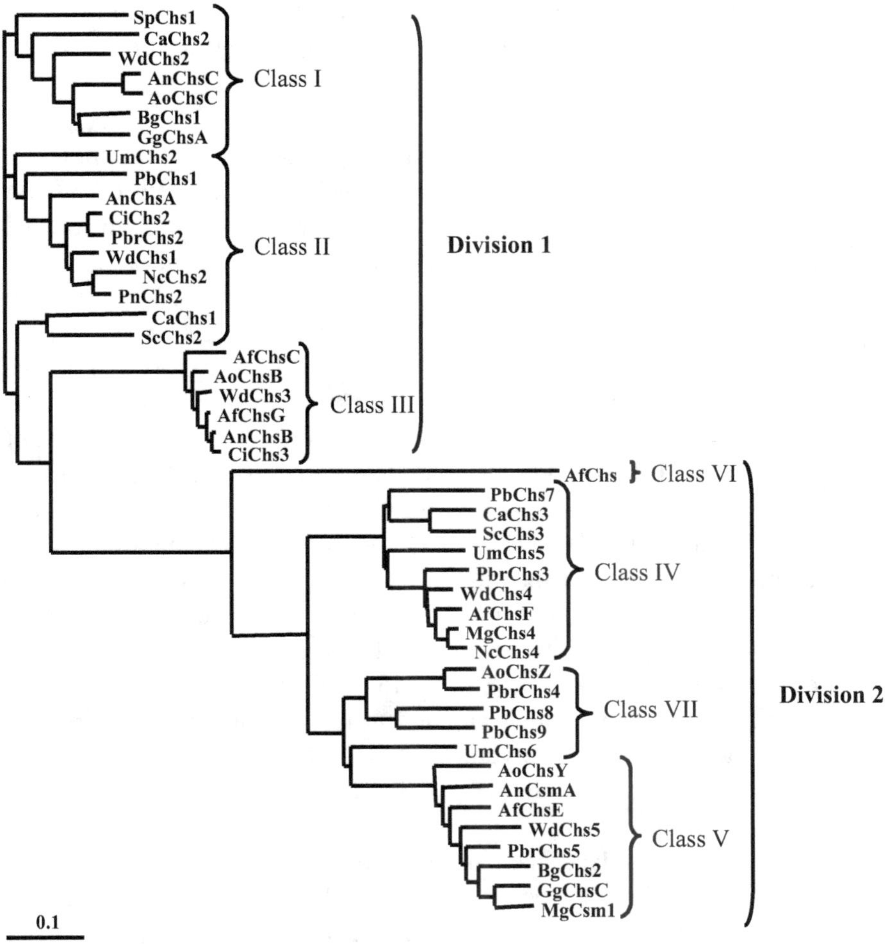

Figure 1.- Tree of relatedness between representative chitin synthases classes IV, V, VI and VII. Sequences were aligned using CLUSTAL W. For the analysis, only the C-terminal chitin synthase domain was taken into account for enzymes classes V and VII. Abbreviations: Ca, *Candida albicans*; Wd, *Wangiella dermatitidis*; An, *Aspergillus nidulans*; Ao, *Aspergillus orizae*; Um, *Ustilago maydis*; Pbr, *Paracoccidioides brasiliensis*; Sc, *Saccharomyces cerevisiae*; Af, *Aspergillus fumigatus*; Mg, *Magnaporthe grisea*. Reproduced with permission from Niño-Vega *et al.* (2004a).

synthesis of chitin in amounts that comprise 43% of the dry weight of the wall of the pathogenic Y form and 13% of the M cell wall (San-Blas and San-Blas, 1994). The nucleotide sequence of *PbrCHS2* contains a 2,540 bp ORF, with three introns, each of which has the characteristic splicing signal (lariat formation) observed in such sequences. The deduced amino acid sequence is 1,043 residues long and predicts a 117 kDa protein that can be classified as a class II chitin synthase. The protein seems to have a three-domain structure that fits within the proposed model of vectorial synthesis and export of chitin from the cytoplasm to its final destination in the cell wall (Niño-Vega *et al.*, 1998).

Expression of *P. brasiliensis* chitin synthase genes throughout the dimorphic shift was explored (Niño-Vega *et al.*, 2000). *PbrCHS1* and *PbrCHS2* have similar transcript sizes at around 3.7 kb, while large transcripts of about 6.5 kb and 6.7 kb were observed for *PbrCHS4* and *PbrCHS5*, respectively. The structures of the last two *CHS*s are similar to *CmsA* from *A. nidulans* (Niño-Vega *et al.*, 2000; 2004a), which has been proposed as a chitin synthase coupled to a myosin motor-like region. In association with cytoskeletal structures, the myosin region might translocate the newly synthesized chitin molecules to its site of deposition. However, PbrChs4p does not possess the P-loop, switch I and switch II domains present in CsmA and conserved among the myosins, a fact that led us to propose a new Class VII to group several chitin synthases with incomplete myosin-motors (Niño-Vega *et al.*, 2004a). Transcripts from these four *PbrCHS* genes were detected in both Y and M cultures where all exhibited similar patterns of expression during the M to Y transition, with a preferential expression in the M form. This enhancement during M growth was unexpected, since the Y form of the organism has a higher chitin content than the M form (San-Blas and San-Blas, 1994).

A soluble galactomannan is also a cell wall polysaccharide component of both morphotypes of *P. brasiliensis*, constituting 2% and 6% of the yeast and mycelial cell walls, respectively (Kanetsuna *et al.*, 1969). Qualitatively, these also differ in structure (Ahrazem *et al.*, 2003). The mycelial galactomannan consists of a trisaccharide repeating unit of →6)-[α-D-Gal*f*-(1→6)-α-D-Man*p*-(1→2)]-α-D-Man*p*-(1→. However the yeastlike galactomannan features a disaccharide repeating unit of [6)-α-D-Man*p*-(1→2)-α-D-Man*p*-(1→], alternating with a trisaccharide repeating block of → 6)-[β-D-Gal*f*-(1→6)-α-D-Man*p*-(1→2)]-α-D-Man*p*-(1→ (see Leal and Bernabé's Chapter 10 of this book for more details). It is not clear whether this polysaccharide has an active role in the dimorphic process.

Cytoskeleton elements, tubulin and actin, have been studied in *P. brasiliensis*. The expression of α-tubulin, as detected by immunoblot assays, is modulated throughout the morphological transition. Levels increased during Y to M shift and decreased during the reverse process, similar to *H. capsulatum* (Silva *et al.*, 2001). Y cells expressed only the α_2 isoform of α-tubulin, whereas isoforms α_1 and α_2 were detected in mycelium, the former probably involved in a specific though unknown role in the morphogenesis of the mycelial phase. The gene that encodes actin, *PbrACT-1*, is present as a single copy (Pérez-Silva, 2003; Niño-Vega *et al.*, unpublished data). Its ORF encodes a putative protein of 375 amino acids, 41.4 kDa, and a very high (>97%) homology with other fungal actins. The N-terminal sequence corresponds to the γ-actin isotype. Other sequences were *PHSE-1* and *PHSE-2* heat shock elements whose presence may be related to the control of expression of the actin message. mRNA expression, in fact, decreased during the first 8 h of the Y to M transition, while a high level of expression after 24 h was recorded in the reverse process (Pérez-Silva, 2003; Niño-Vega *et al.*, unpublished data).

No significant differences were observed in the phospholipid species of both morphological phases of *P. brasiliensis* (San-Blas *et al.*, 1997). However, differences in the degree of unsaturation of fatty acids were recorded. The predominant fatty acid in control Y cells was palmitate (16:0; 56.2 %), followed by linoleate (18:2; 18.2 %), and oleate (18:1; 14.9 %). In the M phase, linoleate was by far the most abundant component (61.1 %), followed by palmitate (26.8 %) and much lower levels of oleate (5 %). Stearic acid (18: 0) was a minor component in both phases (5.5-10.7 %). The ratio of unsaturated to saturated fatty acids (UFA/SFA) was higher in the M form of *P. brasiliensis* (2.10 in the M phase, and 0.49 in the Y phase) (San-Blas *et al.*, 1997). The higher levels of SFA in the Y form are probably required to maintain normal basic permeability properties of the plasma membrane at this relatively high growth temperature (Goyal and Khuller, 1994). As in *P. brasiliensis*, the lowering of the growth temperature increases fatty acid unsaturation in *Aspergillus niger*, *Penicillium chrysogenum*, and *Trichoderma reesei* (Suutari, 1995). So far, this change in UFA:SFA ratio has not been linked to the expression of heat shock proteins, as reported in *H. capsulatum* (Carratù *et al.*, 1996). Sphingolipids, and their structural modifications that occur during the dimorphic process have also been a subject of research in *P. brasiliensis* particularly because of their possible modulation of the activity of cell wall biosynthetic enzymes. Glycosylinositol phosphorylceramides (GIPCs) consist almost exclusively of unbranched (t18:0) 4-hydroxysphinganine (phytosphingosine), in combination with saturated 2-hydroxy fatty acids, predominantly h24:0 (Levery *et al.*, 1998), the mycelial cultures containing more total GPIs by weight than the yeast cells. The major neutral glycosylsphingolipids in both morphological phases were β-glucopyranosylceramides (GlcCer) with (4*E*,8*E*)-9-methyl-4,8-sphingadienine as a long chain base in combination with either *N*-2'-hydroxyoctadecanoate or *N*-2′-hydroxy-(*E*)-3′-octadecenoate, the latter an exclusive modification of fungal cerebrosides (Toledo *et al.*, 1999). The mycelial GlcCer had both fatty acids in a 1:1 ratio, while the yeast GlcCer had only around 15% of the (*E*)-Δ^3-unsaturated fatty acid, a chemical dimorphism that was speculated to work as either a stabilizing factor in the maintenance of proper membrane structure or as a specific messenger in the Y to M conversion in a signalling cascade involving the activation or deactivation of the corresponding desaturase. The possibility that glycosylsphingolipids can modulate the activity of membrane-bound enzymes suggests that they could play a significant role in dimorphism through their interaction with one or more key enzymes affecting fungal morphology, particularly chitin- and glucan-synthases that control cell wall synthesis (Toledo *et al.*, 1999).

The search for factors involved in the expression of fungal dimorphism and morphogenesis has led several authors to explore the differential expression of a wide variety of proteins and genes other than the *PbrCHS*s (Salem-Izaac *et al.*, 1997; Cunha *et al.*, 1999; Silva *et al.*, 1999). Incorporating ^{35}S-methionine in growing cultures of seven *P. brasiliensis* strains, in both morphological phases, Salem-Izaac *et al.* (1997) were able to conclude that some differences among the protein patterns are observable in the strains under analysis, either in each of the morphological phases or during the transition process. There was a greater disparity of protein patterns in the mycelial phase, suggesting a more heterogeneous gene expression profile in the saprobic phase. By two-dimensional electrophoresis of cytosolic proteins from both fungal phases, Cunha *et al.* (1999) were able to detect some differentially expressed proteins associated with either Y or M cells, particularly PbM46 (MW 46 kDa, present in higher amounts in the M phase) and PbY20 (MW 20kDa; present only in the Y phase). The first protein had a sequence similar to enolases as different as *S. cerevisiae* and human. The second had a high degree of similarity with: (a) two 22 kDa allergenic proteins from *Alternaria alternata* and *Cladosporium herbarum*, (b) a 26 kDa protein of unknown function related to a hypothetical gene located in chromosome III of *S. cerevisiae*, (c) a 25 kDa protein related to a thermoregulated gene of *S. pombe*, and (d) a 22.9 kDa protein related to another hypothetical gene from *S. cerevisiae* chromosome II. Silva *et al.* (1999) reported the *hsp70* gene in *P. brasiliensis* which is differentially expressed during transition from the M to the Y form and after heat shock of yeast cells at 42° C. The gene encodes a 649 amino acid protein (predicted MW 70,461 kDa) of high identity with other members of the *hsp70* gene family (73.6% to *Blastocladiella emersonii*, 73.3% to *S. cerevisiae*, 89.2% to *H. capsulatum*), with six conserved sequence motifs characteristic of the Hsp70 family. Another heat shock gene, *hsp60*, was cloned and characterized (Izacc *et al.*, 2001). It is present as a single copy, and is composed of three exons and two introns. The predicted HSP60 protein had two putative N-glycosylation sites, a result that is consistent with the finding of two HSP60 isoforms in *P. brasiliensis* yeast extracts, with differential N-glycosylation attributable to the presence of these two sites. HSP60 also lacks 38 hydrophilic amino acid residues that appear at the NH_2 terminus of the predicted sequence, making it compatible with mitochondrial signal sequences, as have been associated with other HSP60s (Izaac *et al.*, 2001). The higher expression of HSP60 in yeast cells than in mycelium suggests that this protein could be necessary for the survival of *P. brasiliensis* in the host and that it may have a role in morphogenesis.

The signal transduction pathways involved in *P. brasiliensis* dimorphism are relatively unknown. An early report (Paris and Durán, 1985) suggested the possible role of cAMP in the process. More recently, Carvalho *et al.* (2003) found that several inhibitory drugs of the Ca^{2+}/calmodulin signalling pathways led to blockage of the micelium to yeast transition in *P. brasiliensis* strongly suggesting an involvement of this pathway in morphogenesis. The calmodulin gene turned out to be 100% identical to that of *H. capsulatum* and was present as a single copy. Its deduced protein contained 149 amino acid residues with a molecular weight of 16.9 kDa and showed four Ca^{2+} binding sites per molecule.

The synthesis of polyamines appears to be involved in the dimorphic process. Polyamines are micromolecules necessary for cellular growth and differentiation. These are formed during the decarboxylation of ornithine, giving rise to putrescine, the first polyamine in the metabolic pathway. This reaction is catalyzed by ornithine decarboxylase (ODC), one of the most highly regulated enzymes in eukaryotic systems. In the Mucorales (Ruiz-Herrera, 1994) and *C. albicans* (Pfaller *et al.*,

1988), among others, ODC activity is lower in yeasts (Y) and spores than in mycelia (M), increasing significantly during Y to M transition and spore germination. In contrast, Y growth and M to Y transition in *P. brasiliensis* are accompanied by high ODC activity at the onset of budding, while ODC remains at a basal level during vegetative growth of both the M phase and the late stage of Y phase, and also through Y to M transition (San-Blas *et al.*, 1996). The *PbrODC* gene has been cloned and sequenced (Niño-Vega *et al.*, 2004b). It encodes a putative 447 amino acid protein, with high homology to other fungal *ODC* gene products. Its function was tested by transformation of a *S. cerevisiae odc*$^-$ mutant with the *PbrODC*. Expression of *PbrODC* was constant during growth of both the mycelial and yeast forms, as well as during the dimorphic transition of either growth form (Niño-Vega *et al.*, 2004b). These results suggest that, at least during the early stages of yeast growth and M to Y transition, regulation of ODC activity in *P. brasiliensis* might occur at a posttranscriptional level.

Kexin-like proteins are components of the subtilase family of proteinases that are involved in the processing of proproteins to their active forms. Disruption of the *kex2* gene in *C. albicans* (Newport and Agabian, 1997) and *Yarrowia lypolytica* (Richard *et al.*, 2001) affects hyphae production and induces morphological cell defects, strongly suggesting a possible role of kexin-like proteins in dimorphism. Such a gene was reported in *P. brasiliensis* (Venancio *et al.*, 2002). *Pbkex2* has an ORF of 2622 bp, interrupted by only one intron of 93 bp. It contains putative motifs for transcriptional factor binding sites, such as HSE-like motif that functions as a binding site for heat shock proteins with possible involvement in thermo-dependent processes, Xbp1, reported as a transcriptional factor that may control genes involved in cell morphology, and StuAp, which may regulate spore differentiation and pseudohyphal growth in fungi.

Partial transcriptome characterizations have been carried out in *P. brasiliensis* (Felipe *et al.*, 2003a; Goldman *et al.*, 2003). Felipe *et al.*'s goal was to identify genes in the fungal M and Y phases by Expressed Sequence Tag (EST) sequencing in order to generate a partial map of the fungus transcriptome. A total of 3,938 (Y, 1,654; M, 2,274) ESTs were sequenced and clustered into 597 contigs and 1,563 singlets, totaling 2,160 genes, about one quarter of the complete gene repertoire in *P. brasiliensis*. Of these, 1,040 were successfully anotated and 894 could be classified in 18 functional categories: cellular metabolism (44%); information storage and processing

(25%); cellular processes such as cell division or posttranslational modifications (19%); and genes of unknown functions (12%) (Figure 2-A). Computer analysis enabled the authors to identify some genes potentially involved in the dimorphic transition and drug resistance such as genes involved in signal transduction pathways, *hsp* genes, polyubiquitin, and others. Computer substraction analysis revealed several genes possibly expressed in stage-specific forms of *P. brasiliensis*, 12 of these were composed only of yeast-derived ESTs, and 22 from mycelium-derived ESTs. Many others were preferentially expressed in one phase, and poorly in the other. In a recent virtual conference, Felipe *et al.* (2003b) more than duplicated their previous figures so that the number of contigs and singlets increased to 2,307 and 3,189, respectively, totaling 5,496 genes. Of these, 2,707 were successfully anotated and classified into functional groups essentially similar in proportion to the previously reported groups (Felipe *et al.*, 2003a). On the other hand, Goldman *et al.* (2003) surveyed expressed genes in the yeast phase of *P. brasiliensis*. They obtained 13,490 ESTs which, after clustering analysis, yielded the partial sequences of 4,692 expressed genes (about 30-50% of *P. brasiliensis* genome) that were functionally classified by similarity of knowm genes (Figure 2-B). Of these, 8% corresponded to cell metabolism, 4% to protein metabolism, 5% to cell structure, 2% to RNA synthesis, and 60% to hypothetical proteins. The expression of some of these genes during the dimorphic transition was analysed by real-time RT-PCR. Clustering analysis of the M to Y transition revealed three groups: (a) RBT (involved in adhesion), hydrophobin, and isocitrate lyase; (b) malate dehydrogenase, contigs Pb1067 and Pb1145, GPI and alternative oxidase; and (c) ubiquitin, Δ^9 –desaturase, HSP70, HSP82, and HSP104. The first two groups displayed high mRNA expression in the mycelial phase, whereas the third group showed higher mRNA expression in the yeast phase. In a study involving ESTs databases from eight microorganisms belonging to several kingdoms, Faria-Campos *et al.* (2003) established a relationship between available ESTs and protein sequences. One of these microorganisms was *P. brasiliensis*. Using Goldman's data on *P. brasiliensis* ESTs, deposited at the NCBI-USA, Faria-Campos *et al.* (2003) concluded that 553 ESTs matched amino acid sequences deposited there, for a total number of 3,215 candidate ESTs.

Sequencing of 51,022 bp of the whole *P. brasiliensis* genome led to the identification of 18 ORFs, 8 of which were partial ORFs. Of these, 24,452 bp (introns included) represent genes (47.92% of

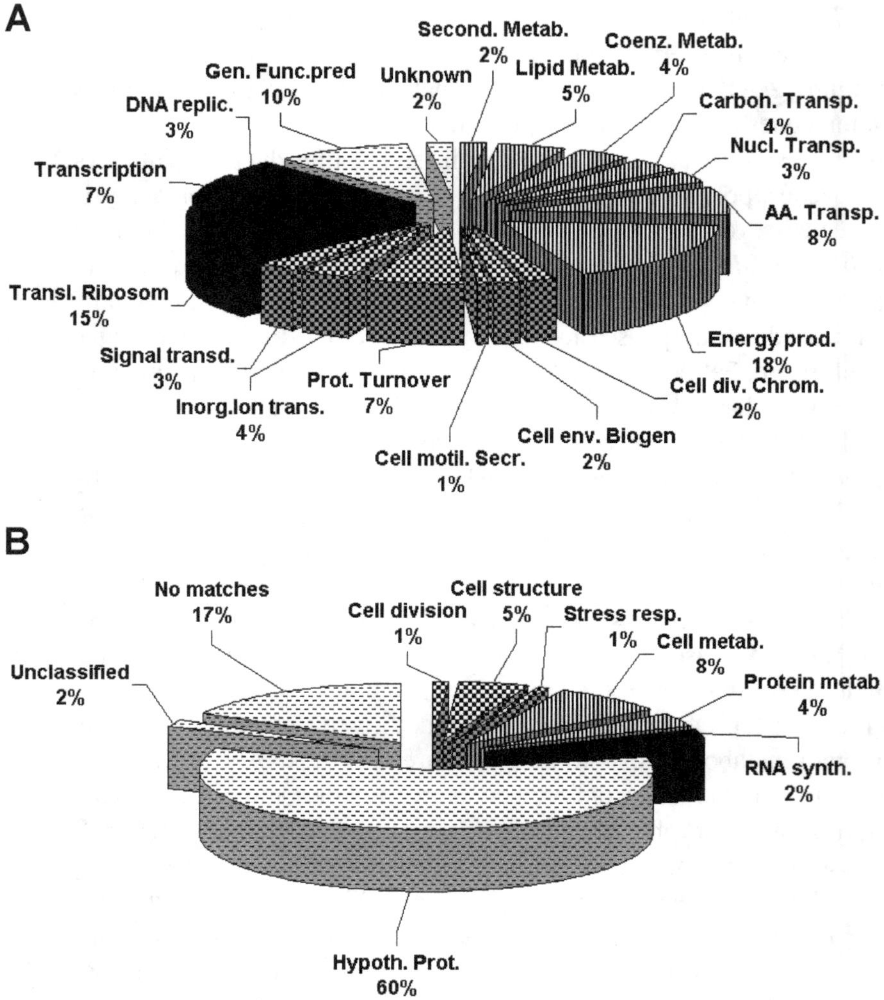

Figure 2. Functional classification of *P. brasiliensis* groups: (A) modified from Felipe *et al.*, 2003; (B) modified from Goldman *et al.*, 2003.

the whole sequence) (Domínguez *et al.*, manuscript in preparation). Fifteen genes were homologous to other fungal genes, of which 10 were homologous to *S. cerevisiae* genes, one was homologous to a protein from *Bacillus halodurans*, and another one to acylCoA dehydrogenase from *Pseudomonas aeruginosa*. Introns were present in 10 such genes; most of these were close to the 5′ end, near to or inmediately after the initiation codon ATG. These were classified into the following functional categories: cellular metabolism, 6 (33%); information storage and processing, 6 (33%); cellular processes, 4 (23%); and genes of unknown function, 2 (11%). The gene density was one gene per 3.5-4.5 kb.

4. *Blastomyces dermatitidis*

This fungus is endemic in the central United States in and around the Ohio and Mississippi River Valleys, although blastomycosis has also been reported in parts of Canada and the southeastern United States. African varieties of *B. dermatitidis* have been isolated but these lack antigens characteristic of the North American strains and produce a variant pattern of blastomycosis, for which it has been suggested that these be considered separate species (Brandhorst *et al.*, 2002). *B. dermatitidis* exists in nature as a sporulating mould, and converts to a large thick-walled yeast forms at 37°C. In a similar fashion as *P. brasiliensis* and *H. capsulatum*, the *B. dermatitidis* cell wall undergoes changes during temperature-dependent morphological transition. Again, α-1,3-glucan is the main neutral sugar polysaccharide in the yeast cell wall, representing 95% of the glucans. By contrast, the proportion of α- and β-glucan is more evenly distributed in the cell walls of the mycelial phase (Kanetsuna and Carbonell, 1971). Like the other two fungal species mentioned above, the loss of the yeast cell wall α-1,3-glucan in *B. dermatitidis* correlates with the loss of virulence in animal models (Hogan and Klein, 1994). Unfortunately, nothing is known about the genetics of these effects.

Medoff *et al.* (1987) demonstrated that the temperature-induced transition is accompanied by a series of biochemical events, similar to *H. capsulatum* (Medoff *et al.*, 1986). After an increase in the temperature, there is a rapid decline in intracellular ATP that follows the uncoupling of oxidative phosphorylation. This is followed by a progressive decrease in respiration that ends in a dormant period of 4-6 days after which, cytochrome components are restored, normal respiration is resumed, yeast-specific cysteine oxidase is induced, and the transition to the yeast morphology is completed. Additionally, the intracellular concentration of cAMP is 1.3-times higher in the mycelial phase than in the yeast form, Y to M transition occurring at the non-permissive temperature of 37°C when the cAMP phosphodiesterase inhibitors theophylline or 3-isobutyl-1-methylxanthine were added to the culture medium (Paris and Garrison, 1983).

A transformation system has been developed for *B. dermatitidis* which has rendered the fungus tractable to genetic manipulation (Hogan and Klein, 1997). Yeast are electro-transformed with linear plasmid DNA expressing dominant selectable markers such as

Escherichia coli hygromycin phosphotransferase resistance, that allows detection of genetically altered expression of determinants in the fungus.

Few phase-specific genes have been identified in *B. dermatitidis*. The most thoroughly studied is *BAD1* (*Blastomyces* adhesin; formerly termed *WI-1*) (Brandhorst and Klein, 2000; Brandhorst *et al.*, 2002; Rooney *et al.*, 2001). Bad1p is released extracellularly and then reassociates to the yeast cell surface by means of disulfide binding to chitin fibrils, that in *B. dermatitidis*, that are accessible in the outermost layers of the fungal cell wall (Brandhorst and Klein, 2000). *BAD1* does not influence morphological transitions, nor is it required for the maintenance of the yeast form. It displays roles as a surface adhesin that promotes the binding of *B. dermatitidis* yeast cells to host cells and tissues that is vital for pathogenicity. This is proven by the fact that deletion of *BAD1* resulted in an avirulent strain that was able to undergo phase transition easily (Brandhorst *et al.*, 1999). However, *BAD1* is upregulated upon conversion to the yeast morphotype that is found in host tissues (Rooney *et al.*, 2001). *BAD1* transcript was apparent in RNA from yeast but was not detectable in RNA from mycelia of North American strains of *B. dermatitidis*. African strains lack the *BAD1* coding region and therefore do not produce the BAD1 adhesin. If such strains are transformed with plasmid pBAD1, containing the *BAD1* locus under the control of its own promoter, the resulting strain conserves the yeast-specific expression typical of the North American *B. dermatitidis* strains (Rooney *et al.*, 2001). The *BAD1* upstream region has two domains which are homologous with the promoter of the *H. capsulatum* yeast-specific gene *yps-3* (Keath and Abidi, 1994). *BAD1* promoter activity increased significantly in the first 12 h after temperature shift, suggesting a role in the initial stages of infection. An epidermal growth factor (EGF)-like cysteine-rich consensus sequence is located at the C-terminus of BAD1 (Hogan *et al.*, 1995). EGF-like domains mediate pleiotropic functions such as maintenance of protein three-dimensional structure and function or adhesiveness of proteins to cells. Deletion of the C-terminal EGF-like domain in BAD1 profoundly affected BAD1 function, leading to non-association with the yeast cell wall chitin, extracellular accumulation and impaired yeast adherence to macrophages (Brandhorst *et al.*, 2003). However, ΔC-term strains and the purified protein itself sharply suppressed tumor necrosis factor (TNF)-α release by phagocytes, and the strains retained pathogenicity in a murine model of blastomycosis.

Therefore, it was concluded that the requirements for host cell binding and immune modulation by BAD1 are dissociated from one another (Brandhorst *et al.*, 2003).

Another yeast-specific gene, *bys1*, has been reported in *B. dermatitidis* by means of differential cDNA library screening (Burg and Smith, 1994). This transcript was expressed at very high levels in the yeast phase but decreased rapidly when cells were transferred to 25°C to promote conversion to the mycelial phase. *Bys1* encoded a protein of 18.6 kDa that contained multiple putative phosphorylation sites, a hydrophobic N terminus, and two 34-amino acid domains with similarly spaced nine-amino acid degenerative repeating motifs. No function has been adscribed to this protein. By using a *bys1* cDNA probe of the coding and untranslated regions, 12 *B. dermatitidis* strains were subjected to Southern blot; two distinct banding patterns were observed, a class I with genomic fragments of 3.17 and 2.81 kb, and class II with fragments of 5.59 and 1.58 kb (Bono *et al.*, 2000). Transcription of both class I and II *bys1* alleles decreased in the mycelial phase, although differences in their regulation was observed. While the class I allele was transcribed 24-48 h after the temperature shift to 37 °C, the class II allele did not transcribe before 120 h.

5. *Coccidioides immitis*

C. immitis is the causative agent of coccidioidomycosis, a human respiratory disease that ranges from mild, influenza-like symptoms to a severe and life-threatening mycosis. It is found in the southwestern United States and in certain arid regions of Mexico, Central America, and South America. Recently, molecular phylogenetic evidence has led to the conclusion that this apparently uniform species should be divided into two: *C. immitis* (geographically restricted to California) and *C. posadasii* (extended to all other regions of North, Central and South America where coccidioidomycosis has been reported) (Fisher *et al.*, 2002) The parasitic cycle of *C. immitis* is characterized by three morphogenetic phases: isotropic growth, segmentation, and sporulation (Figure 3) (Cole and Hung, 2001). Air-dispersed arthroconidia are released from saprobic mycelia and isotropic growth occurs almost immediately after it is inhaled, converting the conidia into spherules. These, in turn, eventually transform into uninucleate endospores (as many as 200 to 300 endospores from a single spherule) which then undergo a period of diametric growth to produce

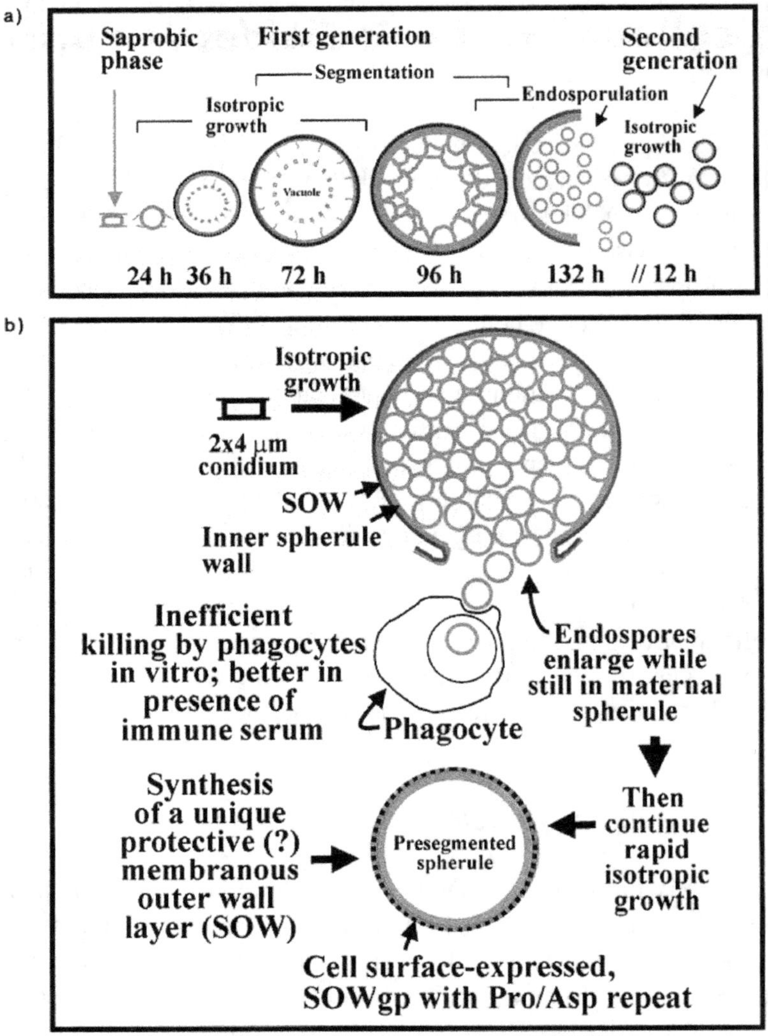

Figure 3. Parasitic cycle of *C. immitis*, showing features of spherule and endospore wall differentiation. (a) Unique parasitic life cycle; (b) High reproductive capacity. Reproduced with permission from Cole and Hung (2001).

a second generation of spherules. Early studies on the *C. immitis* cell wall indicated the presence of insoluble glucans and chitin (Hector and Pappagianis, 1982), while soluble lysates were reported to contain polysaccharides composed of mannose and some galactose (Pappagianis *et al.*, 1961). A third carbohydrate of unusual structure was 3-O-methyl mannose. The lowest lipid levels were found in mycelial walls, followed by arthrospores and spherules (Wheat *et al.*, 1977). Glucosamine, glucose and mannose in variable amounts were the major monosaccharide components of *C. immitis* cell walls. When endospores initiate their isotropic growth phase, a spherule-specific outer wall layer (SOW), composed of lipids and glycolipids, and an antigenic glycoprotein (SOWgp), is deposited at the cell surface (Cole and Hung, 2001). The gene *SOWgp* encodes this immunodominant spherule-specific glycoprotein, which is produced in elevated levels during early spherule development to fulfil probable functions as an adhesin (Hung *et al.*, 2002). Since the principal structural components of the parasitic cell wall of this pathogen are β-glucan and chitin, one would expect that the major hydrolases involved in wall modification during spherule morphogenesis are β-glucanases and chitinases. A spherule wall-associated, 120 kDa β-1,3-glucosidase is secreted from the cell, and is recognized by immunoglobulin M precipitin antibodies (Kruse and Cole, 1992; Hung *et al.*, 2001). Its corresponding gene (*BGL2*) is expressed during isotropic growth of first- and second-generation spherules. A deletion of the *BGL2* gene did not arrest isotropic growth of the parasitic cells, while mycelial cultures of the Δ*bgl2* mutant showed significantly slower growth rate and was less virulent than the parental strain. So far, six additional β-glucosidases have been identified by BLAST searches (Cole and Hung, 2001). Also, the partial cDNA or partial genomic DNA sequences of 5 *C. immitis* glucan synthase genes have been deposited in the GenBank (Cole and Hung, 2001). Initiation of the segmentation phase of spherule development (Figure 3) most likely involves a major upregulation of the chitin biosynthesis machinery, since it is calculated that a threefold increase in total spherule chitin occurs during this process. To date, no identification of chitin synthase genes in *C. immitis* have appeared in the literature, although four complete sequences have been deposited in the GenBank, and two additional sequences have been identified by BLAST analysis of the partial *C. immitis* genome database (Cole and Hung, 2001). The biosynthesis of the endogenous segmentation wall complex is followed by its dissolution during endospore differentiation, a process in which chitinases play a pivotal role. In fact, six genes coding for such enzymes have been revealed. Of these, *CTS1* and *CTS2* have been cloned. Expression of *CTS1* is markedly increased during endospore formation, although its disruption did not affect virulence or endosporulation, probably because of redundancy with other *CTS* genes (Reichard *et al.*, 2000).

Other genes have been studied in *C. immitis*. Some have no proven relationship to either virulence or morphogenesis, i.e., *arg* (Pan *et al.*, 1995), *URE* (Yu *et al.*, 1997) or *URA5* (Yu *et al.*, 1999), while others are related to antigenic functions (Dugger *et al.*, 1996; Johnson *et al.*, 2000; Thomas *et al.*, 1997; Zhu *et al.*, 1996). However, *ODC* and its derived enzyme (ornithine decarboxylase), have been associated with *C. immitis* morphogenetic changes (Guevara-Olvera *et al.*, 2000). The deduced 1341 bp ORF of the cDNA encodes a 447 amino acid protein of 49.5 kDa. The sequence has two predicted sites of glycosylation. Biochemical enzymic assay indicated that ODC activity rises at specific stages of the parasitic cycle, showing a first peak of activity at the initial isotropic growth phase, a second one at the initiation of the spherule segmentation and a third one when the second generation of parasitic cells start its development. However, *ODC* expression remained almost the same throughout the cycle, pointing to a posttranslational regulation of ODC activity.

6. *Wangiella (Exophiala) dermatitidis*

W. dermatitidis is an asexual fungal pathogen of humans, traditionally associated with primary dermatotropic forms of cutaneous and subcutaneous phaeohyphomycosis, but more recently emerging as a systemic pathogen with a marked neurotropic tendency (Szanizlo, 2002). Although it cannot be considered as a public health problem, *W. dermatitidis* is an appropriate model for the study of pathogenic dematiaceous fungi and their polymorphism. It exhibits three distinct vegetative growth modes: blastic, apical and isotropic, which are primarily associated with growth of the yeast, hyphal, and sclerotic-body morphology. Extreme acidity (pH 2.5) regulates cell cycle developmental events in the predominant yeast phase or in the less common hyphal phase, with the formation of an isotropically enlarged, multinucleate phenotype that may become internally partitioned with one or more septa, similar to a sclerotic body (Figure 4). Cell walls of isotropic forms, and to a lesser extent hyphae, are significantly enriched in chitin, while in yeast cells chitin is found primarily in the septal regions. This fact has stimulated the systematic analysis of chitin biosynthesis in *W. dermatitidis* as a structure that contributes to polymorphism and virulence (Szaniszlo, 2002). Chitin biosynthesis in this fungus is also mediated by at least five distinct chitin synthase isozymes, with each one belonging to a different class. As with other fungi, WdChsps are zymogenic, requiring posttranslational cleavage by proteinases to become active. WdChs1p (Class II chitin synthase) seems responsible for the synthesis of chitin that is deposited in the septal region between yeast mother and daughter cells; WdChs2p (Class I) seems redundant with WdChs1p, both being orthologous with similar enzymes in *S. cerevisiae* (Wang *et al.*, 2001). In contrast, WdChs3p (Class III) and WdChs5p (Class V) do not have orthologous

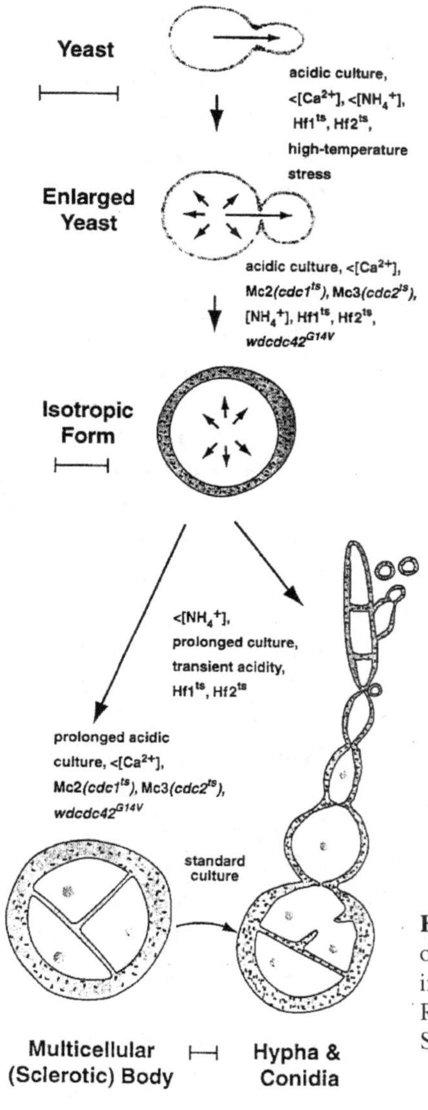

Figure 4. Schematic representation of phenotypes of *W. dermatitidis* induced *in vitro* by various factors Reproduced with permission from Szaniszlo (2002).

genes in yeast (Wang and Szaniszlo, 2000). Finally, WdChs4p (Class IV) synthesizes a large amount of chitin even though its activity is very low, and appears to play a role in the process of cell separation (Wang *et al.*, 1999). No single WdChsp is essential for viability and for yeast budding growth at 25°C, while only WdChs5p is required for viability at 37°C and therefore virulence (Wang and Szaniszlo, 2000; Wang *et al.*, 1999). The structural genes of *WdCHS1* to *WdCHS5* are quite large, ranging from 2756 bp to 5765 bp, with zero (*WdCHS4*), one (*WdCHS1*, *WdCHS2*,) and two (*WdCHS3*, *WdCHS5*) introns (Wang and Szaniszlo, 2000; 2002; Wang *et al.*, 1999; 2001; 2002).

WdCHS5 encodes a protein with a myosin motor-like domain, as reported for other fungi (Ruiz-Herrera *et al.*, 2002; Niño-Vega *et al.*, 2003). Contrary to the essential character of class III chitin synthases for normal growth in *N. crassa* and *Aspergillus* species, disruption of *WdCHS3* (Class III) did not cause growth or morphological defects in any of *W. dermatitidis* forms or in the amounts of chitin in the cell wall (Wang and Szaniszlo, 2000), suggesting a redundancy of this gene. *WdCHS3* is differentially expressed in response to stress, but its expression is not a prerequisite for polymorphic transition. Analysis of the 5′ upstream sequence of *WdCHS3* provides evidence for a negative regulatory element between –780 and –1600 bp (Wang and Szaniszlo, 2000). Surprisingly, heterologous expression of *WdCHS3* in *S. cerevisiae* gave rise to an enzyme that did not require posttranslational proteolysis for activity (Wang and Szaniszlo, 2002). Disruption of *WdCHS2* did not lead to observable defects in yeast vegetative growth or in transition to hyphae or isotropic forms (Wang *et al.*, 2001). Double disruption of *WdCHS2* and *WdCHS3* led to mutant strains that were not affected in morphogenesis. However, these were less virulent in a mouse model of acute infection, an effect that was not seen when single mutations were induced (Wang *et al.*, 2001). Strains with deletions in the *WdCHS4* gene greatly reduce their chitin content at 37°C, indicating that WdChs4p makes more chitin in *W. dermatitidis* than do the other isozymes (Wang *et al.*, 1999). The *wschs4*Δ disruptant strains grew slowly in poor medium and showed abnormal yeast clumping and budding kinetics. These mutants also released higher amounts of melanin, possibly because the lack of an appropriate chitin matrix contributes to an inefficient trapping of the pigment. During the polymorphic transitions of *W. dermatitidis*, the transcription levels of *WdCHS4* did not change significantly. Disruption of this gene did not lead to reduction of virulence in a mouse model (Wang *et al.*, 1999). A compensatory mechanism exists by which increased *WdCHS* expression responds to stress induced by chitin synthase gene disruptions (Wang *et al.*, 2002). Each of the five *WdCHS* genes exhibit a somewhat different expression pattern in response to different growth conditions, suggesting that each one must have a special role in cell growth, viability and virulence. But the general compensatory expression explains why chitin contents are not dramatically changed in any single mutant and suggests its participation in the cell wall integrity pathway (Wang *et al.*, 2002).

The first published genes involved in signal transduction in dematiaceous fungi, controlling yeast bud emergence in *W. dermatitidis*, were the *CDC1* and *CDC2* genes (Cooper and Szaniszlo, 1993), analogous to *S. cerevisiae CDC24*, and *CDC43*. The gene product of *CDC24* is probably a calcium-binding protein bound to the plasma membrane via its interaction with the product of *CDC42*. The *CDC42* gene in *S. cerevisiae* encodes a GTP-binding protein and is most likely prenylated by the *CDC43* gene product, a prenylation that permits the attachment of the *CDC24* gene to the plasma membrane (Cooper and Szaniszlo, 1993). The GTPase function is conserved as it is in yeast. *CDC42* plays crucial roles in the regulation of cell polarity via actin cytoskeleton organization and signal transduction (Ye and Szaniszlo, 2000). *WdCDC42* is conserved, as confirmed by complementation in a *S. cerevisiae cdc42-1*^{ts} mutant. However, disruption of the gene did not result in a lethal phenotype in *W. dermatitidis*, and cellular morphologies were affected only under certain stress conditions, functions that are different from those of *CDC42* in *S. cerevisiae*. Overexpression of the dominant negative alleles of *wdcdc42* in a null mutant did not bring about a cell depolarisation and inhibition of yeastlike bud formation observed in *S. cerevisiae*. In contrast, only expression of the constitutively active allele in *W. dermatitidis* induced transformant cells to exhibit non polarized growth rather than to develop multiple buds or bud elongation (Ye and Szaniszlo, 2000). These results suggest that WdCdc42p negatively regulates cell polarization. Also, this protein cycles between the GTP- and the GDP-binding states; when a sustained activation of the GTP protein exists, *W. dermatitidis* cells initiated isotropic growth, leading to the production of sclerotic bodies and inhibition of hyphal growth.

A calcium/proton exchange mechanism may play an important role in the pH-induced dimorphism of *W. dermatitidis* cultured at 25°C and also in the hyphal-sclerotic bodies dimorphism of three agents of chromoblastomycosis cultured at either 25°C or 37°C (Mendoza *et al.*, 1993). At 25°C and pH 2.5, critical but low calcium concentrations in a defined medium favour non-polarized, multicellular forms in *W. dermatitidis*, whereas even lower calcium concentrations promoted only isotropic growth with no septation or nuclear division. On the contrary, higher concentrations resulted in polarized yeast/hyphal growth (Karuppayil and Szaniszlo, 1997). When the calcium chelator EGTA was added to the medium, yeast cells were arrested in their budding process. These results strongly suggest that at 37°C, isotropic growth, mitosis, septation and polarized yeast budding and hyphal apical extension at pH 2.5 are all Ca^{2+}-dependent processes in *W. dermatitidis*, probably through a calcium-responsive protein.

7. *Sporothrix schenckii*

Sporotrichosis, caused by *S. schenckii*, is a chronic cutaneous infection that affects lymphatic vessels and, in immunocompromised patients may develop into pulmonary and disseminated forms that can be crippling or fatal (Chua *et al.*, 1994). Like other pathogenic fungi, *S. schenckii* grows in a hyphal form in nature, while growing as a yeast at body temperature. Changes in cell wall structure do not seem to accompany this morphogenetic shift, as reported in other dimorphic fungi. Yeast, conidia and hyphae of *S. schenckii* have insoluble glucans in their cell walls whose proportions of β-1,3-, β-1,4-, and β-1,6-linkages were 65-66%, 5-10%, and 25-29%, respectively. These fungi also have soluble glucans of β-1,3-, β-1,4-, and β-1,6-linkages of 44-45%, 28-31%, and 24-26%, respectively (Travassos and Lloyd, 1980). A rhamnomannan, composed of rhamnose and mannose plus traces of glucose and galactose, is also found as a peculiar and exclusive component of *S. schenckii* cell wall. The main repeating unit of rhamnomannan is either a monorhamnosyl- or a dirhamnosyl-structure. The latter is prevalent in polysaccharides obtained at 25°C when the fungus is in the mycelial phase, while monorhamnosyl structures are found in yeast cells grown at 37°C and in conidia. Smaller but similar amounts of chitin (measured as GlucNAc) that range from 7.0 to 8.3% are also present in the cell walls of the yeast, mycelium, and conidia of *S. schencki* (Travassos and Lloyd, 1980). Three chitin synthase genes (*SSCHS1*, *SSCHS2*, and *SSCHS3*), each belonging to a different class, have been identified by PCR by using degenerate PCR primers (Chua *et al.*, 1994), although studies to analyse their expression as a function of morphology were not done. Primers based on *SSCHS1* were designed for the diagnosis of human and animal sporotrichosis (Kano *et al.*, 2001).

Following studies on glycosphingolipids (GSL) in several dimorphic fungi, Takahashi's group (Toledo *et al.*, 2000; 2001a; 2001b) analysed their structure, composition, and dimorphic expression in *S. schenckii*. In lipids extracted from the mycelial phase, a single cerebroside component (GlcCer) was observed, while in the yeast phase a GalCer was also detected. The major long chain core in all three cerebrosides was found to be (4*E*,8*E*)-9-methyl-4,8-sphingadienine, as reported for the majority of fungi (Toledo *et al.*, 2000). Glycosylinositol phosphorylceramides (GIPCs) are a class of GSL that appear to be essential for fungal survival. In *S. schenckii*, GIPC structures were determined to be Manα1→6Ins1-P-1Cer and

Manα1→3Manα1→6Ins1-P-1Cer (where Ins = *myo*inosytol, P = phosphodiester) in the mycelial (Toledo *et al.*, 2001a) and the yeast (Loureiro y Penha *et al.*, 2001) phases. An additional GIPC with the structure Manα1→3Manα1→6GlcNH$_2\alpha$1→2 Ins1-P-1Cer was reported in both phases (Toledo *et al.*, 2001b).

Calcium ions (1 mM) were found to stimulate germ tube formation. The first process to be stimulated was RNA synthesis, particularly rRNA and tRNA, followed by protein synthesis (Rodríguez-del-Valle and Rodríguez-Medina, 1993). Whether the effect of Ca^{2+} on hyphal growth is directly through the actin cytoskeleton is unknown. In fact, differential expression of a 45 kDa actin protein during the dimorphic transition of *S. schenckii* has been reported (Han-Yaku *et al.*, 1996). Actin was marginally detected in yeast cells but predominantly expressed in the intermediate and mycelial cells.

Two kinases that act as regulators of cell cycle progression have been reported in *S. schenckii*. Protein kinase C (PKC) catalyses the transfer of phosphate groups to seryl and threonyl residues in specific target proteins. PKC was found in three isoforms: the γ and ζ isoforms were detected in both the yeast and the mycelial phases, while the β isoform was only found in the yeast form (Aquino-Piñero and Rodríguez-del-Valle, 1997). Two *pkc* genes (*pkcSs-1* and *pkcSs-2*) were reported in *S. schenckii* (Aquino-Piñero and Rodríguez-del-Valle, 2002). The sequence of *pkcSs-2* was completed and revealed an ORF of 3942 bp interrupted by five introns. It encodes a protein of 1194 amino acids and 133 kDa that contains the characteristic domains of other fungal PKCs reported to date. The gene is expressed during the first 6 h after the induction of the yeast to mycelium transition. A second kinase repressed during the transition to the mycelial phase is the Pho85 cyclin-dependent kinase (Berríos and Rodríguez-del-Valle, 2002). Pho85 is a regulator of phosphate metabolism and plays a role in modulating the transcriptional response to nutritional signals. By binding to different cyclins, Pho85 regulates diverse cell cycle processes. The gene encoding a Pho85 homolog, *PhoSs*, was identified and sequenced; it is 990 bp in length, contains one intron, and encodes a protein of 306 amino acids. Its expression decreased 30-fold during the yeast to mycelium transition. The addition of extracellular calcium accelerated the dimorphic transition and restored *phoSs* expression, suggesting a participation of PhoSs in the control of the yeast to mycelium transition in *S. schenckii*.

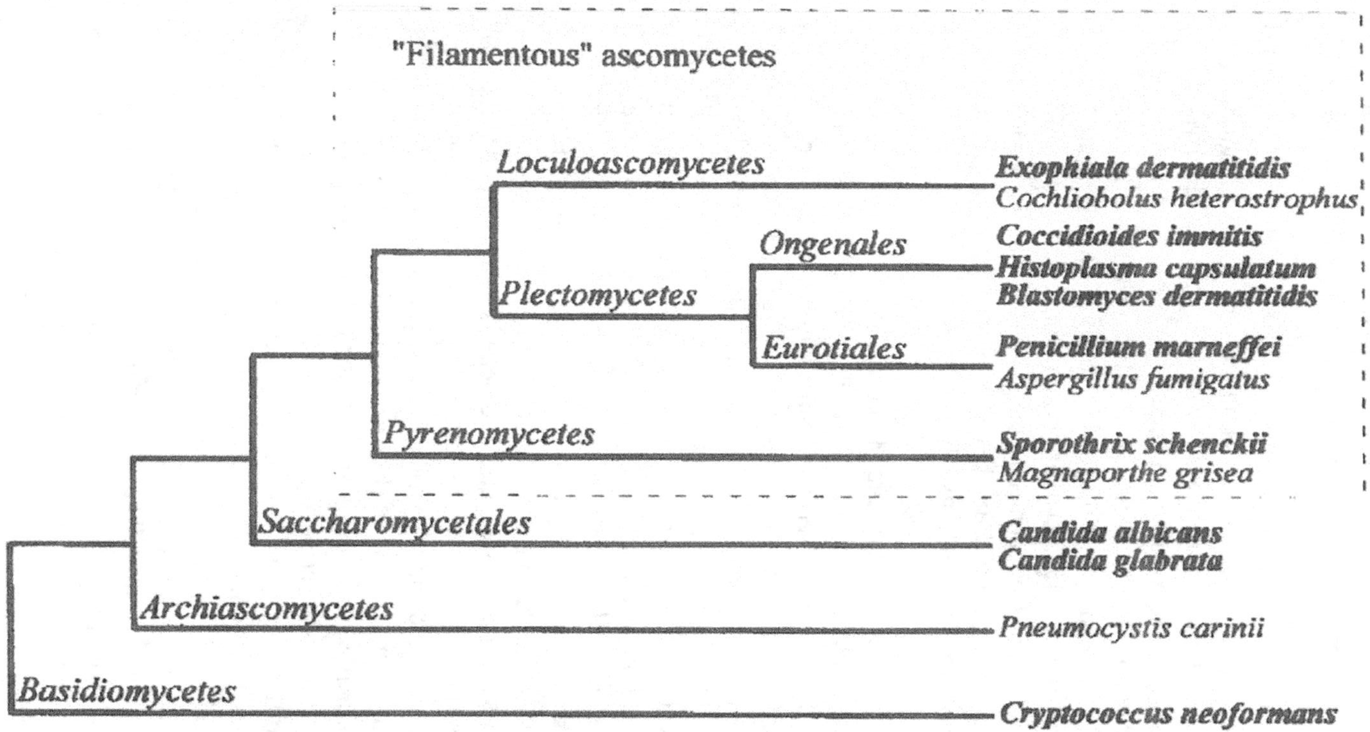

Figure 5. Phylogenetic relationship of *P. marneffei* with other pathogenic dimorphic fungi. Reproduced with permission from Andrianopoulos (2002).

8. *Penicillium marneffei*

The *Penicillium* group of species, comprising the teleomorphic genera *Eupenicillium* and *Talaromyces* and the anamorphic *Penicillium* genus, consists of a number of important industrial and food species. Molecular and morphological phylogenetic analyses suggest that *P. marneffei* is most closely related to *Talaromyces* species with biverticilliate conidiophore morphology. Amongst the *Penicillium* group, *P. marneffei* is the only known dimorphic species, exhibiting all of the filamentous characteristics of this group of organisms at 25°C, in addition to being capable of switching to and growing as a yeast at 37°C (Andrianopoulos, 2002). Other dimorphic pathogenic fungi such as *H. capsulatum*, *B. dermatitidis*, and *C. immitis* are distantly related to *P. marneffei* but closely related to each other (Figure 5; Andrianopoulos, 2002). *P. marneffei* is one of the few medically important species within the *Penicillium* genus; it causes respiratory, skin, and systemic mycosis, geographically prominent in Southeast Asia (Yuen *et al.*, 2003). Morphogenetic changes are multiple in this fungus; at 25°C a uninucleate spore germinates, producing a germ tube and a septate hyphae after a short period of isotropic growth. The vegetative mycelium produces multinucleate aerial stalk cells from which several differentiated cells are derived. When transferred to 37°C, *P. marneffei* produces uninucleate elongate yeast cells which divide by fission, referred to as arthroconidiation (Andrianopoulos, 2002).

Asexual development in *P. marneffei* follows a similar program as that reported for *A. nidulans* (Adams *et al.*, 1998) in which two parallel regulatory pathways, conidiation and dimorphic switching, control morphogenesis (Borneman *et al.*, 2000). This is done through the participation of *abaA*, a gene that encodes a putative protein of 823 amino acids and contains a transcriptional enhancer activator (TEA/ ATTS) DNA-binding motif that is highly conserved with previously identified ATTS proteins from yeast, mouse and human. This motif is found in the N-terminus of certain gene regulatory proteins, such as the *Aspergillus abaA* regulatory gene product (Adams *et al.*, 1998). *P. marneffei abaA* transcript is present at very low levels in vegetative hyphae, while expression during conidiation or yeast growth is 30- or 10-fold higher, respectively. Deletion of the *abaA* gene had no detectable phenotypic effect during vegetatitve growth. During the asexual development, the *abaA* deletion led to the production of aberrant cells with characteristics like those of metulae and phialides, and also to misregulation of arthroconidia. However, *abaA* was not required for budding (Borneman *et al.*, 2000).

The *stuA* gene is a member of the APSES (Asm-1, Phd1p, StuA, Efg1p, and Sok2p) group of transcription factors in fungi which possess a basic helix-loop-helix (bHLH) DNA-binding motif. They regulate processes such as mating, asexual sporulation and dimorphic growth in fungi. Borneman *et al.* (2002) cloned and characterized *stuA* from *P. marneffei*. Like *abaA*, expression of the *stuA* gene was tightly regulated. Its transcript was not detectable during vegetative growth but strongly induced during conidiation. Deletion of the gene resulted in a loss of sterigmata, metulae and phialides, though spores were still produced. Contrary to the participation of APSES proteins in the yeast-mycelium transition in *S. cerevisiae* and *C. albicans*, stuA in *P. marneffei* was not involved in such transition, as *stuA*$^-$ mutants were able to normally undergo yeast-mycelium transition (Borneman *et al.*, 2002).

Genes involved in signal transduction have been reported in *P. marneffei*. *STE12* belongs to the MAPK (mitogen activated protein kinase) pathway, which have been involved in controlling mating and yeast-hyphal transitions in fungi. A homolog of *S. cerevisiae STE12* has been cloned from *P. marneffei* (*stlA*). The *stlA* gene, along with the *A. nidulans steA* and *Cryptococcus neoformans STE12*α genes, form a distinct subclass of *STE12* homologs that have a C_2H_2 zinc-finger motif in addition to the homeobox domain that defines *STE12* genes (Borneman *et al.*, 2001). However, deletion of this gene had no detectable effect on vegetative growth, asexual development, or dimorphic switching in *P. marneffei*, although the gene complemented the sexual defect of an *A. nidulans steA* mutant, suggesting that *P. marneffei* may have an unidentified cryptic sexual cycle.

CDC42, a member of the Rho family of small GTPase-encoding genes is a downstream target of signalling pathways. Cdc42p interacts with Ste20p, being necessary for cell polarization and bud emergence in *S. cerevisiae* (Johnson, 1999). The *P. marneffei CDC42* homologue, *cflA*, is required for correct morphogenesis of yeast cells; deletion of this gene induces the formation of aberrant cells with irregular shapes, branching morphology and poor septation (Boyce *et al.*, 2001). By the generation of dominant-negative and dominant activated *cflA* transformants, these authors showed that CflA initiates polarized growth and extension of the germ tube and subsequently maintains polarized growth in the vegetative mycelium. CflA is also required for polarization and determination of correct cell shape during yeast growth, and for the separation of the yeast cells. However, it is not required for dimorphic switching (Boyce *et al.*, 2001). Another member of the Rho-family GTPases is Rac. Unlike the other Rho GTPases, Rac orthologs are not present in *S. cerevisiae* while in higher eukaryotes they perform act as regulators of cellular morphogenesis (Johnson, 1999). One *RAC* homolog, *cflB*, from *P. marneffei* was cloned. CflB colocalizes with actin at the tips of vegetative hyphal cells and at sites of cell division. Deletion of *cflB* results in cell division and growth defects in both vegetative hyphal and conidiophore cell types such as inappropriate septation, and disruption of the actin cytoskeleton (Boyce *et al.*, 2003).

Signalling pathways also involve repressors. *TUP1* represses yeast to mycelium transition in *C. albicans* (Braun *et al.*, 2000). On the contrary, the *P. marneffei* TupA is a central coordinator of morphogenesis and growth whose primary function is to repress both asexual development during filamentous growth and the mycelium to yeast transition (Todd *et al.*, 2003).

Heterotrimeric guanine nucleotide-binding proteins (G proteins) also act as signal transducers which in fungi play essential roles during pathogenic and sexual development programs (Bolker, 1998). In *P. marneffei*, a *gasA* gene encoding a Gα subunit that is highly similar to *A. nidulans* FadA and other fungal Gα subunits, has been cloned. *P. marneffei* mutants carrying a dominant activating $gasA^{G42R}$ allele is locked in vegetative growth and cannot enter asexual development, suggesting that GasA signalling antagonizes conidiophore development and is a key regulator of asexual development in *P. marneffei* (Zuber *et al.*, 2002). The mutant had no apparent defect in dimorphic switching or yeast-like growth at 37°C. Another G-protein α-subunit with high homology to members of the Class III fungal Gα-subunits, is encoded by *P. marneffei gasC* (Zuber *et al.*, 2003). GasC is a crucial regulator of germination, as demonstrated by Δ*gasC* mutants that are severely delayed in this process; additionally, GasC negatively affects the onset of conidiation and the conidial yield, overlapping with the functions of GasA. However, GasC did not affect yeast to mycelial transition.

Our knowledge of the mechanisms of morphogenetic control and regulation in *P. marneffei* will improve greatly with the exploration of its genome. To this effect, Yuen *et al.* (2003) have initiated an analyses which yielded 2303 random sequence tags (RSTs), corresponding to 9% of the genome. Of the best hits, 269 matched only with yeast-specific sequences, 144 only with non-yeast sequences, and 400 matched with both. The remaining 1,183 sequence tags matched with other sequences in several databanks. Analysis of the RSTs revealed genes for information transfer (ribosomal protein genes, tRNA synthetase subunits, translation initiation, and elongation factors), metabolism (cytochrome P459 enzymes, glucokinase, enzymes of the glycolytic pathway; enzymes involved in synthesis of amino acids and lipids; glucanases and chitinases probably involved in cell wall synthesis), and compartmentalization (permeases and receptors, involved in the transport of carbohydrates, proteins and fatty acids).

9. From Genes to Three-dimensional Proteins to Modifications of Spatial Patterns

While molecular approaches to dimorphism in *Candida* or *Aspergillus* have resulted in significant advances in the dissection of regulatory mechanisms of morphogenesis, particularly signal transduction pathways (see other chapters in this book), such studies are fragmentary in important fungal pathogens like those reviewed in this chapter. Cell wall polysaccharides have major implications in morphological changes and therefore, genes involved in their synthesis are under close scrutiny. However, not all dimorphic fungi (i.e., *S. schenckii*) reveal striking modifications in cell wall structure, despite profound morphological changes. Membrane lipids are also studied to clarify their participation in morphogenesis, although no obvious relationship has so far emerged (i.e., hydroxydecanoate and hydroxydecenoate are components of the mycelial and yeast phases, respectively, of *H. capsulatum* GlcCer; instead, both are present in different molar ratios in *P. brasiliensis*, whereas a GalCer is present in the yeast phase of *S. schenckii*). Metabolic pathways are not always similar during dimorphism. For example, ODC levels are highly elevated in the mycelium to yeast transition in *P. brasiliensis* but the proportions are reversed in other important pathogenic fungi. Furthermore, some data suggest that homologous genes and their deduced proteins may have different functions, as has been reported in more thoroughly studied systems (Sánchez-Martínez and Pérez-Martín, 2001). Examples of this may be the APSES proteins and TUP1, involved in the yeast to mycelium transition in *S. cerevisiae* and *C. albicans*, but not in

P. marneffei, or the *CDC42* gene that plays important roles in the regulation of cell polarity in *S. cerevisiae* but not in *W. dermatitidis*. Recent data suggest that in addition to their amino acid sequences, the three-dimensional structure of proteins is an important determinant of their function. Comparison between all proteins released in 1998 with those known until 1997 indicated that 147 had no significant sequence similarity to any of the pre-1998 proteins, although they had structural similarity to a previously known protein. In two-thirds of these proteins (98), the function was the same (Hegyi and Gerstein, 1999; Koppensteiner *et al.*, 2000). It is then possible that the function of a protein depends upon its form, rather than directly upon its amino acid sequence. This helps one understand the many instances in which three-dimensional structure of proteins is better conserved than the sequence (Harold, 2001).

The molecular paradigm by which morphogenetic information is stored at the level of genes and molecules has been contested by Harold (1999; 2001). The model is not incorrect but is incomplete, since molecules and morphology are loosely coupled. The shapes of chitinous fungal hyphae and those of the Oomycetes, which produce cellulose, differ only in detail; the cell wall composition of yeast and mycelial phases of some fungi (e.g. *S. schenckii*) are very much the same, yet their final morphologies are quite different. Hence, it is not molecular chemistry that determines the final shape of the cell, but how these molecules are put into place. Harold's paradox notes that all heritable information is encoded in the genes; biological forms are inherited; yet form is not encoded in the genome (Harold, 1999; 2001). In looking for solutions to this paradox, the literature of morphogenesis increasingly associates dynamic, self-organizing fields with the generation of spatial order, and fungi can be accommodated in this new trend (Harold, 1999; 2001). In this respect, an interesting field theory of apical growth has been developed by Bartnicki-García and coworkers (see review, Bartnicki-García, 2002). In it, the Spitzenkörper (conceptualised as a vesicle supply center or VSC) is the immediate source of secretory vesicles that travel to the plasma membrane and fuse with it. Growth in isodiametrical or polarized fashion will depend on whether the VSC remain fixed or move alongside, a phenomenon that was mathematically formulated (Bartnicki-García, 2002; see San-Blas and Murgich's Chapter 7 of this book for more details).

In dissecting dimorphic fungi into their molecular constituents, we are making enormous advances towards a better understanding of their morphogenetic shifts. However, this analysis has driven us away from a more integrative view that links the genes (as molecular level of order) with the supramolecular order of the organisms, this being structured in space, regulated and directional, defined by cell boundaries, and shaped by local physical forces. The challenges for future research in fungal morphogenesis is not only continuing the search for a detailed knowledge of genomes but also of the understanding the forces that drive gene products into a given shape from a multitude of alternatives.

10. Acknowledgement

To Fondo Nacional de Investigaciones Científicas y Tecnológicas (FONACIT) for grant G97-000615.

References

Abidi, F.E., Roh, H., and Keath, E.J. 1998. Identification and characterization of a phase-specific, nuclear binding protein from the dimorphic pathogenic fungus *Histoplasma capsulatum*. Infect. Immun. 66: 3867-3873.

Adams, T.H., Wieser, J.K., and Yu, J.H. 1998. Asexual sporulation in *Aspergillus nidulans*. Microbiol. Mol. Biol. Rev. 62: 35-54.

Ahrazem, O., Prieto, A., San-Blas, G., Leal, J.A., Jiménez-Barero, J., and Bernabé, M. 2003. Structural differences between the alkali-extracted water-soluble cell wall polysaccharides from mycelial and yeast phases of the pathogenic dimorphic fungus *Paracoccidioides brasiliensis*. Glycobiology 13: 743-747.

Andrianopoulos, A. 2002. Control of morphogenesis in the human fungal pathogen *Penicillium marneffei*. Int. J. Med. Microbiol. 292: 331-347.

Aquino-Piñero, E.E., and Rodríguez-del-Valle, N. 1997. Different protein kinase C isoforms are present in the yeast and mycelium forms of *Sporothrix schenckii*. Mycopathologia 138: 109-115.

Aquino-Piñero, E., and Rodríguez-del-Valle, N. 2002. Characterization of a protein kinase C gene in *Sporothrix schenckii* and its expression during the yeast to mycelium transition. Med. Mycol. 40: 185-199.

Banuett, F. 1998. Signalling in the yeast: An informational cascade with links to the filamentous fungi. Microbiol. Mol. Biol. Rev. 62: 249-274.

Bartnicki-García, S. 2002. Hyphal growth: Outstanding questions. In: Molecular Biology of Fungal Development. H.D. Osiewacz, ed. Marcel Dekker, New York. p. 29-58.

Batanghari, J.W., Deepe, G.S., Jr., Di Cera, E., and Goldman, W.E. 1998. *Histoplasma* acquisition of calcium and expression of *CBP1* during intracellular parasitism. Mol. Microbiol. 27: 531-539.

Batanghari, J.W., and Goldman, W.E. 1997. Calcium dependence and binding in cultures of *Histoplasma capsulatum*. Infect. Immun. 65: 5257-5261.

Berríos, M.J., and Rodríguez-del-Valle, N. 2002. Expression of a Pho85 cyclin-dependent kinase is repressed during the dimorphic transition in *Sporothrix schenckii*. Fungal Genet. Biol. 37: 39-48.

Braun, B.R., Head, W.S., Wang, M.X., and Johnson, A.D. 2000. Identification and characterization of TUP1-regulated genes in *Candida albicans*. Genetics 156: 31-44.

Bolker, M. 1998. Sex and crime: Heterotrimeric G proteins in fungal mating and pathogenesis. Fungal Genet. Biol. 25: 143-156.

Bono, J.L., Jaber, B., Fisher, M.A., Abuodeh, R.O., O´Leary-Jepson, E., Scalarone, G.M., and Smith, L.H., Jr. 2000. Genetic diversity and transcriptional analysis of the *bys1* gene from *Blastomyces dermatitidis*. Mycopathologia 152: 113-123.

Borneman, A.R., Hynes, M.J., and Andrianopoulos, A. 2000. The *abaA* homologue of *Penicillium marneffei* participates in two developmental programmes: conidiation and dimorphic growth. Mol. Microbiol. 38: 1034-1047.

Borneman, A.R., Hynes, M.J., and Andrianopoulos, A. 2001. An *STE12* homolog from the asexual, dimorphic fungus *Penicillium marneffei* complements the defect in sexual development of an *Aspergillus nidulans steA* mutant. Genetics 157: 1003-1014.

Borneman, A.R., Hynes, M.J., and Andrianopoulos, A. 2002. A basic helix-loop-helix protein with similarity to the fungal morphological regulators, Phd1p, Efg1 and StuA, controls conidiation but not dimorphic growth in *Penicillium marneffei*. Mol. Microbiol. 44: 621-631.

Boyce, K.J., Hynes, M.J., and Andrianopoulos, A. 2001. The *CDC42* homolog of the dimorphic fungus *Penicillium marneffei* is required for correct cell polarization during growth but not development. J. Bacteriol. 183: 3447-3457.

Boyce, K.J., Hynes, M.J., and Andrianopoulos, A. 2003. Control of morphogenesis and actin localization by the *Penicillium marneffei RAC* homolog. J. Cell Sci. 116: 1249-1260.

Brandhorst, T., and Klein, B. 2000. Cell wall biogenesis of *Blastomyces dermatitidis*. Evidence for a novel mechanism of cell surface localization of a virulence-associated adhesin via extracellular release and reassociation with cell wall chitin. J. Biol. Chem. 275: 7925-7934.

Brandhorst, T.T., Rooney, P.J., Sullivan, T.D., and Klein, B.S. 2002. Using new genetic tools to study the pathogenesis of *Blastomyces dermatitidis*. Trends Microbiol. 10: 25-30.

Brandhorst, T., Wüthrich, M., Finkel-Jimenez, B., and Klein, B. 2003. A C-terminal EGF-like domain governs BAD1 localization to the yeast surface and fungal adherence to phagocytes, but is dispensable in immune modulation and pathogenicity of *Blastomyces dermatitidis*. Mol. Microbiol. 48: 53-65.

Brandhorst, T.T., Wüthrich, M., Warner, T., and Klein, B. 1999. Targeted gene disruption reveals an adhesin indispensable for pathogenicity of *Blastomyces dermatitidis*. J. Exp. Med. 189: 1207-1216.

Burg, E.F., III, and Smith, L.S., Jr. 1994. Cloning and characterization of *bys1*, a temperature-dependent cDNA specific to the yeast phase of the pathogenic dimorphic fungus *Blastomyces dermatitidis*. Infect. Immun. 62: 2521-2528.

Carr, J., and Shearer, G., Jr. 1998. Genome size, complexity, and ploidy of the pathogenic fungus *Histoplasma capsulatum*. J. Bacteriol. 180: 6697-6703.

Carratù, L., Franceschelli, S., Pardini, C.L., Kobayashi, G.S., Horvath, I., Vigh, L., and Maresca, B. 1996. Membrane lipid perturbation modifies the set point of the temperature of heat shock response in yeast. Proc. Natl. Acad. Sci. USA, 93: 3870-3875.

Caruso, M., Sacco, M., Medoff, G.S., and Maresca, B. 1987. Heat shock 70 gene is differentially expressed in *Histoplasma capsulatum* strains with different levels of thermotolerance and pathogenicity. Mol. Microbiol. 1: 151-158.

Carvalho, M.J.A., Jesuino, R.S.A., Daher, B.S., Silva-Pereira, I., Freitas, S.M., Soares, C.M.A., and Felipe, M.S.S. 2003. Functional and genetic characterization of calmodulin from the dimorphic and pathogenic fungus *Paracoccidioides brasiliensis*. Fungal Genet. Biol. 39: 204-210.

Chauhan, N., and Calderone, R.. 2004. Adaptation to oxidative, nutrient and pH stress by human pathogenic fungi. In: Pathogenic Fungi: Host Interactions and Emerging Strategies for Control. G. San-Blas, and R.A. Calderone, eds. Caister Academic Press, Wymondham, Norfolk, UK. p129-166.

Chua, S.S., Momany, M., Mendoza, L., and Szaniszlo, P.J. 1994. Identification of three chitin synthase genes in the dimorphic fungal pathogen *Sporothrix schenckii*. Curr. Microbiol. 29: 151-156.

Cole, G.T., and Hung, C.Y. 2001. The parasitic cell wall of *Coccidioides immitis*. Med. Mycol. 39 (Suppl. 1): 31-40.

Colonna-Romano, S., Porta, A., Franco, A., Kobayashi, G.S., and Maresca, B. 1998. Identification and isolation by ddRT-PCR of genes differentially expressed by *Histoplasma capsulatum* during macrophage infection. Microb. Pathog. 25: 55-66.

Cooper, C.R., and Szanniszlo, P.J. 1993. Evidence for two cell division cycle (*CDC*) genes that govern yeast bud emergence in the pathogenic fungus *Wangiella dermatitidis*. Infect. Immun. 61: 2069-2081.

Cunha, A.F., Sousa, M.V., Silva, S.P., Jesuino, R.S.A., Soares, C.M.A., and Felipe, M.S.S. 1999. Identification, N-terminal region sequencing and similarity analysis of differentially expressed proteins in *Paracoccidioides brasiliensis*. Med. Mycol. 37: 115-121.

D'Souza, C.A., and Heitman, J. 2001. Conserved cAMP signalling cascades regulate fungal development and virulence. FEMS Microbiol. Lett. 25: 349-364.

Deepe, G.S.,. Gibbons, R., Brunner, G.D., and Gómez, F.J. 1996. A protective domain of heat-shock protein 60 from *Histoplasma capsulatum*. J. Infect. Dis. 174: 828-834.

Demple, B. 1999. Radical ideas: genetic responses to oxidative stress. Clin. Exp. Pharmacol. Physiol. 26: 64-68.

Denning, D.W. 1997. Echinocandins and pneumocandins – a new antifungal class with a novel mode of action. J. Antimicrob. Chemother. 40: 611-614.

Di Lallo, G., Gargano, S., and Maresca, B. 1994. The *Histoplasma capsulatum cdc2* gene is transcriptionally regulated during the morphologic transition. Gene 140: 51-57.

Dugger, K.O., Villareal, K.M., Ngyuen, A., Zimmermann, C.R., Law, J.H., and Galgiani, J.N. 1996. Cloning and sequence analysis of the cDNA for a protein from *Coccidioides immitis*, with immunogenic potential. Biochem. Biophys. Res. Commun. 218: 485-489.

El-Rady, J., and Shearer, G., Jr. 1996. Isolation and characterization of a calmodulin-encoding cDNA from the pathogenic fungus *Histoplasma capsulatum*. J. Med. Vet. Mycol. 34: 163-169.

El-Rady, J., and Shearer, G., Jr. 1997. Cloning and analysis of an actin-encoding cDNA from the dimorphic pathogenic fungus *Histoplasma capsulatum*. J. Med. Vet. Mycol. 35: 159-166.

Faria-Campos, A.C., Cerqueira, G.C., Anacleto, C., Carvalho, C.M.B., and Ortega, J.M. 2003. Mining microorganisms EST databases in the quest for new proteins. Genet. Mol. Res. 2: 169-177.

Felipe, M.S.S., Andrade, R.V., Petrofeza, S.A., Maranhão, A.Q., Torres, F.A.G., Albuquerque, P., Arraes, F.B.M., Arruda, M., Azevedo, M.O., *et al.* 2003a. Transcriptome characterization of the dimorphic pathogenic fungus *Paracoccidioides brasiliensis* by EST analysis. Yeast 20: 263-271.

Felipe, M.S.S., and Pb Genome Network. 2003b. Functional and differential genome of *Paracoccidioides brasiliensis*. Invited conference. IV Virtual Mycological Congress; June 27-July 14, 2003. http://congresomicologia.ucv.ve.

Fisher, M.C., Koenig, G.L., White, T.J., and Taylor, J.T. 2002. Molecular and phenotypic description of *Coccidioides posadasii* sp. *nov.*, previously recognized as the non-California population of *Coccidioides immitis*. Mycologia 94: 73-84.

Gargano, S., Di Lallo, G., Kobayashi, G.S., and Maresca, B. 1995. A temperature-sensitive strain of *Histoplasma capsulatum* has an altered Δ^9-fatty acid desaturase gene. Lipids 30: 899-906.

Goldman, G.H., Marques, E.R., Ribeiro, D.C.D., Bernardes, L.A.S., Quiapin, A.C., Vitorelli, P.M., Savoldi, M., Semighini, C.P., Oliveira, R.C., *et al.* 2003. Expressed sequence tag analysis of the human pathogen *Paracoccidioides brasiliensis* yeast phase: Identification of putative homologues of *Candida albicans* virulence and pathogenicity genes. Eukaryotic Cell 2: 34-48.

Gómez, F.J., Gómez, A.M. and Deepe, G.S., Jr. 1992. An 80-kilodalton antigen from *Histoplasma capsulatum* that has homology to heat shock protein 70 induces cell-mediated immune responses and protection in mice. Infect. Immun. 60: 2565-2571.

Goyal, S., and Khuller, G. K. 1994. Structural and functional role of lipids in yeast and mycelial forms of *Candida albicans*. Lipids 29: 793-797.

Guevara-Olvera, L., Hung, C.Y., Yu, J.J., and Cole, G.T. 2000. Sequence, expression and functional analysis of the *Coccidioides immitis ODC* (ornithine decarboxylase) gene. Gene 242: 437-448.

Han-Yaku, H., Naka, W., Tajima, S., Harada, T., and Nishikawa, T. 1996. Differential expression of the 45 kDa protein (actin) during the dimorphic transition of *Sporothrix schenckii*. J. Med. Vet. Mycol. 34: 175-180.

Harold, F.M. 1999. In pursuit of the whole hypha. Fungal Genet. Biol. 27: 128-133.

Harold, FM. 2001. The Way of the Cell. Oxford University Press, UK.

Harris, G.S., Keath, E.J., and Medoff, J. 1989a. Characterization of alpha and beta tubulin genes in the dimorphic fungus, *Histoplasma capsulatum*. J. Gen. Microbiol. 135: 1817-1832.

Harris, G.S., Keath, E.J., and Medoff, J. 1989b. Expression of α- and β-tubulin genes during dimorphic-phase transition of *Histoplasma capsulatum*. Mol. Cell Biol. 9: 2042-2049.

Hector, R.F., and Pappagianis, D. 1982. Enzymatic degradation of the wall of spherules of *Coccidioides immitis*. Exp. Mycol. 6: 136-152.

Hegyi, H., and Gerstein, M. 1999. The relationship between protein structure and function: a comprehensive survey with application to the yeast genome. J. Mol. Biol. 288: 147-64.

Hogan, L.H., Josvai, S., and Klein, B.S. 1995. Genomic cloning, characterization and functional analysis of the major surface adhesin WI-1 on *Blastomyces dermatitidis* yeasts. J. Biol. Chem. 270: 30725-30732.

Hogan, L., and Klein, B. 1994. Altered expression of surface alpha-1,3-glucan in genetically related strains of *Blastomyces dermatitidis* that differ in virulence. Infect. Immun. 62: 3543-3546.

Hogan, L.H., and Klein, B.S. 1997. Transforming DNA integrates at multiple sites in the dimorphic fungal pathogen *Blastomyces dermatitidis*. Gene 186: 219-226.

Hung, C.Y., Yu, J.J., Lehmann, P.F., and Cole, G.T. 2001. Cloning and expression of the gene which encodes a tube precipitin antigen and wall-associated β-glucosidase of *Coccidioides immitis*. Infect. Immun. 69: 2211-2222.

Hung, C.Y., Yu, J.J., Seshan, K.R., Reichard, U., and Cole, G.T. 2002. A parasitic phase-specific adhesin of *Coccidioides immitis* contributes to the virulence of this respiratory fungal pathogen. Infect. Immun. 70: 3443-3456.

Hwang, L., Hocking-Murray, D., Bahrami, A.K., Andersson, M., Rine, J., and Sil, A. 2003. Identifying phase-specific genes in the fungal pathogen *Histoplasma capsulatum* using a genomic shotgun microarray. Mol. Biol. Cell 14: 2314-2326.

Ignatov, A., and Keath, E.J. 2002a. Gel shift assay of nuclear extracts from *Histoplasma capsulatum* demonstrates the presence of several DNA binding proteins. Infect. Immun. 70: 2238-2241.

Ignatov, A., and Keath, E.J. 2002b. Molecular cell biology and molecular genetics of *Histoplasma capsulatum*. Int. J. Med. Microbiol. 292: 349-361.

Izacc, S.M.S., Gómez, F.J., Jesuino, R.S.A., Fonseca, C.A., Felipe, M.S.S., Deepe, G.S., and Soares, C.M.A. 2001. Molecular cloning, characterization and expression of the heat shock protein 60 gene from the human pathogenic fungus *Paracoccidioides brasiliensis*. Med. Mycol. 39: 445-455.

Johnson, C.H., Klotz, M.G., York, J.L., Kruft, V., and McEwen, J.E. 2002. Redundancy, phylogeny and differential expression of *Histoplasma capsulatum* catalases. Microbiology 148: 1129-1142.

Johnson, C.H., and McEwen, J.E. 1999. Isolation of a *Histoplasma capsulatum* cDNA that complements a mitochondrial NAD(+)-isocitrate dehydrogenase subunit I-deficient mutant of *Saccharomyces cerevisiae*. Yeast 15: 799-804.

Johnson, C.H., Prigge, J.T., Warren, A.D., and McEwen, J.E. 2003. Characterization of an alternative oxidase activity of *Histoplasma capsulatum*. Yeast 20: 381-388.

Johnson, D.I. 1999. Cdc42: An essential Rho-type GTPase controlling eukaryotic cell polarity. Microbiol. Mol. Biol. Rev. 63: 54-105.

Johnson, S.M., Kerekes, K.M., Zimmermann, C.R., Williams, R.H., and Pappagianis, D. 2000. Identification and cloning of an aspartyl proteinase from *Coccidioides immitis*. Gene 241: 213-222.

Kanetsuna, F., and Carbonell, L.M. 1971. Cell wall composition of the yeastlike and mycelial forms of *Blastomyces dermatitidis*. J. Bacteriol 106: 946-948.

Kanetsuna, F., Carbonell, L.M., Azuma, I., and Yamamura, Y. 1972. Biochemical studies on the thermal dimorphism of *Paracoccidioides brasiliensis*. J. Bacteriol. 110: 208-218.

Kanetsuna, F., Carbonell, L.M., Moreno, R.E., and Rodríguez, J. 1969. Cell wall composition of the yeast and mycelial forms of *Paracoccidioides brasiliensis*. J. Bacteriol. 97: 1036-1041.

Kano, R., Nakamura, Y., Watanabe, S., Tsujimoto, H., and Hasegawa, A. 2001. Identification of *Sporothrix schenckii* based on ssequences of the chitin synthase 1 gene. Mycoses 44: 261-265.

Karuppayil, S.M., and Szaniszlo, P.J. 1997. Importance of calcium to the regulation of polymorphism in *Wangiella* (*Exophiala*) *dermatitidis*. J. Med. Vet. Mycol. 35: 379-388.

Katayama, S., Hirata, D., Arellano, M., Pérez, P., and Toda, T. 1999. Fission yeast α-glucan synthase MOK1 requires the actin cytoskeleton to localize the sites of growth and plays an essential role in cell morphogenesis downstream of protein kinase C function. J. Cell Biol. 144: 1173-1186.

Keath, E.J., and Abidi, F.E. 1994. Molecular cloning and sequence analysis of *yps-3*, a yeast-phase-specific gene in the dimorphic fungal pathogen *Histoplasma capsulatum*. Microbiology 140: 759-767.

Keath, E.J., Kobayashi, G.S., and Medoff, G. 1992. Typing of *Histoplasma capsulatum* by restriction fragment length polymorphisms in a nuclear gene. J. Clin. Microbiol. 30: 2104-2107.

Keath, E.J., Painter, A.A., Kobayashi, G.S., and Medoff, G. 1989. Variable expression of a yeast-phase specific gene in *Histoplasma capsulatum* strains differing in thermotolerance and virulence. Infect. Immun. 57: 1384-1390.

Kersulyte, D., Woods, J.P., Keath, E.J., Goldman, W.E., and Berg, D.E. 1992. Diversity among clinical isolates of *Histoplasma capsulatum* detected by polymerase chain reaction with arbitrary primers. J. Bacteriol. 174: 7075-7079.

Klimpel, K.R., and Goldman, W.E. 1988. Cell walls from avirulent variants of *Histoplasma capsulatum* lack alpha-(1,3)-glucan. Infect. Immun. 56: 2997-3000.

Koppensteiner, W.A., Lackner, P., Wiederstein, M., and Sippl, M.J. 2000. Characterization of novel proteins based on known protein structures. J. Mol. Biol. 296: 1139-52.

Kruse, D., and Cole, G.T. 1992. A seroreactive 120-kilodalton β-1,3-glucanase of *Coccidioides immitis* which may participate in spherule morphogenesis. Infect. Immun. 60: 4350-4363.

Kügler, S., Young, B., Miller, V.L., and Goldman, W.E. 2000. Monitoring phase-specific gene expression in *Histoplasma capsulatum* with telomeric GFP fusion plasmids. Cell. Microbiol. 2: 537-547.

Levery, S.B., Toledo, M.S., Straus, A.H., and Takahashi, H.K. 1998. Structure elucidation of sphingolipids from the mycopathogen *Paracoccidioides brasiliensis*: an immunodominant β-galactofuranose residue is carried by a novel glycosylinositol phosphoceramide antigen. Biochemistry 37: 8764-8775.

Loureiro y Penha, C.V., Todeschini, A.R., Lopes-Bezerra, L.M., Wait, R., Jones, C., Mattos, K.A.,. Heise, N., Mendoça-Previato, L., and Previato, J.O. 2001. Characterization of novel structures of mannosylinositolphosphoryl-ceramides from the yeast forms of *Sporothrix schenckii*. Eur. J. Biochem. 4243-4250.

Magrini, V., and Goldman, W.E. 2001. Molecular mycology: a genetic toolbox for *Histoplasma capsulatum*. Trends Microbiol. 9: 541-546.

Maresca, B., Carratù, L., and Kobayashi, G. 1994. Morphological transition in the human fungal pathogen *Histoplasma capsulatum*. Trends Microbiol. 2: 110-114.

Maresca, B., and Kobayashi, G. 1989. Dimorphism in *Histoplasma capsulatum*: a model for the study of cell differentiation in pathogenic fungi. Microbiol. Rev. 53: 186-209.

Medoff, G., Painter, A., and Kobayashi, G.S. 1987. Mycelial to yeast phase transitions of the dimorphic fungi *Blastomyces dermatitidis* and *Paracoccidioides brasiliensis*. J. Bacteriol. 169: 4055-4060.

Medoff, G., Sacco, M., Maresca, B., Schlessinger, D., Painter, A., Kobayashi, G.S., and Carratù, L. 1986. Irreversible block of the mycelia to yeast transition of *Histoplasma capsulatum*. Science 231: 476-479.

Mendoza, L., Karuppayil, S.M., and Szaniszlo, P.J. 1993. Calcium regulates *in vitro* dimorphism in chromoblastomycotic fungi. Mycoses 36: 157-164.

Newport, G., and Agabian, N. 1997. KEX2 influences *Candida albicans* proteinase secretion and hyphal formation. J. Biol. Chem. 272: 28954-28961.

Niño-Vega, G.A., Buurman, E.T., Gooday, G.W., San-Blas, G., and Gow, N.A.R. 1998. Molecular cloning and sequencing of a chitin synthase gene (*CHS2*) of *Paracoccidioides brasiliensis*. Yeast 14: 181-187.

Niño-Vega, G., Carrero, L., and San-Blas, G. 2004a. Isolation of the CHS4 gene of *Paracoccidioides brasiliensis* and its accommodation in a new class of chitin synthases. Med. Mycol. 41: 51-57.

Nino-Vega, G.A., Sorais, F., Calcagno, A.M., Ruiz-Herrera, J., Martinez-Espinoza, A.D., and San-Blas, G.. 2004b. Cloning and expression analysis of the ornithine decarboxylase gene (PbrODC) of the pathogenic fungus *Paracoccidioides brasiliensis*. Yeast. 21: 211-218.

Niño-Vega, G.A., Munro, C.A., San-Blas, G., Gooday, G.W., and Gow, N.A.R. 2000. Differential expression of chitin synthase genes during temperature-induced dimorphic transition in *Paracoccidioides brasiliensis*. Med. Mycol. 38: 31-39.

Pan, S., Zhang, M., and Cole, G.T. 1995. Isolation and characterization of the arginase-encoding gene (*arg*) from *Coccidioides immitis*. Gene 154: 115-118.

Pappagianis, D., Smith, C.E., Kobayashi, G.S., and Saito, M.T. 1961. Studies on antigens from young mycelia of *Coccidioides immitis*. J. Infect. Dis. 108: 35-44.

Paris, S., and Durán, S. 1985. Cyclic adenosine 3',5'monophosphate (cAMP) and dimorphism in the pathogenic fungus *Paracoccidioides brasiliensis*. Mycopathologia 92: 115-120.

Paris, S., and Garrison, R.G. 1983. Cyclic adenosine 3',5' monophosphate (cAMP) as a factor in phase morphogenesis of *Blastomyces dermatitidis*. Mykosen 27: 340-345.

Patel, J.B., Batanghari, J.W., and Goldman, W.E. 1998. Probing the yeast phase-specific expression of the *CBP1* gene in *Histoplasma capsulatum*. J. Bacteriol. 180: 1786-1792.

Patriarca, E.J., Kobayashi, G.S., and Maresca, B. 1992. Mitochondrial activity and heat-shock response during morphogenesis in the pathogenic fungus *Histoplasma capsulatum*. Biochem. Cell Biol. 70: 207-214.

Patriarca, E., and Maresca, B. 1990. Acquired thermotolerance following heat shock protein synthesis prevents impairment of mitochondrial ATPase activity at elevated temperature in *Saccharomyces cerevisiae*. Exp. Cell Res. 190: 57-64.

Pereira, M., Felipe, M.S.S., Brígido, M.M., Soares, C.M.A., and Azevedo, M.O. 2000. Molecular cloning and characterization of a glucan synthase from the human pathogenic fungus *Paracoccidioides brasiliensis*. Yeast 16: 451-462.

Pérez-Silva, C. 2003. Clonaje del gen de actina del hongo patógeno *Paracoccidioides brasiliensis*. Thesis Licenciado en Biología, Universidad del Zulia, Venezuela.

Pfaller, M.A., Riley, J., and Gerarden, T. 1988. Polyamine depletion and growth inhibition in *Candida albicans* and *Candida tropicalis* by α-difluoromethylornithine and cyclohexylamine. J. Med. Vet. Mycol. 26: 119-126.

Porta, A., Colonna-Romano, S., Callebaut, I., Franco, A., Marzullo, L., Kobayashi, G.S., and Maresca, B. 1999. A homologue of the human 100 kDa protein (p100) is differentially expressed by *Histoplasma capsulatum* during infection of murine macrophages. Biochem. Biophys. Res. Commun. 254: 605-613.

Reichard, U., Hung, C.Y., Thomas, P.W., and Cole, G.T. 2000. Disruption of the gene that encodes a serodiagnostic antigen and chitinase of the human fungal pathogen *Coccidioides immitis*. Infect. Immun. 68: 5830-5838.

Richard, M., Quijano, R.R., Bezzate, S., Bordon-Pallier, F., and Gaillardin, C. 2001. Tagging morphogenetic genes by insertional mutagenesis in the yeast *Yarrowia lypolytica*. J. Bacteriol. 183: 3089-3107.

Rocha, C.R., Schroppel, K., Harcus, D., Marcil, A., Dignard, D., Taylor, B.N., Thomas, D.Y., Whiteway, M., and Leberer, E. 2001. Signaling through adenylyl cyclase is essential for hyphal growth and virulence in the pathogenic fungus *Candida albicans*. Mol. Biol. Cell 12: 3631-3643.

Rodríguez-del-Valle, N., and Rodríguez-Medina, J.R. 1993. Calcium stimulates molecular and cellular events during the yeast to mycelium transition in *Sporothrix schenckii.* J. Med. Vet. Mycol. 31: 43-53.

Rooney, P.J., Sullivan, T.D., and Klein, B.S. 2001. Selective expression of the virulence factor *BAD1* in morphogenesis to the pathogenic yeast form of *Blastomyces dermatitidis*: evidence for transcriptional regulation by a conserved mechanism. Mol. Microbiol. 39: 875-889.

Ruiz-Herrera, J. 1994. Polyamines, DNA methylation, and fungal differentiation. Crit. Rev. Microbiol. 20: 143-150.

Ruiz-Herrera. J., González-Prieto, J.M., and Ruiz-Medrano, R. 2002. Evolution and phylogenetic relationships of chitin synthases from yeast and fungi. FEMS Yeast Res. 1: 247-256.

Ruiz-Herrera, J., and San-Blas, G. 2003. Chitin synthesis as target for antifungal drugs. Curr. Drug Target. Infect. Disorders, 3: 77-91.

Salem-Izaac, S.M., Jesuino, R.S.A., Brito, W.A., Pereira, M., Felipe, M.S.S., and Soares, C.M.A. 1997. Protein synthesis patterns of *Paracoccidioides brasiliensis* isolates in stage-specific forms and during cellular differentiation. J. Med. Vet. Mycol. 35: 205-211.

San-Blas, G. 1979. Biosynthesis of glucans by subcellular fractions of *Paracoccidioides brasiliensis*. Exp. Mycol. 3: 249-258.

San-Blas, G., Niño-Vega, G., and Iturriaga, T. 2002. *Paracoccidioides brasiliensis* and paracoccidioidomycosis: Molecular approaches to morphogenesis, diagnosis, epidemiology, taxonomy and genetics. Med. Mycol. 40: 225-242.

San-Blas, G., and San-Blas, F. 1986. Effect of nucleotides on glucan synthesis in *Paracoccidioides brasiliensis*. Sabouraudia 24: 241-243.

San-Blas, G, and San-Blas, F. 1994. Biochemistry of *Paracoccidioides brasiliensis* dimorphism. In: Paracoccidioidomycosis. M. Franco, C.S. Lacaz, A. Restrepo-Moreno, G. Del Negro, eds. CRC Press, Boca Raton, Florida. p. 49-66.

San-Blas, G., San-Blas, F., and Serrano, L.E. 1977. Host-parasite relationships in the yeastlike form of *Paracoccidioides brasiliensis* strain IVIC Pb9. Infect. Immun. 15: 343-346.

San-Blas, G., Sorais, F., San-Blas, F., and Ruiz-Herrera, J. 1996. Ornithine decarboxylase in *Paracoccidioides brasiliensis*. Arch. Microbiol. 165: 311-316.

San-Blas, G., Urbina, J., Marchán, E., Contreras, L. M., Sorais, F., and San-Blas F. 1997. Inhibition of *Paracoccidioides brasiliensis* by ajoene is associated with blockade of phosphatidylcholine biosynthesis. Microbiology 143: 1583-1586.

Sánchez-Martínez, C., and Pérez-Martín, J. 2001. Dimorphism in fungal pathogens: *Candida albicans* and *Ustilago maydis* – similar inputs, different outputs. Curr. Op. Microbiol. 4: 214-221.

Schmitt, A.P., and McEntee, K. 1996. Msn2p, a zinc finger DNA-binding protein, is the transcriptional activator of the multistress response in *Saccharomyces cerevisiae*. Proc. Natl. Acad. Sci. USA 93: 5777-5782.

Sebghati, T.S., Engle, J.T., and Goldman, W.E. 2000. Intracellular parasitism by *Histoplasma capsulatum*: Fungal virulence and calcium dependence. Science 290: 1368-1372.

Shearer, G., Jr., Birge, C.H., Yuckenberg, P.D., Kobayashi, G.S., and Medoff, G. 1987. Heat-shock proteins induced during the mycelial-to-yeast transition of strains of *Histoplasma capsulatum*. J. Gen. Microbiol. 133: 3375-3382.

Silva, S.P., Borges-Walmsley, M.I., Pereira. I.S., Soares, C.M.A., Walmsley, A.R., and Felipe, M.S.S. 1999. Differential expression of an *hsp70* gene during transition from the mycelial to the infective yeast form of the human pathogenic fungus *Paracoccidioides brasiliensis*. Mol. Microbiol. 31: 1039-1050.

Silva, W.P., Soares, R.B.A., Jesuino, R.S.A., Izacc, S.M.S., Felipe, M.S.S., and Soares, C.M.A. 2001. Expression of alpha tubulin during the dimorphic transition of *Paracoccidioides brasiliensis*. Med. Mycol. 39: 457-462.

Sorais-Landáez, F., and San-Blas, G. 1993. Localization of β-glucan synthase in membranes of *Paracoccidioides brasiliensis*. J. Med. Vet. Mycol. 31: 421-426.

Spitzer, E.D., Lasker, B.A., Travis, S., Kobayashi, G.S., and Medoff, G. 1989. Use of mitochondrial and ribosomal DNA polymorphisms to classify clinical and soil isolates of *Histoplasma capsulatum*. Infect. Immun. 57: 1409-1412.

Suutari, M. 1995. Effect of growth temperature on lipid fatty acids of four fungi (*Aspergillus niger*, *Neurospora crassa*, *Penicillium chrysogenum*, and *Trichoderma reesei*). Arch. Microbiol. 164: 212-216.

Steele, P.E., Carle, G.F., Kobayashi, G.S., and Medoff, G. 1989. Electrophoretic analysis of *Histoplasma capsulatum* chromosomal DNA. Mol. Cell. Biol. 9: 983-987.

Storlazzi, A., Maresca, B., and Gargano, S. 1999. cAMP is involved in transcriptional regulation of Δ^9-desaturase during *Histoplasma capsulatum* morphogenesis. Mol. Cell Biol. Res. Commun. 2: 172-177.

Szaniszlo, P.J. 2002. Molecular genetic studies of the model dematiaceous pathogen *Wangiella dermatitidis*. Int. J. Med. Microbiol. 292: 381-390.

Szaniszlo, P.J., Kang, M.S., and Cabib, E. 1985. Stimulation of β-1,3-glucan synthase of various fungi by nucleoside triphosphates. A generalized regulatory mechanism for cell wall biosynthesis. J. Bacteriol. 161: 1188-1194.

Thomas, P.W., Wyckoff, E.E., Pishko, E.J., Yu, J.J., Kirkland, T.N., and Cole, G.T. 1997. The *hsp60* gene of the human pathogenic fungus *Coccidioides immitis* encodes a T-cell reactive protein. Gene 199: 83-91.

Tian, X., and Shearer, G., Jr. 2001. Cloning and analysis of mold-specific genes in the dimorphic fungus *Histoplasma capsulatum*. Gene 275: 107-114.

Tian, X., and Shearer, G., Jr. 2002. The mold-specific *MS8* gene is required for normal hyphal formation in the dimorphic fungus, *Histoplasma capsulatum*. Eukaryotic Cell 1: 249-256.

Todd, R.B., Greenhalgh, J.R., Hynes, M.J., and Andrianopoulos, A. 2003. TupA, the *Penicillium marneffei* Tup1p homologue, represses both yeast and spore development. Mol. Microbiol. 48: 85-94.

Toledo, M.S., Levery, S.B., Glushka, J., Straus, A.H., and Takahashi, H.K. 2001a. Structure elucidation of sphingolipids from the mycopathogen *Sporothrix schenckii*: Identification of novel glycosylinositol phosphorylceramides with core Manα1→6Ins linkage. Biochem. Biophys. Res. Commun. 280: 19-24.

Toledo, M.S., Levery, S.B., Straus, A.H., and Takahashi, H.K. 2000. Dimorphic expression of cerebrosides in the mycopathogen *Sporothrix schenckii*. J. Lip. Res. 41: 797-806.

Toledo, M.S., Levery, S.B., Straus, A.H., Suzuki, E., Momany, M., Glushka, J., Moulton, J.M., and Takahashi, H.K. 1999. Characterization of sphingolipids from mycopathogens: Factors correlating with expression of 2-hydroxy fatty acyl (*E*)-Δ^3-unsaturation in cerebrosides of *Paracoccidioides brasiliensis* and *Aspergillus fumigatus*. Biochemistry 38: 7294-7306.

Toledo, M.S., Levery, S.B., Straus, A.H., and Takahashi, H.K. 2001b. Sphingolipids of the mycopathogen *Sporothrix schenckii*: Identification of a glycosylinositol phosphorylceramide with novel core $GlcNH_2\alpha1\rightarrow2Ins$ motif. FEBS Lett. 493: 50-56.

Toledo, M.S., Levery, S.B., Suzuki, E., Straus, A.H., and Takahashi, H.K. 2001c. Characterization of cerebrosides from the thermally dimorphic mycopathogen *Histoplasma capsulatum*: expression of 2-hydroxy fatty *N*-acyl (*E*)-Δ^3-unsaturation correlates with the yeast-mycelium phase transition. Glycobiology 11: 113-124.

Tosco, A., Gargano, S., Kobayashi, G.S., and Maresca, B. 1997. An AP1 element is involved in transcriptional regulation of Δ^9-desaturase gene of *Histoplasma capsulatum*. Biochem. Biophys. Res. Commun. 230: 457-461.

Travassos, L.R. and Lloyd, K.O. 1980. *Sporothrix schenckii* and related species of *Ceratocystis*. Microbiol. Rev. 44: 683-721.

Venancio, E.J., Daher, B.S., Andrade, R.V., Soares, C.M.A., Pereira, I.S., and Felipe, M.S.S. 2002. The *kex2* gene from the dimorphic and human pathogenic fungus *Paracoccidioides brasiliensis*. Yeast 19: 1221-1231.

Wang, Q., Liu, H., and Szaniszlo, P.J. 2002. Compensatory expression of five chitin synthase genes, a response to stress stimuli, in *Wangiella* (*Exophiala*) *dermatitidis*, a melanized fungal pathogen of humans. Microbiology 148: 2811-2817.

Wang, Z., and Szaniszlo, P.J. 2000. *WdCHS3*, a gene that encodes a class III chitin synthase in *Wangiella* (*Exophiala*) *dermatitidis*, is expressed differentially under stress conditions. J. Bacteriol. 182: 874-881.

Wang, Z, and Szaniszlo, P.J. 2002. Characterization of WdChs3p, a class III chitin synthase, of *Wangiella* (*Exophiala*) *dermatitidis*, overexpressed in *Saccharomyces cerevisiae*. Med. Mycol. 40: 283-289.

Wang, Z., Zheng, L., Hauser, M., Becker, J.M., and Szaniszlo, P. 1999. WdChs4p, a homolog of chitin synthase 3 in *Saccharomyces cerevisiae*. Alone cannot support growth of *Wangiella* (*Exophiala*) *dermatitidis* at the temperature of infection. Infec. Immun. 67: 6619-6630.

Wang, Z., Zheng, L., Liu, H., Wang, Q., Hauser, M., Kauffman, S., Becker, J.M., and Szaniszlo, P.J. 2001. WdChs2p, a class I chitin synthase, together with WdChs3p (Class III) contributes to virulence in *Wangiella* (*Exophiala*) *dermatitidis*. Infect. Immun. 69: 7517-7526.

Weaver, C.H., Sheenan, K.C.F., and Keath, E.J. 1996. Localization of a yeast-phase-specific gene product to the cell wall in *Histoplasma capsulatum*. Infect. Immun. 64: 3048-3054.

Wheat, R.W., Tritschler, C., Conant, N.F., and Loe, E.P. 1977. Comparison of *Coccidioides immitis* arthrospore, mycelium, and spherule cell walls and influence of growth medium on mycelial cell wall composition. Infect. Immun. 17: 91-97.

Woods, J.P. 2002. *Histoplasma capsulatum* molecular genetics, pathogenesis, and responsiveness to its environment. Fungal Genet. Biol. 35: 81-97.

Ye, X., and Szaniszlo, P.J. 2000. Expression of a constitutively active Cdc42 homologue promotes development of sclerotic bodies but represses hyphal growth in the zoopathogenic fungus *Wangiella (Exophiala) dermatitidis*. J. Bacteriol. 182: 4941-4950.

Yu, J.J., Smithson, S.L., Thomas, P.W., Kirkland, T.N., and Cole, G.T. 1997. Isolation and characterization of the urease gene (*URE*) from the pathogenic fungus *Coccidioides immitis*. Gene 198: 387-391.

Yu, J.J., Zheng, L., Thomas, P.W., Szaniszlo, P.J., and Cole, G.T. 1999. Isolation and confirmation of function of the *Coccidioides immitis URA5* (orotate phosphoribosyl transferase) gene. Gene 226: 233-242.

Yuen, K.Y., Pascal, G., Wong, S.S.Y., Glaser, P., Woo, P.C.Y., Kunst, F., Cai, J.J., Cheung, E.Y.L., Médigue, C., and Danchin, A. 2003. Exploring the *Penicillium marneffei* genome. Arch. Microbiol. 179: 339-353.

Zhu, Y., Yang, C., Magee, D.M., and Cox, R.A. 1996. Molecular cloning and characterization of *Coccidioides immitis* antigen 2 cDNA. Infect. Immun. 64: 2695-2699.

Zuber, S., Hynes, M.J., and Andrianopoulos, A. 2002. G-protein signalling mediates asexual development at 25°C but has no effect on yeast-like growth at 37°C in the dimorphic fungus *Penicillium marneffei*. Eukaryotic Cell 1: 440-447.

From: Pathogenic Fungi: Structural Biology and Taxonomy
Edited by: G. San-Blas and R.A. Calderone

Chapter 6

Regulation of Morphogenesis by Conserved Developmental Pathways in Pathogenic Fungi

Idit Hazan and Haoping Liu

Abstract

The highly conserved pheromone response MAP kinase and nutrient-sensing cAMP/PKA signal pathways are critical for filamentation, mating and virulence in many pathogenic fungi. A comparison of their functions in two human pathogens *Candida albicans* and *Cryptococcus neoformans*, and in two plant pathogens, *Magnaporthe grisea* and *Ustilago maydis,* shows that virulence is tightly associated with filamentation and mating in these fungi, suggesting an evolutionary link between pathogenesis and cellular development.

1. Introduction

Pathogenic fungi are a diverse group of organisms that can infect a variety of hosts including humans and plants, utilizing a range of biological mechanisms. Human pathogenic fungi have become a growing health concern, mainly because of an increase in immunocompromised individuals such as AIDS patients and those undergoing chemotherapy or transplant surgery. Moreover, there is a large increase in such fungal infections seen in surgical and medical wards other than oncology and transplantation, demonstrating that fungal infections are no longer restricted to immunocompromised individuals (Edmond *et al.*, 1999). Over the last two decades, the incidence of nosocomial (hospital acquired) fungal infections has steadily increased (Kao *et al.*, 1999). The problem is compounded by the lack of effective antimycotic drugs, especially for systemic infections, and the increase in pathogen resistance to such drugs (Walsh and Groll, 1999). Plant pathogens are also a growing concern because they cause billions of dollars in damages to crops, especially in third world countries that rely on such crops for their economy and have limited means to deal with such pathogens.

Within the pathogenic fungi, filamentous and dimorphic fungi are a unique group, in which filamentous growth is part of the normal life cycle and filamentation often correlates with pathogenicity. Filamentous fungi grow by apical extension with an appearance of long thin tubes, with occasional branching, called mycelia. Filamentous fungi are also able to form unicellular conidia or spores, which can be released and serve to disseminate the organism, often to a susceptible host. Thus, the spores are usually the infectious agent, while the filamentous form penetrates the host tissues. Dimorphic fungi are a specialized type of filamentous fungi; they grow as filaments, but they also have a vegetative yeast growth form which is round or oval, and these cells reproduce either by budding or fission. In most cases, the disease in the host is marked by some kind of transition from a non-pathogenic infectious form in the environment to a distinct pathogenic form in tissue. This transition is often tightly linked to the morphogenetic transition and in some cases, to mating.

The cues and output morphology for filamentous growth are very different among fungi. Despite their enormous diversity, mounting evidence suggests that two highly conserved receptor-mediated G-protein signaling pathways play a major role in connecting different extracellular signals to cellular development and morphogenesis. G-protein activate a mitogen activated protein (MAP) kinase pathway or control the cAMP regulated / protein kinase A (PKA) pathway. They have been implicated in the regulation of filamentation and mating, as well as virulence in many pathogenic fungi (Lengeler *et al.*, 2000).

In this chapter, we will discuss the two conserved signaling pathways in four fungal pathogens. These include the filamentous fungus *Magnaporthe grisea,* and the polymorphic fungus *Ustilago maydis*, both plant pathogens; the polymorphic ascomycetous fungus *Candida albicans*, a commensal human pathogen, and the basidiomycetous fungus, *Cryptococcus neoformans*, a free-living human pathogenic opportunist. Each organism has adapted distinct ways of using these two signaling pathways to respond to their unique host environment for its survival and pathogenesis. By comparing and contrasting how the two signaling pathways are used to regulate filamentation, mating and virulence, we hope to present a concept of evolutionary conservation. Readers are directed to other reviews on virulence in *Candida* (Chapters 4 of this book; Bignell *et al.*, 2004; Chauhan and Calderone, 2004), *Cryptococcus* (Feldmesser, 2004), morphogenesis in *Candida* (Chapter 3), and morphogenesis in other pathogenic fungi (Chapter 5).

2. *Candida albicans*

2.1. Life Cycle and Polymorphism

Candida albicans is one of the most commonly isolated pathogens in humans, and the fourth leading cause of nosocomial infections in the USA. It is an opportunistic pathogen, which most frequently causes irritating superficial infections of the oral and urological tracts. Around 90% of AIDS patients will experience at least one episode of oral candidiasis. Even in the healthy host, 70% of all women will experience *Candida* related vaginitis. Moreover, it can cause deep tissue or systemic infections, mostly in neutropenia (low neutrophil count) patients. The attributable mortality rate for invasive candidiasis is over 35% (Wenzel, 1995).

C. albicans is polymorphic and undergoes reversible morphological transitions between budding, pseudohyphal growth, and hyphal growth forms (Odds, 1985) (Figure 1A). In the yeast form, cells grow like other budding yeast such as *Saccharomyces cerevisiae*, as an oval-shaped single-nucleated cell that multiplies by budding, followed by cell separation, each cell containing a nucleus. In pseudohyphal growth, daughter cells form by budding, but the cells are elongated and form a chain with constricted septa. In the hyphal form, cells grow by apical extension to form germ tubes that are extended into hyphal filaments, forming chains of cells. Cells are separated by septa, but there are no constrictions at the septa (Figure 1A). Many media

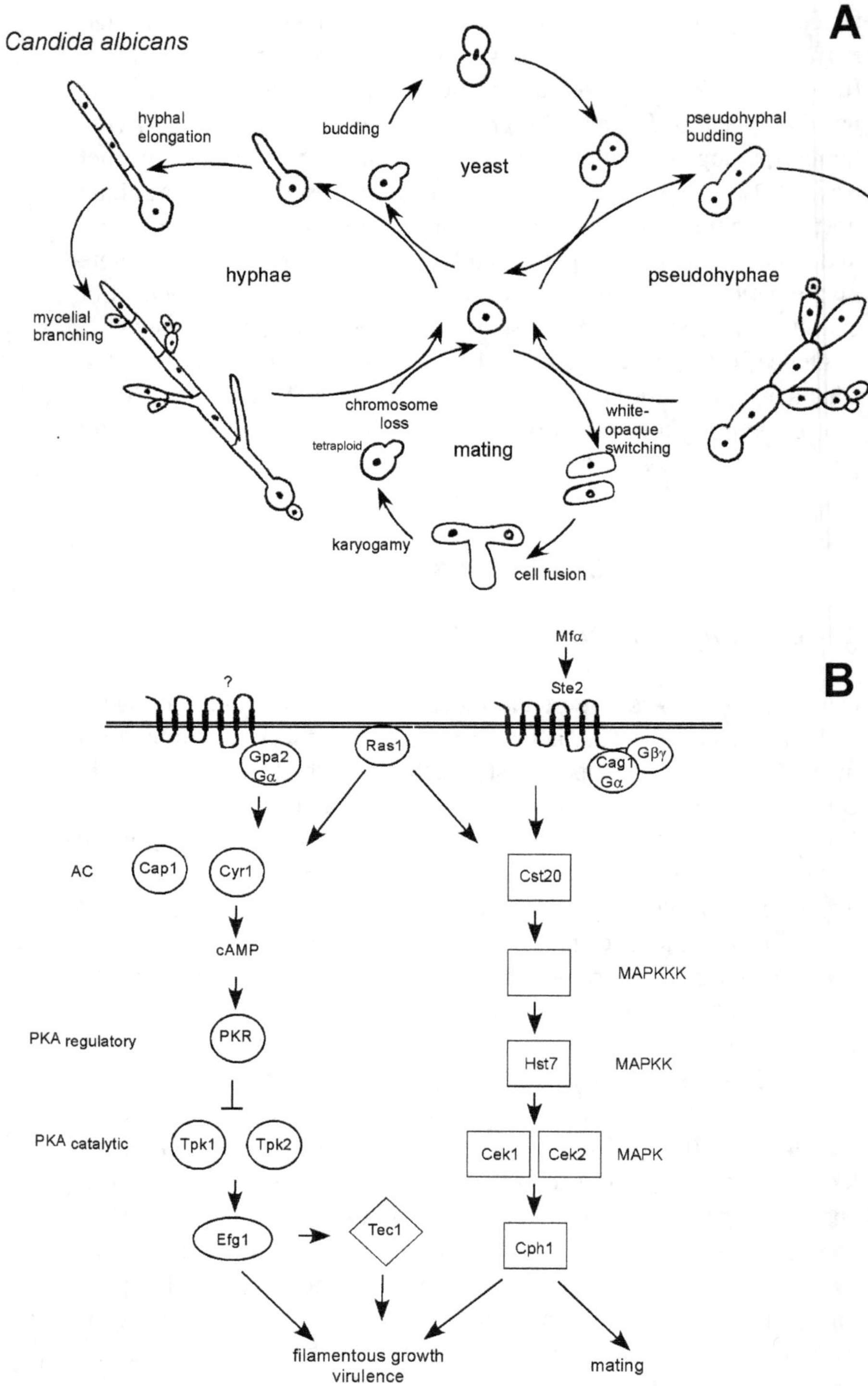

Figure 1. (A) Life cycle of *Candida albicans*. (B) The conserved cAMP/PKA and MAP kinase pathways (see text for details).

and growth conditions can induce the yeast-to-hyphal transition *in vitro*. These include serum, an increase in temperature, pH 7, nutrient starvation and other morphogens related to the host environment. It is clear that some of the signals that promote hyphal morphogenesis may be detected by cell-surface receptors, and other signals may be detected by intracellular mechanisms (Brown and Gow, 1999).

The ability of *C. albicans* to adapt to numerous environments and conditions makes it a successful opportunistic pathogen, able to survive as commensal in many sites in the body. Likewise, its ability to switch between yeast and hyphal growth forms is related to its virulence. *C. albicans* mutants that are unable to form hyphae are less virulent in mouse models than their wild-type counterparts (Lo *et al.*, 1997). Hyphae, which have the ability to adhere and penetrate tissues, may be suited to breach barriers in the host, but the yeast form is more easily disseminated. It also allows it to escape the immune system by puncturing holes in macrophages (Lo *et al.*, 1997). Formation of hyphal filaments is influenced strongly by contact or touch (thigmotropism), and hyphae are able to grow through grooves in agar or ridges on polystyrene (Sherwood *et al.*, 1992; Gow *et al.*, 1994). Hyphae of *C. albicans* also exhibit a tropic responses to oxygen (aerotropism) and applied electrical fields (galvanotropism), but chemotropism has not been shown (Gooday and Adams, 1993). In addition to dimorphic switching, phenotypic switching in *C. albicans* provides phenotypic variability, which leads to the necessary plasticity for adaptation and infection at vastly different host sites (Soll, 2002). One example of phenotypic switching is white-opaque switching, where cells in the white phase switch to the opaque phase which *in vitro* occurs at a frequency of 10^{-4}. Cells in the white phase have a morphology of budding yeast cells and produce white colonies, whereas those in the opaque phase are elongated with prominent pimples on the cell wall that form opaque colonies (Soll, 2002). The ability of *C. albicans* to adhere to host cells through adhesins on the cell surface and to secrete proteinases and phospholipases has been shown to be important for its pathogenicity (Calderone and Fonzi, 2001). Interestingly, the signaling pathways that are responsible for hyphal morphogenesis and the white-opaque switching also regulate the expression of some of the cell surface adhesins and secreted proteinases (Liu, 2002). The co-regulation of virulence genes with the yeast-to-hypha transition and white-opaque switching may be important for *C. albicans* to exist both as a harmless commensal and as a pathogen in various host environments.

C. albicans exists as a diploid and was considered to be asexual until recently, when the mating type locus (*MTL*) of *C. albicans* was discovered (Hull and Johnson, 1999). *MTL* of *C. albicans* contains **a**1 and **a**2 in the *MTL***a** locus, α1 and α2 in the *MTL*α locus on the other homologous chromosome of **a**/α cells (Tsong et al., 2003). In **a**/α cells, **a1** and α2 work together to repress mating (Hull and Johnson, 1999). **a**/α cells can become homozygous at the *MTL* at a high frequency, as about 3% clinical isolates are **a/a** or α/α (Lockhart *et al.*, 2002). **a** cells are able to mate with α cells *in vitro* and *in vivo* (Hull *et al.*, 2000; Magee and Magee, 2000) (Figure 1A). **a**2 positively regulates **a**-type mating, and α1 positively regulates α-type mating (Tsong *et al.*, 2003). Mating in *C. albicans* is tightly linked to white-opaque switching (Miller and Johnson, 2002). First, white-opaque switching is inhibited by **a**1- α2 in **a**/α cells, and only occurs in **a** and α cells (Miller and Johnson, 2002). Furthermore, opaque cells mate 10^6 times more efficiently than white-phase cells and the mating frequency between opaque cells is comparable to that of *S. cerevisiae*. So, *C. albicans* **a** and α cells must first undergo the switch from the white phase to the opaque phase before they are competent to mate (Miller and Johnson, 2002). The resulting tetraploid strain does not undergo meiosis but can be induced to loose chromosomes to give rise to progeny that has a variety of ploidity (Bennett and Johnson, 2003). Approximately, one-third of the progeny become diploid again. The mating and induced chromosome loss of the mated tetraploid constitute a parasexual cycle in *C. albicans* (Bennett and Johnson, 2003) (Figure 1A).

2.2. The cAMP/PKA Pathway

In eukaryotic cells, the second messenger cAMP, is produced in response to extracellular nutrients in order to regulate a variety of growth and development processes. Cellular levels of cAMP depend upon biosynthesis by the enzyme adenylate cyclase (AC) and degradation by phosphodiesterases. AC is regulated by highly conserved receptor-mediated heterotrimeric G-proteins. In *S. cerevisiae*, the G-protein that regulates the adenylate cyclase is the Gα Gpa2. The G-protein couple receptor for Gpa2 is Gpr1, which acts as a sensor of extracellular glucose (Xue *et al.*, 1998). In addition to the Gpr1-coupled Gpa2, AC activity is also regulated by Ras2. Both Gpa2 and Ras2 are important for pseudohyphal growth of *S. cerevisiae* (Pan and Heitman, 1999).

The target of cAMP is the cAMP-dependent protein kinase (PKA), which mediates many processes in the cell. When the cAMP level is low, the PKA holoenzyme is an inactive tetramer comprised of two highly-conserved regulatory and two highly conserved catalytic subunits. When the cAMP level increases, cAMP binds to the regulatory subunits and induces a conformational change that causes dissociation of the tetramer into dimeric regulatory subunits and active monomeric catalytic subunits. The liberated catalytic subunits are enzymatically active and phosphorylate target substrates that include metabolic enzymes and transcription factors. *S. cerevisiae* has three catalytic subunits of overlapping and distinctive functions (Robertson and Fink, 1998). Active PKA activates enzymes involved in the utilization of storage carbohydrates, represses stress-activated genes, inhibits entry to stationary phase, and promotes pseudohyphal growth in *S. cerevisiae*.

The cAMP pathway is essential for hyphal development and virulence in *C. albicans* (Rocha et al., 2001). Exogenously added cAMP and its analog have been reported to induce filamentation (Sabie and Gadd, 1992). An increase in the intracellular cAMP level precedes germ tube emergence in *C. albicans* (Niimi *et al.*, 1980). Many homologues of the components in the cAMP pathway of *S. cerevisiae* have also been identified in *C. albicans* and are thought to function in the same positions in the cAMP pathway as their *S. cerevisiae* counterparts (Figure 1B).

RAS1 is important for hyphal formation in *C. albicans* (Feng *et al.*, 1999). It is the only *RAS* gene in the *C. albicans* genome, but *ras1*/*ras1* strains are viable, although the mutants are defective for hyphal formation under induction by dialyzed serum. Strains expressing the dominant active *RAS1* are hyperfilamentous and cells with a dominant negative *RAS1* are reduced for hyphal growth (Feng *et al.*, 1999). Gpa2 is also important for hyphal development in the response of *C. albicans* to some environmental inducers but is not required for induction by serum (Sanchez-Martinez and Perez-Martin, 2002).

C. albicans has a single gene for the adenylate cyclase, *CYR1*, which is not essential for growth but is essential for hyphal development under all hyphal inducing conditions investigated, even when induced by serum (Rocha *et al.*, 2001). The cyclase mutants are unable to cause vaginal and systemic infections in mice. The AC- associated protein Cap1 also plays an essential function in hyphal development.

The *cap1/cap1* mutant is defective in germ tube formation in all media tested, and is avirulent. This defect can be suppressed by exogenous cAMP (Bahn and Sundstrom, 2001). Mutant phenotypes of *RAS1*, *GPA1*, *CYR1* and *CAP1* suggest that cAMP is necessary for *C. albicans* to undergo hyphal development. On the other hand, deletion of *PDE2*, which encodes the high affinity cAMP phosphodiesterase, causes elevated cAMP levels, allows pseudohyphal development, but inhibits normal hyphal development in hyphal inducing liquid media. Interestingly, *pde2/pde2* mutants form aberrant hyphae with fewer braches and almost no lateral buds on solid media (Jung and Stateva, 2003). Therefore, cAMP levels are tightly regulated during hyphal development in *C. albicans*, and appropriate levels of cAMP are important for normal hyphal development.

C. albicans has two catalytic subunits of the PKA, Tpk1 and Tpk2 (Bockmuhl *et al.*, 2001). The two kinases have redundant roles in growth and both are regulators of hyphal formation, although they also have distinct functions in hyphal development because their mutants have differential effects in different media.

One potential target of the PKA is Efg1, a basic helix-loop-helix protein similar to StuA of *Aspergillus nidulans* and Sok2 and Phd1 of *S. cerevisiae* (Stoldt *et al.*, 1997). Sok2 and Phd1 are an inhibitor and activator of pseudohyphal growth in *S. cerevisiae*, respectively. StuA is a regulator of conidia formation in *A. nidulans*. *efg1/efg1* mutants in *C. albicans* do not form hyphae under most hypha inducing conditions, including induction by serum (Lo *et al.*, 1997), but the mutants show enhanced hyphal formation under embedded growth conditions (Giusani *et al.*, 2002). Efg1 is downstream of the PKA because *TPK2* overexpression cannot suppress the *efg1/efg1* defect in hyphal development, whereas overexpression of *EFG1* can suppress the filamentation defect in *tpk2/tpk2* (Sonneborn *et al.*, 2000) (Figure 1B). The activity of *EFG1* depends on the potential PKA phosphorylation site threonine-206 (Bockmuhl and Ernst, 2001)).

Efg1 is essential for the expression of genes induced during the yeast-to-hypha transition. These include *ECE1*, *HWP1*, *HYR1*, *ALS3*, *RBT1*, and *RBT4*. They encode either cell wall or secreted proteins. Although none of them are required for hyphal morphogenesis, many have been shown to be important for virulence in systemic infections. *HWP1* encodes a cell wall protein that can serve as a substrate for mammalian transglutaminases to form covalent attachments between

C. albicans and host epithelial cells (Staab *et al.*, 1999). *RBT1* encodes a cell wall protein, and *RBT4* encodes a secreted protein similar to the plant pathogenesis-related protein (*PRY1*) (Braun *et al.*, 2000). Both genes are necessary for the full virulence of *C. albicans* in a systemic mouse model. Efg1 is also necessary for the expression of three secreted aspartyl proteinase genes, *SAPs4-6* when hyphal development is induced with serum or media containing polypeptides as the sole nitrogen source. *SAP* genes have been shown to contribute to virulence in both systemic and mucosal *Candida* infections (Monod *et al.*, 1994; Hube, 1996; Sanglard *et al.*, 1997). Therefore, hyphal morphogenesis and expression of virulence factors are co-regulated by the cAMP/PKA pathway in *C. albicans*.

One of the Efg1-regulated genes is *TEC* (Lane *et al.*, 2001) (Figure 1B). Tec1 is a member of the TEA/ATTS family of transcription factors. Its homologues AbaA in *A. nidulans* and *TEC1* in *S. cerevisiae* are involved in the regulation of conidiophore formation and filamentous growth, respectively (Boylan *et al.*, 1987; Gavrias *et al.*, 1996). *tec1/tec1* mutants show a medium-specific defect in hyphal growth, are less virulent (Schweizer *et al.*, 2000), and overexpression of *TEC1* causes hyperfilamentation. Part of the Efg1 function in hyphal development is believed to be mediated through Tec1 because *TEC1* overexpression can partially suppress the *efg1/efg1* mutant phenotype (Lane *et al.*, 2001). Efg1 also negatively regulates its own transcription (Tebarth *et al.*, 2003). Efg1 can *complement sok2* in the regulation of a Sok2 repressed element in *S. cerevisiae,* where Sok2 functions as a negative regulator of meiosis (Shenhar and Kassir, 2001). How Efg1 functions as both a positive regulator of some genes while a negative regulator of others remains to be addressed.

2.3. The Pheromone-Responsive MAPK Pathway

MAP kinases are a family of serine/threonine protein kinases known as mitogen activated protein (MAP) kinases. They are involved in sensing extracellular signals, including growth factors, pheromones, and other changes to regulate growth and differentiation (Dickman and Yarden, 1999). A MAP kinase is activated by a MAP kinase kinase (MAPKK), which itself is activated by a MAP kinase kinase kinase (MAPKKK). The cascades of kinases are highly conserved among organisms from yeast to humans. In multicellular organisms, the MAP kinase cascades are regulated by receptor-tyrosine kinases.

In *S. cerevisiae*, there are five MAP kinase pathways. They regulate mating, pseudohyphal growth, osmotic response, cell wall integrity, and meiosis (Herskowitz, 1995). Among them, the pheromone-responsive MAP kinase pathway of mating is regulated by a G-protein coupled receptor. Activation of the pheromone receptors Ste2 or Ste3 by pheromone causes dissociation of a heterotrimeric G protein subunit G_{α} (Gpa1) from the $G_{\beta\gamma}$ subunits. The $G_{\beta\gamma}$ subunits then activate the cascade of kinases, which consists of Ste20 (a homologue of the mammalian P21-activated kinase, PAK), Ste11 (a MAPKKK homologue), Ste7 (a MAPKK homologue), and the MAP kinases Fus3 and Kss1. The activated MAP kinases induce the activation of the transcription factor Ste12 through the release of the inhibitory activity of Dig1,2. Active Ste12 then induces the transcription of pheromone-responsive genes through the Ste12 homodimer or in cooperation with Mcm1. Many components of the pheromone-responsive MAP kinase pathway, including Ste20, Ste11, Ste7, Kss1 and Ste12, are also required for invasive and filamentous growth in *S. cerevisiae*. The upstream activator of the filamentation MAP kinase is not known. Ras2 has been suggested to activate the pathway via Cdc42 in *S. cerevisiae*. One of the outputs of the pathway is *TEC1*. *TEC1* transcription is regulated by Ste12 and by itself. Together with Ste12, it regulates the expression of genes in pseudohyphal growth in *S. cerevisiae* (Madhani and Fink, 1998).

Many homologs of the filamentation MAP kinase pathway have been identified in *C. albicans*. These include Cst20 (Ste20), Hst7 (Ste7), Cek1 (Kss1), and Cph1 (Ste12) (Figure 1B). Their regulation and functions are believed to be similar to their *S. cerevisiae* homologs, because they can complement the *S. cerevisiae* mutants in mating and pseudohyphal growth. While the MAPK pathway is essential for pseudohyphal morphogenesis in *S. cerevisiae*, it seems to have limited effect on hyphal morphogenesis in *C. albicans*. Mutants of the MAP kinase pathway show suppressed hyphal formation on certain solid media but can still form hyphae in liquid inducing media and in response to serum (Whiteway *et al.*, 1992; Liu *et al.*, 1994; Kohler and Fink, 1996; Leberer *et al.*, 1997a; 1997b; Csank *et al.*, 1998). The upstream regulators for the MAP kinase pathway in filamentation have not been identified. Part of Ras1 function has been suggested to act through the MAP kinase pathway based on epistasis analysis (Sanchez-Martinez and Perez-Martin, 2002).

C. albicans has a pheromone-responsive MAP kinase pathway highly homologous to that of *S. cerevisiae*, and the pathway is essential for mating in *C. albicans*. The α mating pheromone produced by *C. albicans* α cells has been identified, and the gene (Mfα) is required for α cells, but not **a** cells, to mate (Bennett *et al.*, 2003; Lockhart *et al.*, 2003). *C. albicans* also has **a** mating pheromone because disruption of *HST6*, the orthologue of the *STE6* gene which encodes an ABC transporter responsible for secretion of the Mfa pheromone, prevents mating in **a** but not in α cells (Magee *et al.*, 2002). The α pheromone receptor encoded by *STE2* is required for the mating of **a** cells but not α cells (Bennett *et al.*, 2003). The Gα protein for mating is likely Cag1 (Sadhu *et al.*, 1992). Genes encoding for the putative receptor for Mfa and the G protein βγ subunits are also present in *C. albicans*, as revealed by the *C. albicans* genome-sequencing project (Tzung *et al.*, 2001). *HST7* and *CPH1* are essential for mating in *C. albicans* (Chen *et al.*, 2002; Magee *et al.*, 2002). The two MAP kinases, Cek1 and Cek2, have redundant functions in mating, as *cek1/cek1* or *cek2/cek2* single mutants are only partially defectively in mating, while the *cek1 cek2* double mutant is completely blocked in mating (Chen *et al.*, 2002). Pheromone treatment of *C. albicans* cells induces the expression of genes previously implicated in mating in *S. cerevisiae* as well as genes specific to *C. albicans* (Bennett *et al.*, 2003; Tsong *et al.*, 2003). Interestingly, pheromone-induced genes also include some cell surface and secreted proteins (*HWP1, RBT1, RBT4, SAP4-6*) that are known to be induced during the yeast-to-hypha transition and are required for virulence in a mouse model of disseminated candidiasis (Bennett *et al.*, 2003; Lockhart *et al.*, 2003). This indicates that there is considerable overlap among mating, filamentation and pathogenesis in *C. albicans*, and suggests an evolutionary link among these three processes. The pheromone response MAP kinase pathway regulates all three.

3. *Magnaporthe grisea*

3.1. Life Cycle

The heterothallic, haploid, ascomycete *M. grisea* is pathogenic to crops such as barley, wheat, rice and millet (Talbot, 1995). Each year this fungus causes losses of 10-30% of the rice harvest, an important food in many parts of the world (Talbot, 1995). *M. grisea*'s pathogenic life cycle starts from an asexual spore dispersed in moist

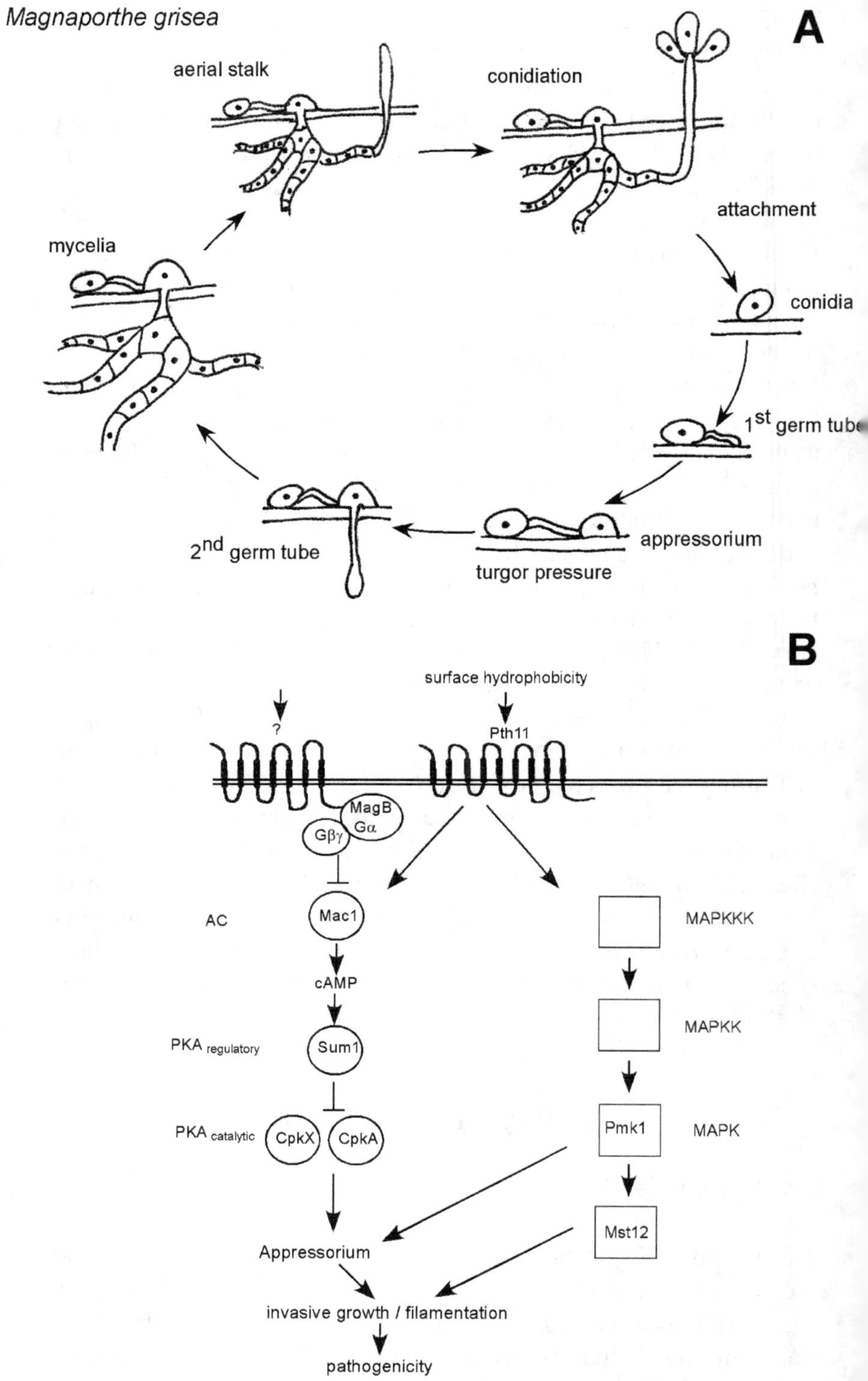

Figure 2. (A) Life cycle of *Magnaporthe grisea*. (B) The conserved cAMP/PKA and MAP kinase pathways (see text for details).

air (Figure 2A). When the spore lands on a leaf, it attaches itself to the cuticle by releasing an adhesive molecule. The spore then germinates, and its short germ tube swells at the tip and becomes flat on the leaf, a process known as hooking (Talbot, 2003). The swelling coincides with a mitotic division, one of the nuclei migrates to the incipient appressorium and the other migrates back to the conidium, a septum forming between them (Talbot, 2003). The formation of the appressorium requires a hydrophobic surface with no nutrients, the exact conditions on a rice leaf. The appressorium wall thickens and turgor pressure is generated by accumulating high concentrations of glycerol, which causes an influx of water from the plant surface (Howard *et al.*, 1991), followed by growth of an invasive hypha, a process called peg penetration for invasion of the plant epidermis (Howard and Valent, 1996). The invading hypha is narrow, but widens as it penetrates before developing into a branched and bulbous mycelium. The secondary hypha infects neighboring cells. Small lesions appear on the plant, accompanied by local chlorosis. This process, as well as production of phytotoxic substances, is thought to be the mechanism of plant death (Howard and Valent, 1996). The mycelia form thousands of conidia, which are disseminated in the air to attach to new hosts (Figure 2A).

There are two mating types in *M. grisea*, *MAT1-1* and *MAT1-2* (Kang *et al.*, 1994). Cells of opposite mating types are able to mate in the proper medium conditions, forming sexual fruiting bodies that contain asci with ascospores, the products of meiosis and mitosis arranged as unordered octads (Valent *et al.*, 1991). The pathogenic isolates found in nature are mostly infertile. Thus, it appears that mating and virulence are not tightly linked in *M. grisea* (Valent *et al.*, 1991).

3.2. The cAMP/PKA Pathway

cAMP is involved in the process of appressorium formation as early as during initial germination. Many mutants in the cAMP/PKA pathway are unable to form appressoria and are therefore avirulent. In addition, high levels of cAMP can cause appressorium formation on non-hydrophobic surfaces (Lee and Dean, 1993; Choi and Dean, 1997) (Figure 2B).

A heterotrimeric G-protein α subunit encoded by *MAGB* has been shown to affect morphogenesis. Its deletion causes reduced vegetative growth and mutant cells are unable to make appressoria and instead make undifferentiated germ tubes that are unable to hook or respond to surface contact (Liu and Dean, 1997). Δ*magB* mutants can be complemented by exogenous cAMP, indicating that *MAGB* operates upstream of the generation of a cAMP signal for appressorium formation (Liu and Dean, 1997). Interestingly, a mutation in *MAGB* that abolishes its GTPase activity causes production of normal appressoria. Furthermore, a mutation in *MAGB* that causes constitutive association of the Gα to the Gβγ complex has little effect on *M. grisea*, and rescues the phenotype of a Δ*magb*. Thus, it is assumed that the Gβγ operates as a negative regulator of appressorium morphogenesis, and deletion of *MAGB* releases the Gβγ which then represses adenylate cyclase and prevents appressorium formation (Fang and Dean, 2000). The receptor and upstream signal for MAGB have not been identified. *M. grisea* also has two more Gα genes *MAGA* and *MAGC*, which are involved in the sexual development of *M. grisea* (Liu and Dean, 1997).

The adenylate cyclase of *M. grisea* is encoded by *MAC1*, and its mutation causes reduced conidiation, decreased vegetative growth, and an inability to form appressoria (Choi and Dean, 1997) (Figure 2B). This phenotype is rescued by the addition of cAMP. In a screen to suppress the Δ*mac1* phenotype, the regulatory subunit of PKA, *SUM1*, was discovered. Δ*mac1 sum1* has accelerated germination and appressorium formation but is still impaired in virulence (Choi and Dean, 1997). Disruption of the catalytic PKA subunit, encoded by *CPKA*, causes *M. grisea* to be defective in appressorium formation, resulting in small nonfunctional appressoria (Mitchell and Dean, 1995). The mutant is unable to penetrate plant tissue but is still invasive once host tissue barriers have been removed. The partial defects associated with Δ*cpka* suggest the existence of an additional PKA catalytic gene that is involved in earlier stage of appressoria development.

A putative membrane-bound sensor or receptor encoded by the *PTH11* gene has been identified in *M. grisea* (Figure 2B). *PTH11* encodes a nine membrane-spanning protein in the plasma membrane that responds to the presence of hydrophobic surfaces to transmit a cue for appressorium formation (DeZwaan *et al.*, 1999). *pth11* mutants fail to make appressoria efficiently on hydrophobic surfaces and are avirulent. The virulence and appressorium formation defects in *pth11* can be suppressed by the addition of cAMP. Therefore, *PTH11* is thought to be upstream of the accumulation of cAMP during appressorium development (DeZwaan *et al.*, 1999).

3.3. The PMK1 MAPK Pathway

The MAP kinase pathway of *M. grisea* that is an equivalent of the pheromone-response MAPK pathway in *S. cerevisiae* is involved in appressoria development and subsequent invasive growth of *M. grisea*. *PMK1* was the first MAPK gene identified from *M. grisea*. It rescues the mating defect in a *fus3 kss*1 mutant of *S. cerevisiae* (Xu and Hamer, 1996). Δ*pmk1* mutants are unable to make appressoria on any surface or in response to cAMP and are nonpathogenic. However, they are still able to respond to cAMP for early stages of appressoria formation (Xu and Hamer, 1996). Hence, *PMK1* is thought to act downstream of or parallel to the initial cAMP-mediated signal for appressorium morphogenesis (Figure 2B). In addition to the defect in appressoria formation, the Δ*pmk1* mutants are also defective in invading the plants even when applied to wounds. However, *PMK1* is not essential for the sexual and asexual cycles in culture.

Because of the similarity of *PMK1* to Fus3/Kss1, the *M. grisea* homolog of Ste12, *MST12*, was characterized. Δ*mst12* mutants can still form appressoria (Park *et al.*, 2002). However, they are unable to invade plants from this appressoria and are thus non-pathogenic. The mutants are defective in spreading disease lesions and infective growth. Thus, *MST12* appears to be required for the later stages of infection.

Although the PMK1 MAP kinase pathway is involved in appressorium formation, both Δ*pmk1* and Δ*mst12* are not defective in mating. In addition, two G*a* genes *MAGA* and *MAGC* involved in mating are not required for appressorium formation. Therefore, mating is not required for pathogenesis. It is possible that PMK1 has a functionally redundant MAPK and only deletion of both can block mating, as in the case of *S. cerevisiae* and *C. albicans*. In this case, the MAGA and MAGC should be upstream of the PMK1 pathway in the mating pathway. The upstream components for the *PMK1* MAP kinase pathway in morphogenesis are not identified yet, but the *PTH11* sensor is a potential candidate. Therefore, both a MAP kinase pathway and the cAMP/PKA pathway regulate morphogenesis, which is directly linked to pathogenesis in *M. grisea*.

4. *Ustilago maydis*

4.1. Life Cycle

U. maydis is the casual agent of corn smut disease, which can cause severe economical losses in agriculture worldwide, with the exception of Mexico, where "huitlacoche", *U. maydis*-infected corn, is an expensive delicacy going back to Aztec times. *U. maydis* is a basidiomycete fungus with a life cycle that includes a dimorphic switch between a saprophytic (= able to survive in the environment on refuse), nonpathogenic, haploid yeast phase and a pathogenic, dikaryotic, filamentous form after mating of two compatible haloids in the host. Therefore, mating and infection are intricately connected in *U. maydis*.

During the saprophytic stage, cigar shaped haploid *U. maydis* cells grow by budding and divide by mitosis, like *S. cerevisiae* (Figure 3A). Haploid cells are unable to invade the plant host and are not pathogenic. To form filaments, two haploid cells must mate by forming conjugation tubes or mating hyphae (Kahmann *et al.*, 1995). Fusing of two conjugation tubes at their tips give rise to infectious hypha with two nuclei, which grows into dikaryotic filaments in the host. Mating compatibility of haploids is determined by allelic differences in the *a* and *b* mating-type loci. The *a* locus encodes both the pheromones (*mfa1* and *mfa2* genes) and pheromone receptors (*pra1* and *pra2* genes on *a1* and *a2*, respectively) (Bolker *et al.*, 1992). The *a* locus controls fusion of haploid cells and maintenance of dikaryotic filamentous growth. The *b* locus has multiple alleles that encode two homeodomain proteins (e.g. bE1 and bW1 for *b1*) (Gillissen *et al.*, 1992). Formation of a heterodimer bE/bW from different *b* alleles is required for the expression of genes necessary for filamentation and pathogenicity. The dikaryotic filaments are pathogenic to corn by causing neoplastic growth which give rise to tumor-like structures. Other symptoms include chlorosis, growth stunting, and pigmentation of the plants. The establishment of the infectious dikaryotic filaments requires not only the appropriate mating loci and mating pheromone but also signals from the host. Without a host, karyogamy occurs in a dikaryon with the formation of diploid cells, which grow as a haploid cells like yeasts (Ruiz-Herrera and Martinez-Espinoza, 1998).

In the host, karyogamy occurs in some dikaryotic mycelia which septate to form teliospores. The teliospores are initially round with a thick cell wall, but the spores mature and become pigmented and

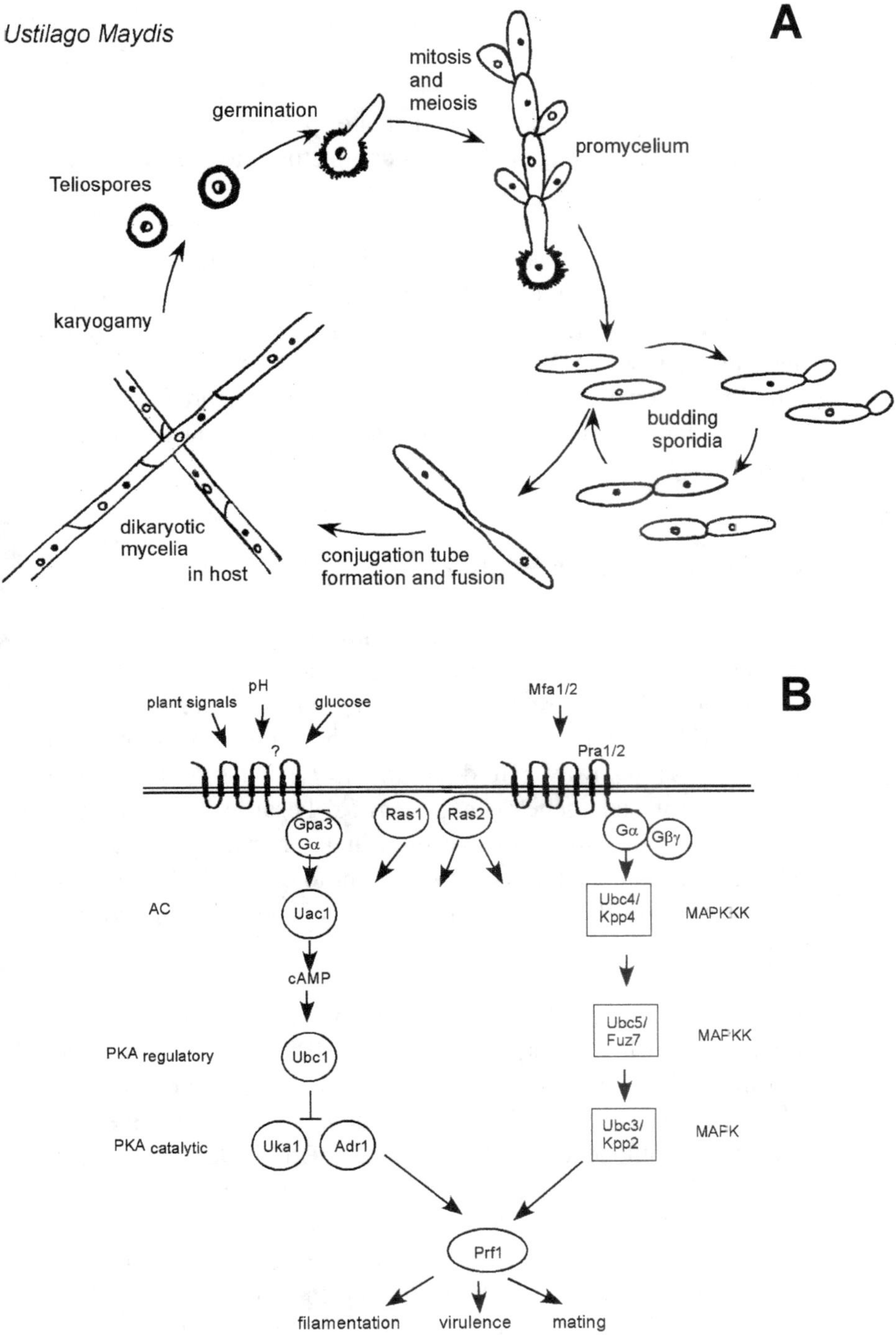

Figure 3. (A) Life cycle of *Ustilago maydis*. (B) The conserved cAMP/PKA and MAP kinase pathways (see text for details).

rough. When the tumors burst, many tiny spores are released that are spread by wind and rain. Germination of the teliospores creates promycelium, and meiosis and mitosis occurs to form basidiospores, which are then able to bud off and create sporidia. These sporidia are capable of yeast-like budding, thus completing the life cycle (Figure 3A).

The signals for pathogenicity and morphogenesis are largely unknown. Cell fusion, normally occurring in the leaf whorl, requires starvation or activated charcoal on plates (Day *et al.*, 1971; Banuett and Herskowitz, 1994). Only after penetration of the host tissue can mitotic division of the dikaryon be observed. Teliospore formation has never been induced *in vitro*. Thus, these processes must require cues provided by the plant.

4.2. The cAMP/PKA Pathway

In *U. maydis*, the cAMP pathway is required for pheromone-responsive transcription, mating and pathogenesis, and negatively regulates filamentation. Therefore, the yeast to hypha transition and mating are controlled both by the cAMP pathway and the pheromone-responsive MAP kinase pathway (Gold *et al.*, 1994). Mating is especially important for pathogenesis in *U. maydis*, because it is required for formation of the invasive mycelium and for pathogenesis. On the other hand, filamentation is not sufficient to cause virulence because some constitutively filamentous mutants of the cAMP pathway are avirulent.

U. maydis is amenable to genetic analysis and thus has been a model system for dimorphism (Banuett, 1995). Characterization of constitutively filamentous mutants led to the identification of the *uac1* adenylate cyclase gene (Figure 3B). The *uac1* mutant is defective in mating and pathogenesis. Addition of cAMP to the mutant suppresses filamentation and produces a budding phenotype. The *ubc1* regulatory subunit of PKA (*ubc* = *Ustilago* bypass of cyclase) was isolated as a suppressor of the *uac1* mutation. *ubc1* strains grow as yeast (Gold *et al.*, 1994). Two catalytic subunits of the PKA were identified: *adr1* and *uka*1. Deletion of *adr1* causes constitutive filamentation similar to that of *uac1*, as well as impairment in mating and pathogenicity. The *adr1* mutant also suppresses the yeast-like colony morphology of *ubc1*, causing filamentous growth consistent with the Adr1 catalytic subunit acting downstream of Ubc1 regulatory subunit. In contrast to *adr1*, deletion of *uka1* has little effect, suggesting that uka1 plays a minor role (Durrenberger *et al.*, 1998).

Four Gα subunits are identified in *U. maydis* by reverse genetics. Among the four Gαs, only *gpa3* is required for mating and pathogenesis. A *gpa3* mutant is also constitutively filamentous similar to *uac1*. Although Gpa3 was originally thought to function in the pheromone-induced MAP kinase pathway, cAMP suppression of its filamentous morphology and sterility subsequently revealed that Gpa3 acts as the G-protein in the cAMP pathway (Kruger *et al.*, 1998) (Figure 3B). The ability to form filaments *in vitro* is not sufficient for pathogenesis as several mutants of the cAMP pathway, *uac1*, *adr1*, and *gpa3*, are constitutively filamentous and avirulent.

One target of the PKA in *U. maydis* is the transcription factor Prf1 (pheromone response factor) (Hartmann *et al.*, 1996). *prf1* mutants do not express the *a* and *b* genes and are sterile. The disruption of *prf1* also results in a loss of pathogenicity and filamentation. Prf1 recognizes pheromone response elements (PREs) present in the regulatory regions of the *a* and *b* mating type genes. Constitutive expression of the *b* genes restores pathogenicity and induces filamentous growth in the absence of pheromone signal. Prf1 is regulated on the transcriptional level through its upstream activating sequences by nutrient signaling and PREs. Prf1 protein is also regulated post-transcriptionally through phosphorylation by the Adr1 PKA and the MAP kinase Kpp2 (Hartmann *et al.*, 1999). The PKA phosphorylation sites in Prf1 are essential for the induction of *a* and *b* mating type genes, whereas the MAPK phosphorylation sites are required for pheromone responsive *b* genes (Kaffarnik *et al.*, 2003). This illustrates how mating in *U. maydis* is regulated by both the cAMP pathway and the pheromone-responsive MAP kinase pathway.

4.3. The Pheromone-Responsive MAP Kinase Pathway

The mating pheromones and receptors are encoded by the mating type *a* loci and the binding of a pheromone to a G-coupled receptor will activate the pheromone responsive MAP kinase pathway, although the G-protein for the pathway has yet to be determined (Figure 3B). The MAPKK homologue *fuz7/ubc5* is involved in conjugation tube formation, filament formation, and maintenance as well as tumor formation (Banuett and Herskowitz, 1994). Deletion of *ubc3/kpp2*, a MAP kinase homologue, causes reduced production of mating pheromone and a reduced level of mating and pathogenicity (Muller *et al.*, 1999). It also has a strong tendency to form dikaryotic hyphae (Mayorga and Gold, 1999; Muller *et al.*, 1999). Because *ubc3* mutants are not entirely sterile, additional MAP kinases may function in parallel with the *ubc3* to regulate mating and pathogenesis (Mayorga and Gold, 1999; Muller *et al.*, 1999). These two genes and several additional genes in the MAP kinase cascade were also identified as suppressors of *uac1*. *ubc4/kpp4* encodes the MAPKKK and is required for mating, filamentous growth and pathogenicity. *ubc2* is required for pheromone responsive morphogenesis and has been shown to genetically interact with ubc4 (Figure 3B). Since it has homology to SH3 protein, it is thought to be an important adaptor protein acting directly upstream of the MAPK cascade (Mayorga and Gold, 2001). The fact that components of the MAPK pathway are discovered in a screen for suppressors of the adenylate cyclase mutant supports the integrated regulation of mating and filamentation by both the cAMP pathway and the MAPK pathway. Both pathways are required for mating and pathogenicity via Prf1, but the cAMP pathway inhibits filamentation, while the MAPK pathway activates filamentation.

Another point of cross-talk between the mating MAPK and cAMP pathways in *U. maydis* is at the level of two Ras proteins, Ras1 and Ras2. Ras2 is identified in a screen for suppressors of the filamentous phenotype of *adr1* mutant (Lee and Kronstad, 2002). Expression of dominant active alleles of *ras1* and *ras2* showed that Ras2 induces filamentous growth through the MAPK cascade involving Fuz7 and Ubc3, while Ras1 does not affect cell morphology but elevates pheromone gene expression. (Lee and Kronstad, 2002; Muller *et al.*, 2003). Deletion of *ras2* causes altered morphology of yeast cells and reduced pathogenicity and mating (Lee and Kronstad, 2002). Thus, it appears that Ras1 and Ras2 have different roles in morphogenesis in *U. maydis*, but both play a role in mating (Figure 3B).

Both the cAMP pathway and the pheromone responsive MAP kinase pathway regulate mating. Mating is required for subsequent dikaryon mycelium formation, which causes diseases in plants. Therefore, mating, dikaryon mycelia growth, and pathogenesis are regulated tightly by both pathways.

5. *Cryptococcus neoformans*

5.1. Life Cycle

C. neoformans is a human fungal pathogen. It infects the central nervous system and causes meningoencephalitis, which is fatal if not treated. It is largely opportunistic, occurring most frequently in patients with AIDS or other immunocompromising conditions (Hull and Heitman, 2002). *C. neoformans* is a basidiomycete, and, therefore, is more similar to *U. maydis* than to the ascomycete *S. cerevisiae* (Hull and Heitman, 2002). *C. neoformans* is ubiquitous in the environment and is most commonly found in pigeon guano. Humans are exposed to it by inhalation of small round spores, which are deposited in the lung alveoli. The infection spreads to the blood stream and then to the central nervous system, the most common site of clinical presentation (Hull and Heitman, 2002).

There are 5 serotypes: A, B, C, D and AD. Seroytpe A is the most common cause of human disease. It exists as a haploid with two mating types: **a** and α. The environmental and clinical isolates are usually α, and *MAT*α cells are innately more virulent than ***MATa*** cells (Kwon-Chung and Bennett, 1978). Its life cycle includes vegetative growth by budding that can undergo a morphogenetic change to filamentous form by mating or haploid fruiting. During mating, partners form conjugation tubes and fuse in response to opposite mating pheromones (Figure 4A). In contrast to *S. cerevisiae*, karyogamy is delayed and the heterokaryon becomes filamentous. Ultimately, basidia are formed at the termini of the dikaryotic hyphae, and nuclear fusion and meiosis occur in the basidia. Spores are produced external to the basidium, typical of basidiomycetes. In haploid fruiting, haploid, α-cells only form filaments under nitrogen starvation and desiccation conditions, independently of mating pheromone (Kwon-Chung, 1975; 1976). *C. neoformans* produces two virulence factors in response to host conditions - the polysaccharide capsule and the pigment melanin (Hull and Heitman, 2002). The production of virulence factors shares the same signaling pathways that control mating and filamentation in *C. neoformans*. The defined sexual cycle, existence of haploid growth phase, and efficient homologous recombination make *C. neoformans* an attractive genetic system for studying dimorphic pathogenic fungi (Hull and Heitman, 2002).

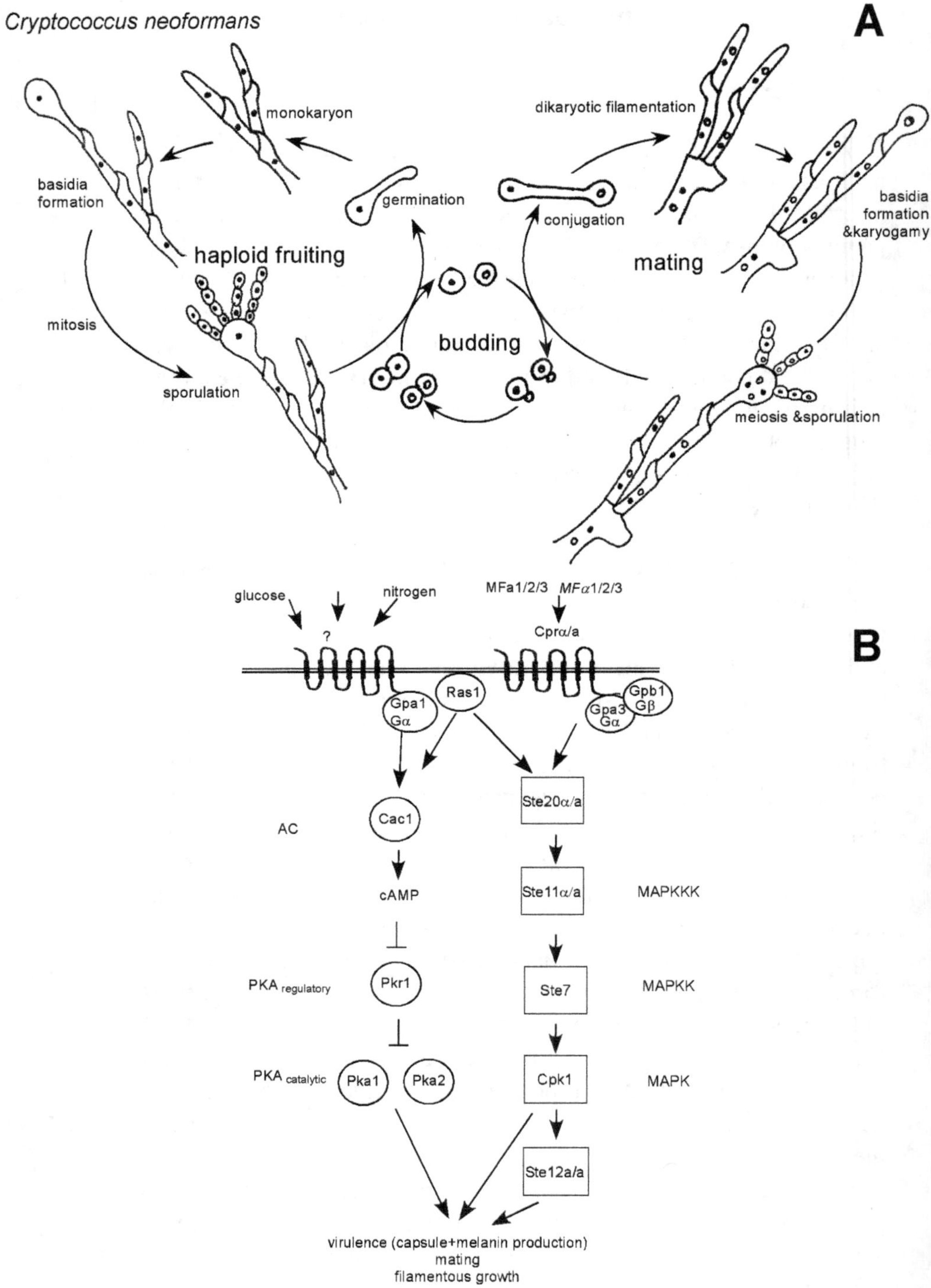

Figure 4. (A) Life cycle of *Cryptococcus neoformans*. (B) The conserved cAMP/PKA and MAP kinase pathways (see text for details).

5.2. The cAMP/PKA Pathway

The cAMP signaling pathway is important in the production of the virulence factors capsule and melanin, and mating. The *GPA1* gene of *C. neoformans* is highly homologous to *U. maydis gpa3* and *S. cerevisiae* Gpa2 (Gα) (Kubler *et al.*, 1997; Lorenz and Heitman, 1997; Pan and Heitman, 1999) (Figure 4B). The *gpa1* mutant is defective in mating and the production of melanin and capsule and is avirulent in a rabbit model (Alspaugh *et al.*, 1997). The defects can be rescued by the addition of cAMP. The adenylate cyclase gene *CAC1* of *C. neoformans* is 51% identical to *U. maydis uac1* gene (Alspaugh *et al.*, 2002). *cac1* mutant strains are viable, have no vegetative growth defect, but are unable to mate or produce capsule and melanin and are thus avirulent (Alspaugh *et al.*, 2002). The *cac1* mutant strains maintain the yeast-like morphology during haploid fruiting, and therefore, are defective in morphogenesis, as the cyclase mutant in *C. albicans*. The phenotypes of *cac1* can be suppressed by the addition of cAMP (Alspaugh *et al.*, 2002).

C. neoformans has two PKA catalytic subunits *PKA1* and *PKA2* and one regulatory subunit *PKR1* (D'Souza *et al.*, 2001). In serotype A, *pka1* mutants are similar to *gpa1* and *cac*1 mutants in that they are sterile, cannot produce capsule and melanin, and are avirulent (D'Souza *et al.*, 2001). In serotype D, *pka1* mutants have no phenotype, can produce capsule and melanin, and are virulent. However, *pka2* mutants in serotype D have similar phenotypes as *pka1* mutants in serotype A (D'Souza *et al.*, 2001). Deletion of the regulatory subunit leads to constitutive activation of the PKA. As expected, *pkr1* mutants overproduce capsule, and are hypervirulent (Hull and Heitman, 2002). The *pkr1 pka1* double mutant has similar phenotypes as the *pka1* mutants, thus Pka1 functions downstream of the Pkr1 regulatory subunit (D'Souza *et al.*, 2001).

5.3. The Pheromone-Responsive MAPK Pathway

The pheromone-responsive MAPK pathway is essential for mating and haploid fruiting in *C. neoformans*. Its role in virulence, however, is less clear. Many components of the MAP kinase pathway are encoded by the mating type **a** and α loci. Three genes, *MFα1, 2* and *3*, encode the α pheromones in α cells (Figure 4B). Deletion of these genes results in a pheromone-less α strain that is impaired in mating

(Shen *et al.*, 2002). Three related genes encoding the Mf**a** pheromones in **a** cells have also been discovered (McClelland *et al.*, 2002). Two putative seven-membrane receptor proteins, CPRα and CPR**a,** have been identified, and their deletion exhibits a drastic reduction in mating efficiency and virulence (Chung *et al.*, 2002; Chang *et al.*, 2003). Two G-protein coupled receptors, that likely function in pheromone sensing, are encoded by *MAT***a** and *MAT*α loci. They are similar to Ste3 of *S. cerevisiae* in sequence. The Gβ unit of the receptor-coupled G-protein, Gpb1, is critical for mating. The Gpb1 mutant is sterile, and the phenotype cannot be suppressed by the addition of cAMP but can be rescued by activating the MAP kinase pathway (Wang *et al.*, 2000). The kinases in the MAP kinase cascade include the Ste20α (Ste20**a**), Ste11α (Ste11**a**), Ste7 and Cpk1. Deletions of *ste20*, *ste11*, *ste7* and *cpk1* are sterile, unable to respond to pheromones, and defective in haploid fruiting, but are nearly fully virulent (Clarke *et al.*, 2001; Davidson *et al.*, 2003). Epistasis studies have shown that they function in analogous positions to their *S. cerevisiae* homologues in the pathway (Davidson *et al.*, 2003) (Figure 4B).

A potential target of the MAPK pathway is the transcription factor Ste12, which exists as Ste12**a** and Ste12α (Wickes *et al.*, 1997; Yue *et al.*, 1999; Karos *et al.*, 2000; Chang *et al.*, 2001). A *ste12*α mutant only displays a slight defect in mating, but has reduced virulence and is defective in haploid fruiting. The defect in haploid fruiting can be suppressed by overexpression of *MF*α, dominant active Ste11α, or overexpression of Cpk1. Although Ste12 is not required for mating, overexpression of *STE12* can bypass mutations in *ste11, ste7* and *cpk1*. Based on these findings, Ste12 is suggested to be either one of two redundant targets of the MAPK pathway in mating (Davidson *et al.*, 2003) or acts in a pathway parallel to the MAPK pathway.

RAS1 has been implicated in the functions of both the cAMP pathway and the pheromone-responsive MAPK pathways (Alspaugh *et al.*, 2000) (Figure 4B). A *ras1* mutant is viable, but unable to grow at 37°C and is deficient for mating. Addition of cAMP partially suppresses the mating defect. Overexpression of *GPB1* of the MAPK pathway fully suppresses the mating defect. In addition, a dominant active *RAS1* can cause haploid fruiting, and the phenotype is blocked by a Ste12α mutation (Alspaugh *et al.*, 2000). Recently, Ras1 was shown to be required for the transcriptional induction of elements of the pheromone response pathway, and its role in the mating response is through the pheromone responsive MAP kinase pathway

(Waugh *et al.*, 2003). In addition to Ras1, *C. neoformans* has another functionally redundant *RAS* gene, *RAS2*. RAS2 has limited functions in *C. neoformans*. A *ras2* mutation shows no alterations in vegetative growth rate, mating differentiation or virulence factor expression (Waugh *et al.*, 2002). *A ras1 ras2* double mutant strain is viable, but grows slower than the *ras1* single mutant.

C. neoformans is very much like *U. maydis* in that both the cAMP pathway and the pheromone-responsive MAP kinase pathway are required for mating and the MAP kinase pathway also regulates filamentation. Virulence factors, however, are only regulated by the cAMP pathway in *C. neoformans*.

6. Closing Remarks

A comparison of the two highly conserved signaling pathways in four fungi shows that ubiquitous signaling elements are used in diverse ways in the regulation of filamentation, mating and pathogenesis. In *C. albicans*, both pathways positively regulate hyphal development and virulence, and the MAP kinase also controls mating. Some virulence genes are expressed in hyphae and during mating. In *M. grisea*, both pathways regulate appressorium morphogenesis and virulence. Virulence is tightly associated with morphogenesis, but not with mating. In *U. maydis*, both pathways converge on a transcription factor to regulate mating, virulence and filamentation. However, the cAMP pathway inhibits filamentation, and the MAP kinase pathway is required for filamentation. In *C. neoformans*, the cAMP pathway is critical for mating and production of virulence factors, whereas the MAP kinase pathway is required for mating and filamentation. Interactions between the two conserved signaling pathways incellular development seem to be a common theme in these fungi.

Signaling pathways that regulate filamentation and mating are closely coupled in filamentous fungi. In *C. albicans*, components of the MAP kinase pathway are important for both mating and filamentation. In *U. maydis*, the entire MAP kinase pathway is involved in mating and formation of dikaryotic mycelia. In *C. neoformans*, the MAP kinase pathway is essential for filamentation during mating or haploid fruiting. The shared signaling components between filamentation and mating may reflect evolutionary origins, given that mating and filamentation both involve polarized growth. In fact, the mechanism of polarity establishment seems similar between the formation of mating projections and apical expansion in hyphae (Hazan and Liu, 2002). Identification of the targets of the signaling pathways responsible for the polarity establishment in these two processes will bring this into sharper focus.

Finally, virulence is controlled by the same signaling pathways that regulate filamentation and mating in two human pathogens and two plant pathogens. This suggests an evolutionary link between pathogenesis and cellular development (mating and filamentation) in these fungi. It is possible that these pathogens have evolved to use these ubiquitous signaling pathways to sense diverse cues in their unique host environments for their survival and pathogenesis.

7. Acknowledgement

We thank Heather Chen for her help in drawing the fungal life cycles. The research from authors' laboratory is supported by National Institute of Health Grant GM/AI 55155.

References

Alspaugh, J.A., Cavallo, L.M., Perfect, J.R., and Heitman, J. 2000. *RAS1* regulates filamentation, mating and growth at high temperature of *Cryptococcus neoformans*. Mol. Microbiol. 36: 352-365.

Alspaugh, J.A., Perfect, J.R., and Heitman, J. 1997. *Cryptococcus neoformans* mating and virulence are regulated by the G-protein alpha subunit *GPA1* and cAMP. Genes Dev. 11: 3206-3217.

Alspaugh, J.A., Pukkila-Worley, R., Harashima, T., Cavallo, L.M., Funnell, D., Cox, G.M., Perfect, J.R., Kronstad, J.W., and Heitman, J. 2002. Adenylyl cyclase functions downstream of the Galpha protein Gpa1 and controls mating and pathogenicity of *Cryptococcus neoformans*. Eukaryot. Cell 1: 75-84.

Bahn, Y. S., and Sundstrom, P. 2001. *CAP1*, an adenylate cyclase-associated protein gene, regulates bud-hypha transitions, filamentous growth, and cyclic AMP levels and is required for virulence of *Candida albicans*. J. Bacteriol. 183: 3211-3223.

Banuett, F. 1995. Genetics of *Ustilago maydis*, a fungal pathogen that induces tumors in maize. Annu. Rev. Genet. 29: 179-208.

Banuett, F., and Herskowitz, I. 1994. Identification of *fuz7*: a *Ustilago maydis* MEK/MAPKK homolog required for a-locus-dependent and -independent steps in the fungal life cycle. Genes Dev. 8: 1367-1378.

Bennett, R.J., and Johnson, A.D. 2003. Completion of a parasexual cycle in *Candida albicans* by induced chromosome loss in tetraploid strains. EMBO J. 22: 2505-2515.

Bennett, R.J., Uhl, M.A., Miller, M.G., and Johnson, A.D. 2003. Identification and characterization of a *Candida albicans* mating pheromone. Mol. Cell. Biol. 23: 8189-8201.

Bignell, E., Rogers, T., and haynes, K. 2004. Host recognition by fungal pathogens. In: Pathogenic Fungi: Host Interactions and Emerging Strategies for Control. G. San-Blas, and R.A. Calderone, eds. Caister Academic Press, Wymondham, Norfolk, UK. p. 3-48.

Bockmuhl, D.P., and Ernst, J.F. 2001. A potential phosphorylation site for an A-type kinase in the Efg1 regulator protein contributes to hyphal morphogenesis of *Candida albicans*. Genetics 157: 1523-1530.

Bockmuhl, D.P., Krishnamurthy, S., Gerads, M., Sonneborn, A., and Ernst, J.F. 2001. Distinct and redundant roles of the two protein kinase A isoforms Tpk1p and Tpk2p in morphogenesis and growth of *Candida albicans*. Mol. Microbiol. 42: 1243-1257.

Bolker, M., Urban, M., and Kahmann, R. 1992. The a mating type locus of *Ustilago maydis* specifies cell signaling components. Cell 68: 441-450.

Boylan, M.T., Mirabito, P.M., Willett, C.E., Zimmerman, C.R., and Timberlake, W.E. 1987. Isolation and physical characterization of three essential conidiation genes from *Aspergillus nidulans*. Mol. Cell. Biol. 7: 3113-3118.

Braun, B.R., Head, W.S., Wang, M.X., and Johnson, A.D. 2000. Identification and characterization of *TUP1*-regulated genes in *Candida albicans*. Genetics 156: 31-44.

Brown, A. J., and Gow, N. A. 1999. Regulatory networks controlling *Candida albicans* morphogenesis. Trends Microbiol. 7: 333-338.

Calderone, R.A., and Fonzi, W.A. 2001. Virulence factors of *Candida albicans*. Trends Microbiol. 9: 327-335.

Chang, Y.C., Miller, G.F., and Kwon-Chung, K.J. 2003. Importance of a developmentally regulated pheromone receptor of *Cryptococcus neoformans* for virulence. Infect. Immun. 71: 4953-4960.

Chang, Y.C., Penoyer, L.A., and Kwon-Chung, K.J. 2001. The second *STE12* homologue of *Cryptococcus neoformans* is *MATa*-specific and plays an important role in virulence. Proc. Natl. Acad. Sci. USA 98: 3258-3263.

Chauhan, N., and Calderone. 2004. Adaption to oxidative, nutrient and pH stress by human pathogenic fungi. In: Pathogenic Fungi: Host Interactions and Emerging Strategies for Control. G. San-Blas, and R.A. Calderone, eds. Caister Academic Press, Wymondham, Norfolk, UK. p. 129-166.

Chen, J., Lane, S., and Liu, H. 2002. A conserved mitogen-activated protein kinase pathway is required for mating in *Candida albicans*. Mol. Microbiol. 46: 1335-1344.

Choi, W., and Dean, R.A. 1997. The adenylate cyclase gene *MAC1* of *Magnaporthe grisea* controls appressorium formation and other aspects of growth and development. Plant Cell 9: 1973-1983.

Chung, S., Karos, M., Chang, Y.C., Lukszo, J., Wickes, B.L., and Kwon-Chung, K.J. 2002. Molecular analysis of *CPR*alpha, a *MAT*alpha-specific pheromone receptor gene of *Cryptococcus neoformans*. Eukaryot. Cell 1: 432-439.

Clarke, D.L., Woodlee, G.L., McClelland, C.M., Seymour, T.S., and Wickes, B.L. 2001. The *Cryptococcus neoformans STE11*alpha gene is similar to other fungal mitogen-activated protein kinase kinase kinase (MAPKKK) genes but is mating type specific. Mol. Microbiol. 40: 200-213.

Csank, C., Schroppel, K., Leberer, E., Harcus, D., Mohamed, O., Meloche, S., Thomas, D.Y., and Whiteway, M. 1998. Roles of the *Candida albicans* mitogen-activated protein kinase homolog, Cek1p, in hyphal development and systemic candidiasis. Infect. Immun. 66: 2713-2721.

Davidson, R.C., Nichols, C.B., Cox, G.M., Perfect, J.R., and Heitman, J. 2003. A MAP kinase cascade composed of cell type specific and non-specific elements controls mating and differentiation of the fungal pathogen *Cryptococcus neoformans*. Mol. Microbiol. 49: 469-485.

Day, P.R., Anagnostakis, S.L., and Puhalla, J.E. 1971. Pathogenicity resulting from mutation at the *b* locus of *Ustilago maydis*. Proc. Natl. Acad. Sci. USA 68: 533-535.

DeZwaan, T.M., Carroll, A.M., Valent, B., and Sweigard, J.A. 1999. *Magnaporthe grisea* Pth11p is a novel plasma membrane protein that mediates appressorium differentiation in response to inductive substrate cues. Plant Cell 11: 2013-2030.

Dickman, M.B., and Yarden, O. 1999. Serine/threonine protein kinases and phosphatases in filamentious fungi. Fungal Genet. Biol. 26: 99-117.

D'Souza, C.A., Alspaugh, J.A., Yue, C., Harashima, T., Cox, G.M., Perfect, J.R., and Heitman, J. 2001. Cyclic AMP-dependent protein kinase controls virulence of the fungal pathogen *Cryptococcus neoformans*. Mol. Cell. Biol. 21: 3179-3191.

Durrenberger, F., Wong, K., and Kronstad, J.W. 1998. Identification of a cAMP-dependent protein kinase catalytic subunit required for virulence and morphogenesis in *Ustilago maydis*. Proc. Natl. Acad. Sci. USA 95: 5684-5689.

Edmond, M.B., Wallace, S.E., and McClish, D.K. 1999. Nosocomial bloodstream infections in United States hospitals: A three year analysis. Clin. Infect. Dis. 29: 239-244.

Fang, E.G., and Dean, R.A. 2000. Site-directed mutagenesis of the *magB* gene affects growth and development in *Magnaporthe grisea*. Mol. Plant Microbe Interact. 13: 1214-1227.

Feldmesser, M. 2004. Virulence factors that protect the fungus from elimination. In: Pathogenic Fungi: Host Interactions and Emerging Strategies for Control. G. San-Blas, and R.A. Calderone, eds. Caister Academic Press, Wymondham, Norfolk, UK. p. 49-96

Feng, Q., Summers, E., Guo, B., and Fink, G. 1999. Ras signaling is required for serum-induced hyphal differentiation in *Candida albicans*. J. Bacteriol. 181: 6339-6346.

Gavrias, V., Andrianopoulos, A., Gimeno, C.J., and Timberlake, W.E. 1996. *Saccharomyces cerevisiae TEC1* is required for pseudohyphal growth. Mol. Microbiol. 19: 1255-1263.

Gillissen, B., Bergemann, J., Sandmann, C., Schroeer, B., Bolker, M., and Kahmann, R. 1992. A two-component regulatory system for self/non-self recognition in *Ustilago maydis*. Cell 68: 647-657.

Giusani, A.D., Vinces, M., and Kumamoto, C.A. 2002. Invasive filamentous growth of *Candida albicans* is promoted by Czf1p-dependent relief of Efg1p-mediated repression. Genetics 160: 1749-1753.

Gold, S., Duncan, G., Barrett, K., and Kronstad, J. 1994. cAMP regulates morphogenesis in the fungal pathogen *Ustilago maydis*. Genes Dev. 8: 2805-2816.

Gooday, G.W., and Adams, D.J. 1993. Sex hormones and fungi. Adv. Microb. Physiol. 34: 69-145.

Gow, N.A., Perera, T.H., Sherwood-Higham, J., Gooday, G.W., Gregory, D.W., and Marshall, D. 1994. Investigation of touch-sensitive responses by hyphae of the human pathogenic fungus *Candida albicans*. Scanning Microsc. 8: 705-710.

Hartmann, H.A., Kahmann, R., and Bolker, M. 1996. The pheromone response factor coordinates filamentous growth and pathogenicity in *Ustilago maydis*. EMBO J. 15: 1632-1641.

Hartmann, H.A., Kruger, J., Lottspeich, F., and Kahmann, R. 1999. Environmental signals controlling sexual development of the corn smut fungus *Ustilago maydis* through the transcriptional regulator Prf1. Plant Cell 11: 1293-1306.

Hazan, I., and Liu, H. 2002. Hyphal tip-associated localization of Cdc42 is F-actin dependent in *Candida albicans*. Eukaryot. Cell 1: 856-864.

Herskowitz, I. 1995. MAP kinase pathways in yeast: for mating and more. Cell 80: 187-197.

Howard, R.J., Ferrari, M.A., Roach, D.H., and Money, N.P. 1991. Penetration of hard substrates by a fungus employing enormous turgor pressures. Proc. Natl. Acad. Sci. USA 88: 11281-11284.

Howard, R.J., and Valent, B. 1996. Breaking and entering: host penetration by the fungal rice blast pathogen *Magnaporthe grisea*. Annu. Rev. Microbiol. 50: 491-512.

Hube, B. 1996. *Candida albicans* secreted aspartyl proteinases. Curr. Top. Med. Mycol. 7: 55-69.

Hull, C.M., and Heitman, J. 2002. Genetics of *Cryptococcus neoformans*. Annu. Rev. Genet. 36: 557-615.

Hull, C.M., and Johnson, A.D. 1999. Identification of a mating type-like locus in the asexual pathogenic yeast *Candida albicans*. Science 285: 1271-1275.

Hull, C.M., Raisner, R.M., and Johnson, A.D. 2000. Evidence for mating of the "asexual" yeast *Candida albicans* in a mammalian host. Science 289: 307-310.

Jung, W.H., and Stateva, L.I. 2003. The cAMP phosphodiesterase encoded by Ca*PDE2* is required for hyphal development in *Candida albicans*. Microbiology 149: 2961-2976.

Kaffarnik, F., Muller, P., Leibundgut, M., Kahmann, R., and Feldbrugge, M. 2003. PKA and MAPK phosphorylation of Prf1 allows promoter discrimination in *Ustilago maydis*. EMBO J. 22: 5817-5826.

Kahmann, R., Romeis, T., Bolker, M., and Kamper, J. 1995. Control of mating and development in *Ustilago maydis*. Curr. Opin. Genet. Dev. 5: 559-564.

Kang, S., Chumley, F.G., and Valent, B. 1994. Isolation of the mating-type genes of the phytopathogenic fungus *Magnaporthe grisea* using genomic subtraction. Genetics 138: 289-296.

Kao, A.S., Brandt, M.E., Pruitt, W.R., Conn, L.A., Perkins, B.A., Stephens, D.S., Baughman, W.S., Reingold, A.L., Rothrock, G.A., Pfaller, M.A., *et al.* 1999. The epidemiology of candidemia in two United States cities: results of a population-based active surveillance. Clin. Infect. Dis. 29: 1164-1170.

Karos, M., Chang, Y.C., McClelland, C.M., Clarke, D.L., Fu, J., Wickes, B.L., and Kwon-Chung, K.J. 2000. Mapping of the *Cryptococcus neoformans MAT*alpha locus: presence of mating type-specific mitogen-activated protein kinase cascade homologs. J. Bacteriol. 182: 6222-6227.

Kohler, J.R., and Fink, G.R. 1996. *Candida albicans* strains heterozygous and homozygous for mutations in mitogen-activated protein kinase signaling components have defects in hyphal development. Proc. Natl. Acad. Sci. USA 93: 13223-13228.

Kruger, J., Loubradou, G., Regenfelder, E., Hartmann, A., and Kahmann, R. 1998. Crosstalk between cAMP and pheromone signalling pathways in *Ustilago maydis*. Mol. Gen. Genet. 260: 193-198.

Kubler, E., Mosch, H.U., Rupp, S., and Lisanti, M.P. 1997. Gpa2p, a G-protein alpha-subunit, regulates growth and pseudohyphal development in *Saccharomyces cerevisiae* via a cAMP-dependent mechanism. J. Biol. Chem. 272: 20321-20323.

Kwon-Chung, K.J. 1975. A new genus, *Filobasidiella*, the perfect state of *Cryptococcus neoformans*. Mycologia 67: 1197-1200.

Kwon-Chung, K.J. 1976. Morphogenesis of *Filobasidiella neoformans*, the sexual state of *Cryptococcus neoformans*. Mycologia 68: 821-833.

Kwon-Chung, K.J., and Bennett, J.E. 1978. Distribution of alpha and alpha mating types of *Cryptococcus neoformans* among natural and clinical isolates. Am. J. Epidemiol. 108: 337-340.

Lane, S., Zhou, S., Pan, T., Dai, Q., and Liu, H. 2001. The basic helix-loop-helix transcription factor Cph2 regulates hyphal development in *Candida albicans* partly via *TEC1*. Mol. Cell. Biol. 21: 6418-6428.

Leberer, E., Thomas, D.Y., and Whiteway, M. (1997a). Pheromone signalling and polarized morphogenesis in yeast. Curr. Opin. Genet. Dev. 7: 59-66.

Leberer, E., Ziegelbauer, K., Schmidt, A., Harcus, D., Dignard, D., Ash, J., Johnson, L., and Thomas, D.Y. (1997b). Virulence and hyphal formation of *Candida albicans* require the Ste20p- like protein kinase CaCla4p. Curr. Biol. 7: 539-546.

Lee, N., and Kronstad, J.W. 2002. ras2 controls morphogenesis, pheromone response, and pathogenicity in the fungal pathogen *Ustilago maydis*. Eukaryot. Cell 1: 954-966.

Lee, Y.H., and Dean, R.A. 1993. cAMP regulates infection structure formation in the plant pathogenic fungus *Magnaporthe grisea*. Plant Cell 5: 693-700.

Lengeler, K.B., Davidson, R.C., D'Souza, C., Harashima, T., Shen, W.C., Wang, P., Pan, X., Waugh, M., and Heitman, J. 2000. Signal transduction cascades regulating fungal development and virulence. Microbiol. Mol. Biol. Rev. 64: 746-785.

Liu, H., Kohler, J., and Fink, G.R. 1994. Suppression of hyphal formation in *Candida albicans* by mutation of a *STE12* homolog [published erratum appears in Science 1995 267: 17]. Science 266: 1723-1726.

Liu, H. 2002. Co-regulation of pathogenesis with dimorphism and phenotypic switching in *Candida albicans*, a commensal and a pathogen. Int. J. Med. Microbiol. 292: 299-311.

Liu, S., and Dean, R.A. 1997. G protein alpha subunit genes control growth, development, and pathogenicity of *Magnaporthe grisea*. Mol. Plant Microbe Interact. 10: 1075-1086.

Lo, H.J., Kohler, J.R., DiDomenico, B., Loebenberg, D., Cacciapuoti, A., and Fink, G.R. 1997. Nonfilamentous *Candida albicans* mutants are avirulent. Cell 90: 939-949.

Lockhart, S.R., Pujol, C., Daniels, K.J., Miller, M.G., Johnson, A.D., Pfaller, M.A., and Soll, D.R. 2002. In *Candida albicans*, white-opaque switchers are homozygous for mating type. Genetics 162: 737-745.

Lockhart, S.R., Zhao, R., Daniels, K.J., and Soll, D.R. 2003. Alpha-pheromone-induced "shmooing" and gene regulation require white-opaque switching during *Candida albicans* mating. Eukaryot. Cell 2: 847-855.

Lorenz, M.C., and Heitman, J. 1997. Yeast pseudohyphal growth is regulated by *GPA*2: a G protein alpha homolog. EMBO J. 16: 7008-7018.

Madhani, H.D., and Fink, G.R. 1998. The riddle of MAP kinase signaling specificity. Trends Genet. 14: 151-155.

Magee, B.B., Legrand, M., Alarco, A.M., Raymond, M., and Magee, P.T. 2002. Many of the genes required for mating in *Saccharomyces cerevisiae* are also required for mating in *Candida albicans*. Mol. Microbiol. 46: 1345-1351.

Magee, B.B., and Magee, P.T. 2000. Induction of mating in *Candida albicans* by construction of *MTL*a and *MTL*alpha strains. Science 289: 310-313.

Mayorga, M.E., and Gold, S.E. 1999. A MAP kinase encoded by the ubc3 gene of *Ustilago maydis* is required for filamentous growth and full virulence. Mol. Microbiol. 34: 485-497.

Mayorga, M.E., and Gold, S.E. 2001. The ubc2 gene of *Ustilago maydis* encodes a putative novel adaptor protein required for filamentous growth, pheromone response and virulence. Mol. Microbiol. 41: 1365-1379.

McClelland, C.M., Fu, J., Woodlee, G.L., Seymour, T.S., and Wickes, B.L. 2002. Isolation and characterization of the *Cryptococcus neoformans MATa* pheromone gene. Genetics 160: 935-947.

Miller, M.G., and Johnson, A.D. 2002. White-opaque switching in *Candida albicans* is controlled by mating-type locus homeodomain proteins and allows efficient mating. Cell 110: 293-302.

Mitchell, T.K., and Dean, R.A. 1995. The cAMP-dependent protein kinase catalytic subunit is required for appressorium formation and pathogenesis by the rice blast pathogen *Magnaporthe grisea*. Plant Cell 7: 1869-1878.

Monod, M., Togni, G., Hube, B., and Sanglard, D. 1994. Multiplicity of genes encoding secreted aspartic proteinases in *Candida* species. Mol. Microbiol. 13: 357-368.

Muller, P., Aichinger, C., Feldbrugge, M., and Kahmann, R. 1999. The MAP kinase kpp2 regulates mating and pathogenic development in *Ustilago maydis*. Mol. Microbiol. 34: 1007-1017.

Muller, P., Katzenberger, J.D., Loubradou, G., and Kahmann, R. 2003. Guanyl nucleotide exchange factor Sql2 and Ras2 regulate filamentous growth in *Ustilago maydis*. Eukaryot. Cell 2: 609-617.

Niimi, M., Niimi, K., Tokunaga, J., and Nakayama, H. 1980. Changes in cyclic nucleotide levels and dimorphic transition in *Candida albicans*. J. Bacteriol. 142: 1010-1014.

Odds, F. C. 1985. Morphogenesis in *Candida albicans*. Crit. Rev. Microbiol. 12: 45-93.

Pan, X., and Heitman, J. 1999. Cyclic AMP-dependent protein kinase regulates pseudohyphal differentiation in *Saccharomyces cerevisiae*. Mol. Cell. Biol. 19: 4874-4887.

Park, G., Xue, C., Zheng, L., Lam, S., and Xu, J.R. 2002. *MST12* regulates infectious growth but not appressorium formation in the rice blast fungus *Magnaporthe grisea*. Mol. Plant Microbe Interact. 15: 183-192.

Robertson, L.S., and Fink, G.R. 1998. The three yeast A kinases have specific signaling functions in pseudohyphal growth. Proc. Natl. Acad. Sci. USA 95: 13783-13787.

Rocha, C.R., Schroppel, K., Harcus, D., Marcil, A., Dignard, D., Taylor, B.N., Thomas, D.Y., Whiteway, M., and Leberer, E. 2001. Signaling through adenylyl cyclase is essential for hyphal growth and virulence in the pathogenic fungus *Candida albicans*. Mol. Biol. Cell 12: 3631-3643.

Ruiz-Herrera, J., and Martinez-Espinoza, A.D. 1998. The fungus *Ustilago maydis*, from the aztec cuisine to the research laboratory. Int. Microbiol. 1: 149-158.

Sabie, F.T., and Gadd, G.M. 1992. Effect of nucleosides and nucleotides and the relationship between cellular adenosine 3':5'-cyclic monophosphate (cyclic AMP) and germ tube formation in *Candida albicans*. Mycopathologia 119: 147-156.

Sadhu, C., Hoekstra, D., McEachern, M.J., Reed, S.I., and Hicks, J.B. 1992. A G-protein alpha subunit from asexual *Candida albicans* functions in the mating signal transduction pathway of *Saccharomyces cerevisiae* and is regulated by the a1-alpha 2 repressor. Mol. Cell. Biol. 12: 1977-1985.

Sanchez-Martinez, C., and Perez-Martin, J. 2002. Gpa2: a G-protein alpha subunit required for hyphal development in *Candida albicans*. Eukaryot. Cell 1: 865-874.

Sanglard, D., Hube, B., Monod, M., Odds, F.C., and Gow, N.A. 1997. A triple deletion of the secreted aspartyl proteinase genes *SAP4*: *SAP5*: and *SAP6* of *Candida albicans* causes attenuated virulence. Infect. Immun. 65: 3539-3546.

Schweizer, A., Rupp, S., Taylor, B.N., Rollinghoff, M., and Schroppel, K. 2000. The TEA/ATTS transcription factor CaTec1p regulates hyphal development and virulence in *Candida albicans*. Mol. Microbiol. 38: 435-445.

Shen, W.C., Davidson, R.C., Cox, G.M., and Heitman, J. 2002. Pheromones stimulate mating and differentiation via paracrine and autocrine signaling in *Cryptococcus neoformans*. Eukaryot. Cell 1: 366-377.

Shenhar, G., and Kassir, Y. 2001. A positive regulator of mitosis, Sok2: functions as a negative regulator of meiosis in *Saccharomyces cerevisiae*. Mol. Cell. Biol. 21: 1603-1612.

Sherwood, J., Gow, N.A., Gooday, G.W., Gregory, D.W., and Marshall, D. 1992. Contact sensing in *Candida albicans*: a possible aid to epithelial penetration. J. Med. Vet. Mycol. 30: 461-469.

Soll, D. R. 2002. *Candida* commensalism and virulence: the evolution of phenotypic plasticity. Acta Trop. 81: 101-110.

Sonneborn, A., Bockmuhl, D.P., Gerads, M., Kurpanek, K., Sanglard, D., and Ernst, J.F. 2000. Protein kinase A encoded by *TPK2* regulates dimorphism of *Candida albicans*. Mol. Microbiol. 35: 386-396.

Staab, J.F., Bradway, S.D., Fidel, P.L., and Sundstrom, P. 1999. Adhesive and mammalian transglutaminase substrate properties of *Candida albicans* Hwp1. Science 283: 1535-1538.

Stoldt, V.R., Sonneborn, A., Leuker, C.E., and Ernst, J.F. 1997. Efg1p, an essential regulator of morphogenesis of the human pathogen *Candida albicans*, is a member of a conserved class of bHLH proteins regulating morphogenetic processes in fungi. EMBO J. 16: 1982-1991.

Talbot, N.J. 1995. Having a blast: exploring the pathogenicity of *Magnaporthe grisea*. Trends Microbiol. 3: 9-16.

Talbot, N.J. 2003. On the trail of a cereal killer: Exploring the biology of *Magnaporthe grisea*. Annu. Rev. Microbiol. 57: 177-202.

Tebarth, B., Doedt, T., Krishnamurthy, S., Weide, M., Monterola, F., Dominguez, A., and Ernst, J.F. 2003. Adaptation of the Efg1p morphogenetic pathway in *Candida albicans* by negative autoregulation and PKA-dependent repression of the EFG1 gene. J. Mol. Biol. 329: 949-962.

Tsong, A.E., Miller, M.G., Raisner, R.M., and Johnson, A.D. 2003. Evolution of a combinatorial transcriptional circuit: a case study in yeasts. Cell 115: 389-399.

Tzung, K.W., Williams, R.M., Scherer, S., Federspiel, N., Jones, T., Hansen, N., Bivolarevic, V., Huizar, L., Komp, C., Surzycki, R., *et al.* 2001. Genomic evidence for a complete sexual cycle in *Candida albicans*. Proc. Natl. Acad. Sci. USA 98: 3249-3253.

Valent, B., Farrall, L., and Chumley, F. G. 1991. *Magnaporthe grisea* genes for pathogenicity and virulence identified through a series of backcrosses. Genetics 127: 87-101.

Walsh, T.J., and Groll, A.H. 1999. Emerging fungal pathogens: evolving challenges to immunocompromised patients for the twenty-first century. Transpl. Infect. Dis. 1: 247-261.

Wang, P., Perfect, J.R., and Heitman, J. 2000. The G-protein beta subunit *GPB1* is required for mating and haploid fruiting in *Cryptococcus neoformans*. Mol. Cell. Biol. 20: 352-362.

Waugh, M.S., Nichols, C.B., DeCesare, C.M., Cox, G.M., Heitman, J., and Alspaugh, J.A. 2002. Ras1 and Ras2 contribute shared and unique roles in physiology and virulence of *Cryptococcus neoformans*. Microbiology 148: 191-201.

Waugh, M.S., Vallim, M.A., Heitman, J., and Alspaugh, J.A. 2003. Ras1 controls pheromone expression and response during mating in *Cryptococcus neoformans*. Fungal Genet. Biol. 38: 110-121.

Wenzel, R. P. 1995. Nosocomial candidemia: risk factors and attributable mortality. Clin. Infect. Dis. 20: 1531-1534.

Whiteway, M., Dignard, D., and Thomas, D.Y. 1992. Dominant negative selection of heterologous genes: isolation of *Candida albicans* genes that interfere with *Saccharomyces cerevisiae* mating factor-induced cell cycle arrest. Proc. Natl. Acad. Sci. USA 89: 9410-9414.

Wickes, B.L., Edman, U., and Edman, J.C. 1997. The *Cryptococcus neoformans STE12*alpha gene: a putative *Saccharomyces cerevisiae STE12* homologue that is mating type specific. Mol. Microbiol. 26: 951-960.

Xu, J.R., and Hamer, J.E. 1996. MAP kinase and cAMP signaling regulate infection structure formation and pathogenic growth in the rice blast fungus *Magnaporthe grisea*. Genes Dev. 10: 2696-2706.

Xue, Y., Batlle, M., and Hirsch, J.P. 1998. *GPR1* encodes a putative G protein-coupled receptor that associates with the Gpa2p Galpha subunit and functions in a Ras-independent pathway. EMBO J. 17: 1996-2007.

Yue, C., Cavallo, L.M., Alspaugh, J.A., Wang, P., Cox, G.M., Perfect, J.R., and Heitman, J. 1999. The *STE12*alpha homolog is required for haploid filamentation but largely dispensable for mating and virulence in *Cryptococcus neoformans*. Genetics 153: 1601-1615.

From: Pathogenic Fungi: Structural Biology and Taxonomy
Edited by: G. San-Blas and R.A. Calderone

Chapter 7

Beyond Molecular Biology: Fungal Morphology as a Mathematical, Biophysical and Computational Subject

Gioconda San-Blas and Juan Murgich

Abstract

Morphogenesis at the level of the cell requires details of molecular biology, but that is not sufficient to determine spatial patterns that exist at a scale that is larger than three to five orders of magnitude. Thus, the molecular paradigm, while not incorrect, is incomplete. Therefore, it follows that principles of spatial organization must exist at the level of the cell that influence the morphogenetic process. Physical forces and fields, applied to biological systems, must be at work to explain morphogenesis, once the molecules have been synthesized. Biophysical, mathematical, and computational approaches have been developed in order to explain morphogenetic events, in an attempt to raise the molecular data to a higher level of understanding. In this chapter we summarize a mathematical morphological index, the steady-state, biomechanical and vesicle-supply centre models, and molecular modelling of cell wall glucans.

1. Introduction

In addition to the importance of fungal morphogenesis in explaining fungal relationships with its environment and its host, the process of morphogenesis and morphogenetic switch in fungi has been fruitful in bringing about new ideas on the problems of spatial biological patterns in general.

While molecular biologists believe the premise that morphogenetic information is wholly stored at the level of genes and molecules, other scientists assert that morphogenesis at the cellular level is not completely reduced to the molecular level (Harold, 1999). Both schools of study are correct, but molecular biological information while necessary is not sufficient to determine spatial patterns on a scale three to five orders of magnitude larger. The fact that molecules and morphology are loosely coupled can be illustrated by comparing, for example, the shapes of chitinous fungal hyphae and those of cellulolytic oomycetes whose shape differ only in detail or, for that matter, the radically different chemical structures that shape the similar hyphae of fungi and actinomycetes. Therefore, the molecular paradigm is not incorrect, but incomplete, for which it follows that principles of spatial organization must exist that operate on the cellular scale to lead the morphogenetic process (Harold, 1999; 2001).

To explain spatial patterns, the concept of morphogenetic fields assumes a major role in explaining processes such as morphogenesis. A dynamic physiological field, coordinating the molecular activities underlying growth, may be the obligatory intermediate between what is encoded in the gene and the supramolecular order of the organism. Once gene products are synthesized, they are released into a cellular context that is already structured in space, regulated and directional, a process that will require the production of physical forces and its localized application (Harold, 1999).

Therefore, physical forces and fields, applied to biological systems, must be at work to explain morphogenesis, once the molecules have been synthesized. Biophysical, mathematical, and computational approaches have been developed in order to explain morphogenetic events, in an attempt to raise the molecular data to a higher level of understanding. Some of these concepts are summarized in this chapter.

2. A Morphological Index for Characterization of Cell Shape and Morphogenetic Transition in Fungi

The morphogenetic switch between the extreme yeast and mycelial shapes moves along a progressive pathway of intermediate cell shapes. The latter can also be found as the end point for particular wild or mutant strains that are loosely described as pseudohyphae, blurring the morphological descriptions in the scientific literature. To give a more precise, numerical description of extreme and intermediate fungal shapes, Merson-Davies and Odds (1989) expressed the morphology of *Candida albicans* cells as a mathematical ratio, the morphology index (Mi), which was derived from the cell's maximum length, maximum diameter and septal diameter. For spherical yeast cells, Mi = 1.0 and for true hyphae, Mi = 4.0, while elongated yeast cells and pseudohyphae have intermediate values. Mi is mathematically defined in the equation:

$$\mathrm{Mi} = 2 + 1.78\ \log_{10}(ls/d^2)$$

where l = length, d = maximum diameter, s = diameter at septal junction. The ratio l/d increases with the length of the cell, while the ratio d/s increases with the degree of ballooning of the cell (also see Chapter 4 for morphology index).

Mi was determined with high reproducibility for *C. albicans* grown in a variety of environments and was found to be directly related to the chitin content of the cells. In this way, the subjective description of cell morphologies was substituted by objective parameters aimed at facilitating comparison of inter-laboratory data while creating a better way to distinguish phenotypes.

In 1997, San-Blas *et al.* used Merson-Davies and Odd's equation to define a Mi for *Paracoccidioides brasiliensis*, another dimorphic fungal pathogen. To account for particular characteristics of this species, the Mi equation had to be adjusted slightly:

$$\mathrm{Mi} = 2.13 + 1.13\ \log_{10}(ls/d^2)$$

in order to set the intercept and slope in such a way as to derive Mi = 1 for yeast cells and Mi = 4 for true hyphae. The discriminatory power of the equation was further used to quantify morphological population mixtures through Mi histograms (Figure 1). During the temperature-

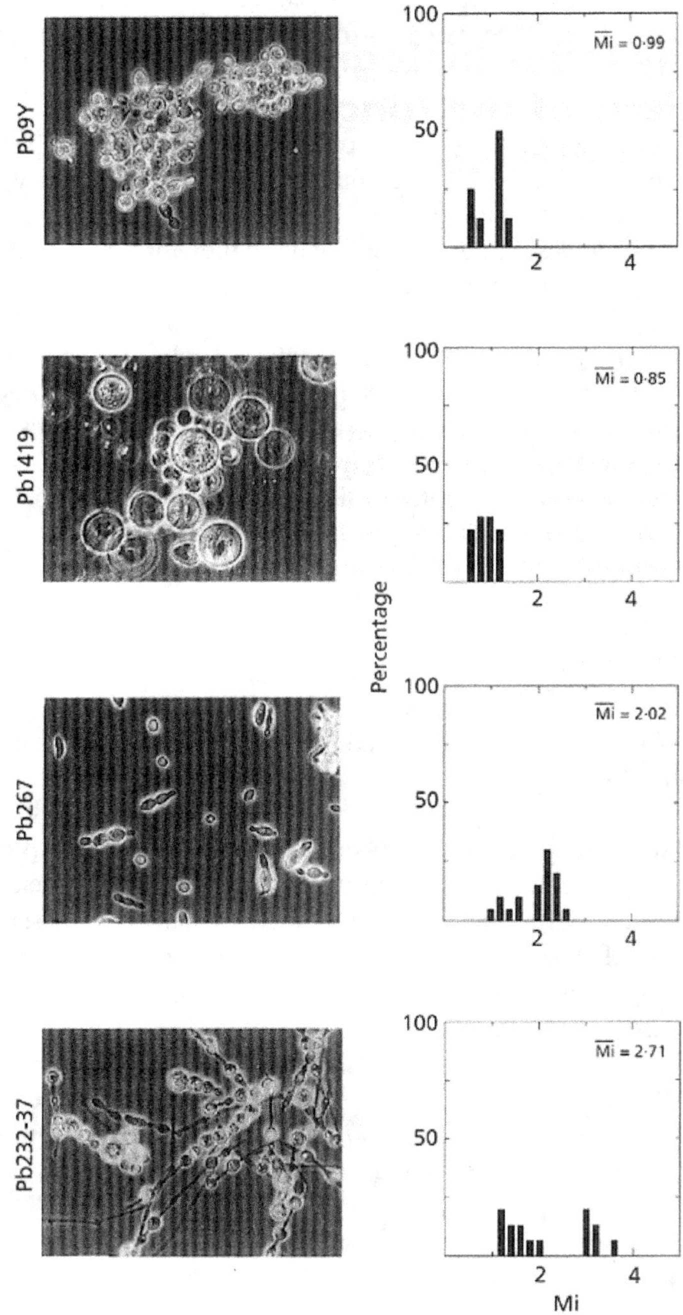
Pb9Y
Pb1419
Pb267
Pb232-37
Percentage
100
50
2
4
Mi
$\overline{Mi}$ = 0·99
$\overline{Mi}$ = 0·85
$\overline{Mi}$ = 2·02
$\overline{Mi}$ = 2·71

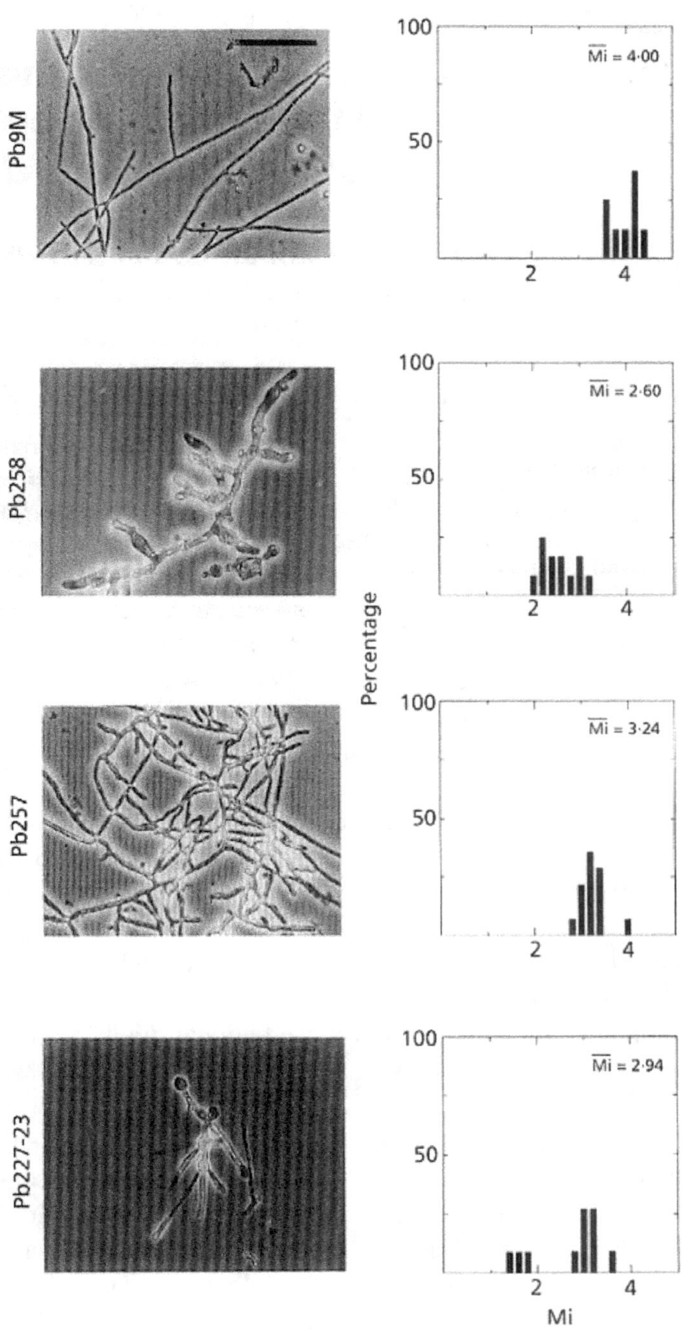

Figure 1. Mi values and histograms of *P. brasiliensis*: wild strain Pb9, yeast and mycelial phases; mutant strains ex-Pb9: Pb257, Pb258, and Pb267; mutant strains ex-Pb73: Pb1419, Pb232-37, and Pb227-23. Frequencies of Mi values are expressed as percentages. Bar, 20 μm. Reproduced with permission from San-Blas *et al.* (1997).

induced reversible dimorphic transition, mean Mi varied linearly with time, suggesting a continuity in the process. Also, in wild strains and their mutants, an inverse relationship between mean Mi and content of both cell wall chitin and α-1,3-glucan was found (San-Blas *et al.*, 1997).

3. Modelling the Morphology of the Hyphal Growth: Steady-State, Biomechanical and Vesicle-Supply Centre Models

The manner in which hyphae, the apically-growing tubular cells typical of fungi, attain their characteristic morphology has been an appealing subject for experimental and theoretical studies. The steady-state model of apical growth was developed by Sietsma and Wessels (1994) with experimental data suggesting that a plastic wall material was first synthesized and deposited at the apical dome, after which it converted into a rigid lateral wall with the apex containing a steady-state amount of plastic wall. This increasing rigidity was the result of progressive cross-linking of wall polymers by covalent bonds between β-1,3-glucan and chitin that led to changes in the physical properties of the wall, as the apical wall became subapical.

The idea that changes in wall properties from plastic to rigid would explain the origin of shape in hyphae was challenged by Bartnicki-García (1999) who argued that the wall of the apical dome, even at its potentially most plastic point, must be strong enough to resist deformation by the high turgor pressure of the cytoplasm; otherwise the cell would bulge and explode under normal growth conditions. Therefore, it is difficult to see how any additional rigidification of the apical wall would limit wall expansion to create a cylindrical wall.

The view that a hypha is a polarized secretory system was suggested in numerous studies. Vesicles carrying precursors and enzymes for wall synthesis are exocytosed at the apex (Harold, 1990; Gooday, 1995; Gow and Gadd, 1995). As such, a fungal hypha can be described as an organized territory, a physiological and morphogenetic field that is associated with the generation of spatial order (Harrison, 1993; Lumsden *et al.*, 1997). With this line of thought, Bartnicki-García and coworkers (Bartnicki-García *et al.*, 1989; Gierz and Bartnicki-García, 2001) developed a field theory of apical growth, based on the Spitzenkörper as the immediate source of secretory

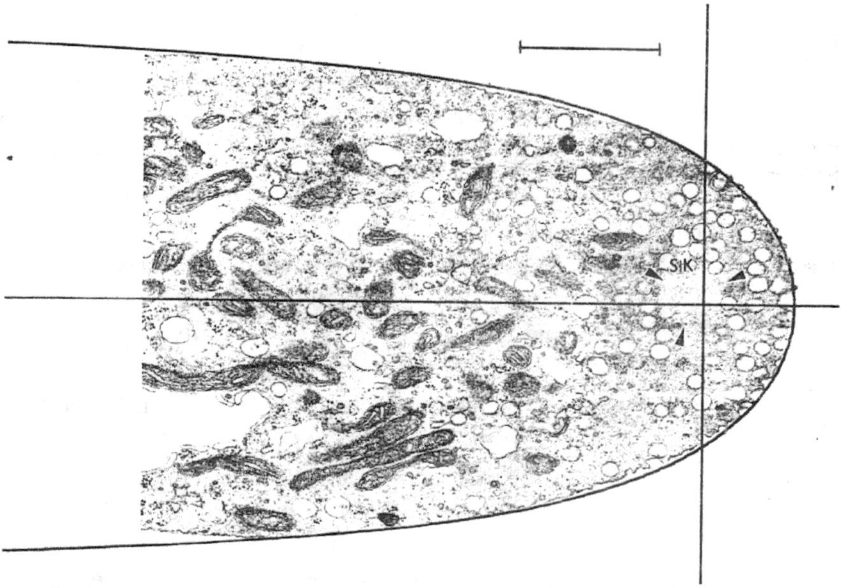

Figure 2. Correspondence between a curve predicted by the hyphoid equation and the electron microscopic profile of a hypha of *Armillaria mellea*. SK= Spitzenkörper. Reproduced with permission from Bartnicki-García *et al.* (1989).

vesicles that travel to the plasma membrane and increase its size by a fusion process. Spitzenkörper are characteristic accumulations of wall-building vesicles in the hyphal apex (Brunswik, 1924) that organize and direct vesicle traffic for hyphal elongation (Bartnicki-García, 2002). A vesicle-based computer simulation of fungal morphogenesis led Bartnicki-García *et al.* (1989) and Gierz and Bartnicki-García (2001) to realize that different cell shapes could be generated by displacing the immediate source of wall-making vesicles, called the vesicle-supply centre (VSC) (Bartnicki-García *et al.*, 1989). This shape-building process was described by a simple equation, called the hyphoid equation:

$$y = x \cot (xV/N)$$

which defined both the shape and diameter of a hypha by two parameters, to which physiological significance can be adscribed: N, the amount of wall-destined vesicles released from the VSC per unit time, and, V, the rate of displacement of the VSC. The two-dimensional shape described by the plotted hyphoid curve was almost identical to the shape of wall-preserved hyphae in longitudinal median sections (Figure 2) (Bartnicki-García *et al.*, 1989).

A three-dimensional model of hyphal morphogenesis was later developed by Gierz and Bartnicki-García (2001) under the same basic assumption used for the construction of their two-dimensional model, namely, that the polarized growth of tubular cells arises from a gradient of wall-building vesicles generated by a VSC. Contrary to the simple two-dimensional mathematical formulation, the three-dimensional derivation led to a mathematical indetermination whose solution required defining *a priori* the patterns of expansion of the wall. These were described by experimentally tracing the movement of marker points on the cell surface of a growing hypha (Bartnicki-García *et al.*, 2000), points that followed orthogonal trajectories, i.e., the newly synthesized material moved perpendicular to the wall surface. An orthogonal pattern of growth is consistent with the long-held view that turgor, rather than cytoskeleton, is the physical force driving cell expansion (Bartnicki-García *et al.*, 2000). Therefore, Gierz and Bartnicki-García (2001) could solve the mathematical indetermination of their three-dimensional equation. Once solved, the resulting equation bore a close similarity in profile to the two-dimensional hyphoid equation, validating in this way its use on this biological field model.

Elaborating on mathematical models based on biomechanical principles, rather than the geometrical principles of Bartnicki-Garcia and coworkers, Goriely's group (Goriely and Tabor, 2003a; 2003b; Goriely *et al.*, 2004) has proposed sophisticated mathematical formulations for biomechanical models of hyphal growth in actinomycetes and fungi. By means of turgor pressure-driven apical growth, and using large deformation elasticity theory, these authors developed a self-consistent, three-dimensional, biomechanical model in which the cell wall is modelled as a stretchable and growing elastic membrane with geometric-dependent moduli. Their formulations are sufficiently general to accommodate many different effects such as external stress, varying elastic moduli, and strain-energy functions describing the elastic response of the cell walls. This model is capable of producing realistic tip shapes and demonstrates that there is not entirely a special tip shape but rather that the tip expansion is essentially self-similar; that is, at each step the new shape is a duplicate of the previous one. The idea of a geometry-dependent elastic modulus is biophysically plausible, as classical results in polymer physics indicate that the modules increase with the degree of polymeric cross-linking, consistent with the process of wall rigidification.

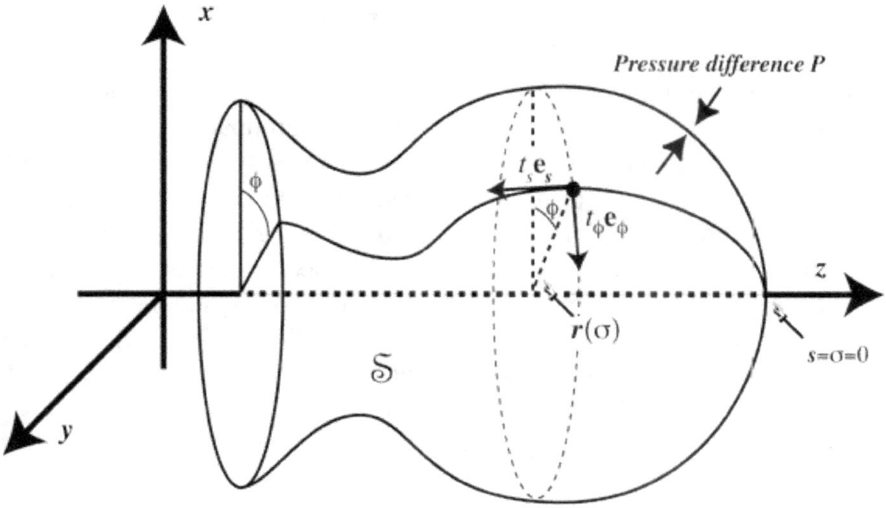

Figure 3. The surface S with stresses (t_s , t_φ). Reproduced with permission from Goriely and Tabor (2003b).

A natural question is whether this type of theoretical framework can be used to model morphologically similar, but biochemically very different process of hyphal growth in hyphal fungi. Although the role of turgor pressure and cytoskeletal structure in the expansion process of fungal hyphae is not completely clear (Money, 1997), at the biomechanical level, all the normal stresses are balanced in the equation:

$$P = \kappa_s\, t_s + \kappa_\varphi\, t_\varphi$$

in which P is pressure, t_s is the tension on the surface along the tangent $\mathbf{e}_s$ (Figure 3), in the direction of increasing archlength, t_φ is the tension along the unit vector $\mathbf{e}_\varphi$, normal to $\mathbf{e}_s$ in a plane tangent to S, and in the direction of increasing azimuthal angle φ (Goriely and Tabor, 2003b). This equation represents a generalized form of the Young-Laplace law for a membrane with principal curvatures supporting a pressure difference P. Therefore, P can be equated to the sum of all net forces generated by external and internal turgor pressure, and cytoskeleton on the cell wall. In this way, the biomechanical model of Goriely and coworkers (Goriely and Tabor, 2003a; 2003b; Goriely *et al*., 2004) can still capture the essence of a continuous wall-building process.

4. Molecular Modelling of Cell Wall Glucans From *Paracoccidioides brasiliensis*

Computer simulation techniques have been used for over 30 years primarily to look at the motion of atoms or groups of atoms within a protein or other biopolymers, and more recently, in the modelling of protein structures or macromolecular complexes. It has also been used in calculating free energy differences arising from binding drugs or from small changes in sequences, and as refinement technique in determining macromolecular conformations from X-ray crystallography or NMR data (Goodfellow, 1990). Molecular modelling of man-made polymers has also grown notably over the past few years (Rigby and Eichinger, 2001). This technique has been used in the study of biopolymers that shape the fungal cell wall (Murgich *et al.*, unpublished results).

In a simplified fungal cell wall model, molecular modelling of amorphous α-1,3- and β-1,3-glucans from *P. brasiliensis* was undertaken to obtain properties such as the most stable molecular conformations of the strands in solid state. Also, the cohesive energy density for both biopolymers was calculated at 21 °C and 37 °C. *P. brasiliensis* was taken as the subject of study because in this fungus, the cell wall varies in composition according to the morphological phase, particularly with regards to the carbohydrate constituents, which account for >80% of its dry weight (San-Blas and San-Blas, 1994; also see Chapter 5). Of their biochemical structures, glucose polymers are arranged mainly as α-1,3-glucan in the cell wall of the pathogenic yeastlike phase (95%) plus a small amount (5%) of β-1,3-glucan. The latter is the only glucose polymer found in the mycelial cell wall. α-1,3-Glucan plays a role as a protective layer against host defences, becoming a virulence factor (see review, San-Blas and Niño-Vega, 2001), while a decisive role of glucans in the establishment of either morphology in the dimorphic process of *P. brasiliensis* has been proposed (San-Blas and San-Blas, 1994).

To this cell wall model of *P. brasiliensis*, molecular mechanics was applied. This technique provides information about the most stable conformation of molecules and their aggregates, with the use of analytical functions that represent bond stretchings, binding and torsional as well as non-bonded energies of molecules (i.e., electrostatic interactions, dispersion attraction and exchange repulsion). The functions and the different parameters constitute a force field. The procedure of searching a minimum energy for the system is such that an initial configuration is specified and the interatomic distances and bond angles are adjusted, using an iterative computational method, until the minimum energy configuration is obtained. The algorithms used in this work were part of the InsightII and Discover programs (Accelrys Inc., San Diego, CA). To construct a cell wall model, the Amorphous Cell Module (Accelrys Inc., San Diego, CA) was used. This module generates strands of the polymer that are tightly packed in a cube until a predetermined density is achieved, and static and dynamic properties can be derived from these simulations. Once the cube is built as close as possible to the chosen physical state, molecular dynamic trajectories (Leach, 2001) and energy minimizations may be performed. Minimizations will probe the detailed structure of the low energy configurations while trajectories can be used to examine paths between physically significant system configurations. If a sufficiently large volume of phase space is searched, one can then calculate equilibrium properties. Both static and dynamic properties may be obtained from these simulations (Leach, 2001). In particular, properties such as cohesive energy density, radial plus pair distribution functions and elastic constants of the material can be obtained. To avoid the problems generated by the boundaries of the cube, periodic boundary conditions were applied (Leach, 2001). The number of non-bonded interactions in a dense system is such that it is customary to use summation cutoffs. In this way, the atoms that only generate small or negligible contributions are ignored.

When simulating polymers, it is necessary to start with a configuration in which the backbone bond conformer populations and other characteristics of the chains, are representative of those more likely to be found in the material. A combination of the method developed by Theodorou and Suter (1985) with scanning (Meirovitch, 1983), was employed with some improvements (Polymer Users Guide, Accelrys Inc., San Diego, CA). In them, an initial guess structure is generated and allowed to evolve toward a minimum potential energy structure. The first three backbone atoms of the starting structure are placed in the box in a random orientation. The location of the next atom is determined from the estimated conditional probability of the rotational bond states. Once the cube is filled, an equilibration procedure encompassing minimizations and dynamics follows (Polymer Users Guide, Accelrys Inc., San Diego, CA).

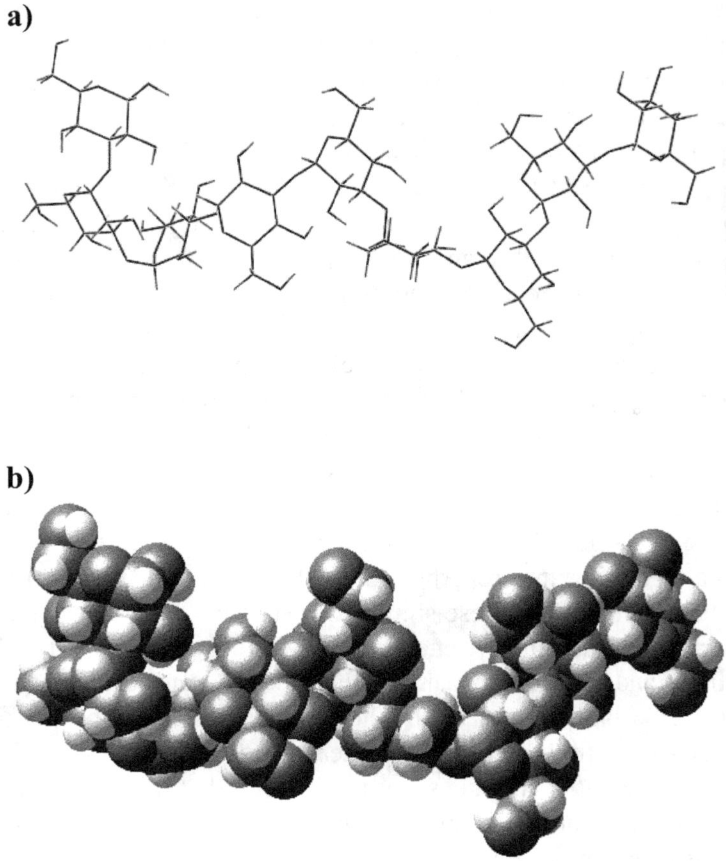

Figure 4. Conformation of α-glucan strands at 21°C. (a) Wire frame drawing; (b) Space filling drawing (Murgich, Abanero and San-Blas, unpublished results).

The strands of the α- and β-glucan polysaccharides were built with the Amorphous Cell program. The results obtained for strands containing an increasing number of monomers leveled off around 40. For this reason, the strands employed for the calculations contained 50 monomers. The CVFF force field has been used to describe the intra- and intermolecular interactions. Since the density of the α- and β-glucans is not known, the value of glucose was used instead. All the calculations were made using a summation cutoff distance of 9.5Å.

Figures 4 and 5 show the α-and β-glucan strands generated by means of the Amorphous Cell module. A significant difference between the strand morphologies (chain conformations) of both types of polymers can be observed, implying the existence of different cohesive energy densities in the solid phase. From the minimization calculations, Murgich *et al.* (unpublished results) concluded that α- and

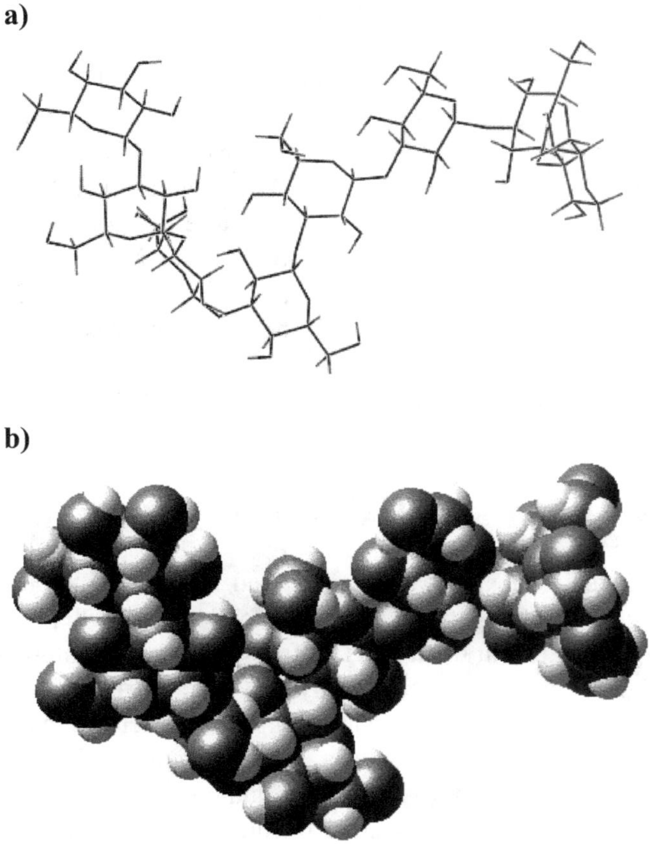

Figure 5. Conformation of β-glucan strands at 21°C. (a) Wire frame drawing; (b) Space filling drawing (Murgich, Abanero and San-Blas, unpublished results).

β-glucans have a sort of helical conformation with a different number of turns per unit length. This could be a consequence of differences in the location of the single bond found between the monomers of the α- and β-glucans and also from some steric interference between the different atoms of the monomers. In the cell wall, the polymer chains try to minimize the energy by packing themselves in such a way that the number of attracting interatomic interactions is maximized. This means that the different chains will try to change their conformation in such a way that the density of the system is at a maximum (Wunderlich, 1973). The dissimilarities in chain conformations of α-and β-glucans are reflected in their mechanical properties. The more dense packing of the chains produces "stronger" material than lesser dense ones, as a result of the steric interference of the molecules in their chains. This correlates with the density of the system and is reflected in the mechanical properties of the polymers (Wunderlich, 1973). Cohesive energy density results (not shown) indicated that their values decreased with increasing temperature. In β-glucan, intermolecular interactions decreased with temperature, giving rise to weaker interatomic contacts at 37°C. On the other hand, data from molecular mechanics indicate that α-glucan is in a state of higher energy (i.e., less stable) at 21°C. Therefore, β-glucan will be the preferred polymer at 21°C and α-glucan at 37°C. This is in accordance with real-life facts, where the mycelial cell wall of *P. brasiliensis*, generated at 21-23°C, is structured exclusively by β-glucan as the sole glucose polymer, while α-glucan is the preferred glucan when *P. brasiliensis* is transferred to 37°C to induce the pathogenic yeast phase.

5. Final Remarks

The unconventional approaches discussed above are not meant to invalidate data arising from molecular biology or cell physiology, but rather complement them in a fruitful way. Thus, a comprehensive model should include changes in physical properties of the wall, the cytoskeleton to provide the scaffolding for vesicle dinamics, the VSC to orchestrate the exocytosis gradient, and the mathematical and computational analyses to provide numerical logic to biological problems, each of them accounting for different features of the wall-building process, both in the spatial control of wall synthesis and the temporal control of wall extensibility. As remarked by Harold (1999), "holistic and reductionist viewpoints are complementary rather than antagonistic in biology, having made strides by dissecting organisms into their molecular constituents, has paid a price for doing so, and now the time has come to rediscover the organism".

6. Acknowledgement

To Fondo Nacional de Investigaciones Científicas y Tecnológicas (FONACIT), Caracas, Venezuela, for grant G97-000615.

References

Bartnicki-García, S. 1999. Glucans, walls, and morphogenesis: On the contributions of J.G.H. Wessels to the golden decades of fungal physiology and beyond. Fungal Genet. Biol. 27: 119-127.

Bartnicki-García, S. 2002. Hyphal tip growth: Outstanding questions. In: Molecular Biology of Fungal Development. H.D. Osiewacz, ed. Marcel Dekker, New York, p. 29-58.

Bartnicki-García, S., Bracker, C.E., Gierz, G., López-Franco, R., and Lu, H. 2000. Mapping the growth of fungal hyphae: orthogonal cell wall expansion during tip growth and the role of turgor. Biophys. J. 79: 2382-2390.

Bartnicki-García, S., Hergert, F., and Gierz, G. 1989. Computer simulation of fungal morphogenesis and the mathematical basis for hyphal (tip) growth. Protoplasma 153: 46-57.

Brunswik, H. 1924. Untersuchungen über Geschlechts- und Kernverhältnisse bei der Hymenomyzetengattung *Coprinus*. In: Botanische Abhandlungen. K. Goebel, ed. Gustav Fischer, Jena. p. 1-152.

Gierz, G., and Bartnicki-García, S. 2001. A three-dimensional model of fungal morphogenesis based on the vesicle supply center concept. J. Theor. Biol. 208: 151-164.

Gooday, G.W. 1995. The dynamics of hyphal growth. Mycol. Res. 99: 385-394.

Goodfellow, J.M. 1990. Computer simulation in molecular biology. Chem. Britain 1066-1068.

Goriely, A., Károly, G., and Tabor, M. 2004. Growth induced curve dynamics for filamentary micro-organisms. J. Mathem. Biol. In press.

Goriely, A., and Tabor, M. 2003a. Self-similar tip growth in filamentary organisms. Phys. Rev. Lett. 90: 108101-1 –108101-4.

Goriely, A., and Tabor, M. 2003b. Biomechanical models of hyphal growth in actinomycetes. J. Theor. Biol. 222: 211-218.

Gow, N.A.R., and Gadd, G.M. 1995. The Growing Fungus. Chapman and Hall, London.

Harold, F.M. 1990. To shape a cell: an inquiry into the causes of morphogenesis of microorganisms. Microbiol. Rev. 54: 381-431.

Harold, F.M. 1999. In pursuit of the whole hypha. Fungal Genet. Biol. 27: 128-133.

Harold, F.M. 2001. The Way of the Cell. Molecules, Organisms and the Order of Life. Oxford Univ. Press., New York.

Harrison, L.A. 1993. Kinetic theory of living pattern. Cambridge Univ. Press, Cambridge.

Leach, A.R. 2001. Molecular Modelling: Principles and Applications, 2nd. Ed., Prentice Hall, London.

Lumsden, C.J., Brandts, W.A., and Trainor, L.E.H. eds. 1997. Physical Theory in Biology: Foundations and Explorations. World Scientific, Singapore.

Meirovitch, H.J. 1983. Computer simulation technique for calculating the entropy of polymer chains, based on the scanning method with a mean field parameter. Macromolecules 18: 569-574.

Merson-Davies, L.A., and Odds, F.C. 1989. A morphology index for characterization of cell shape in *Candida albicans*. J. Gen. Microbiol. 135: 3143-3152.

Money, N.P. 1997. Wishful thinking of turgor revisited: The mechanics of fungal growth. Fungal Genet. Biol. 21: 173-187.

Rigby, D., and Eichinger, B.E. 2001. Polymer modeling. Curr. Opin. Sol. State Mat. Sci. 5: 445-450.

San-Blas, G., and Niño-Vega, G. 2001. *Paracoccidioides brasiliensis*: virulence and host response. In: Fungal Pathogenesis: Principles and Clinical Applications. R.L. Cihlar, and R.A. Calderone, eds. Marcel Dekker, New York. p. 205-226.

San-Blas, G., Padrón, R., Álamo, L., and San-Blas, F. 1997. Use of morphology index histograms to quantify populations of the fungal pathogen *Paracoccidioides brasiliensis*. Microbiology 143: 197-202.

San-Blas, G., and San-Blas, F. 1994. Biochemistry of *Paracoccidioides brasiliensis* dimorphism. In: Paracoccidioidomycosis. M. Franco, C.S. Lacaz, A. Restrepo-Moreno, and G. Del Negro, eds. CRC Press, Boca Raton, Florida. p. 49-66.

Siestma, J.H., and Wessels, J.G.H. 1994. Apical wall biogenesis. In: The Mycota. Vol. I. Growth, Differentiation and Sexuality. J.G.H. Wessels, and F. Meinhardt, eds. Springer-Verlag, Berlin Heidelberg. p. 125-141.

Theodorou, D.N., and Suter, U.W. 1985. Detailed molecular structure of polymeric glasses. Macromolecules 18: 1467-1477.

Wunderlich, B. 1973. Macromolecular Physics, Vol. 1, Academic Press, New York.

New Taxonomic Tools

From: Pathogenic Fungi: Structural Biology and Taxonomy
Edited by: G. San-Blas and R.A. Calderone

Chapter 8

The Use of Phylogenetic Analysis to Investigate Uncultivated Microbes in Medical Mycology

Leonel Mendoza and Victor Silva

Abstract

The classical way to investigate the relationship between organisms includes, but is not limited to, taxonomic macro- and microscopic morphological changes within individuals of the same group. One drawback to this approach has been that the uncultured microbes are difficult to study in fine detail in the absence of cultures. With the advent of phylogenetic analysis, however, several taxonomic mysteries, including the phylogenetic connections of uncultivated microbes, have been revisited. In medical mycology in particular, two enigmatic microbes that resisted cultivation were investigated: *Lacazia loboi* and *Rhinosporidium seeberi*. Despite ultra-structural, serological and chemical studies, the relationship of these two pathogens remained a mystery for more that 100 years. Using 18S SSU rDNA, ITS sequences and chitin synthase genes, however, we found that *L. loboi* formed a sister clade to *Paracoccidioides brasiliensis* in the dimorphic Onygenales, and that *R. seeberi* was a protistal Mesomycetozoea located at the point where the fungi first diverged from animals about 1.6 billion year ago. This article will discuss in detail some of the aspects that made these studies possible and future directions to investigate other traits in these unusual microbes.

1. Introduction

Based on the classical taxonomic classification of the living things, first Copeland (1956) and then Whittaker (1969) introduced the concept of multiple kingdoms to properly place all living things in a tree of life. According to this categorization, the prokaryotes remain at the base and the eukaryotes in four other kingdoms. The latter group comprised one unicellular protistal eukaryote kingdom and three kingdoms formed by the multicellular animals, fungi and plants, now known as the "crown eukaryotes". With the advent of molecular phylogeny some of these schemes were validated and others have been challenged (Baldauf and Palmer, 1993; Cavalier-Smith, 1998; Katz, 1999; Wainright *et al.*, 1993). Molecular phylogeny using either 18S small subunit (SSU), 28S large subunit rDNA, internal transcriber spacers (ITS) sequences or the deduced amino acid sequences of conserve proteins such as actin, β-tubulin, and elongation factor 1α, have been an invaluable tools to study the phylogenetic connections of the three multicellular kingdoms and other organisms (Baldauf and Palmer, 1993; Baldauf *et al.*, 2000; Katz, 1999; Patterson, 1999).

One of the three kingdoms of multicellular eukaryotes, the fungi, has been referred as the fifth kingdom (Kendrick, 1985). This probably was because, before the proposal of five kingdoms and for a long time, fungi were studied within the kingdom Plantae, a proposal presented earlier in the 20th century by Haeckel (see Margulis and Schewartz, 1998). Molecular data, however, have in part validated Whittaker's proposal. Using this technology it was found that the fungi constitute a monophyletic group of organism that share phylogenetic characteristics and several conserved morphological and ultrastructural features (Alexopolus *et al.*, 1996; Cavalier-Smith, 1998). Interestingly, phylogenetic analysis also revealed that the animals and fungi might have had a recent common ancestor different from the one shared with the alveolates, plants and straminipilans[1] (Berbee and Taylor 2001; Wainright *et al.*, 1993). Using DNA, RNA or protein sequences to determine the origin of Metazoans and Fungi proved to be difficult (Katz, 1999; Lang *et al.*, 2002; Mendoza *et al.*, 2002; Patterson, 1999). Recently, however, Burger *et al.* (2003) and Lang *et al.* (2002), using complete genomes of mtDNA, resolved

[1]The term Stramenopila proposed by Patterson (1989) has been widely used. However, Dick (2001) recently called the attention that the term is a misspelling of the Latin world Stramini, which means hair. The correct spelling of the term should be Straminipila.

the controversial phylogenetic position of the Mesomycetozoea, the Choanoflagellata and the Metazoa (they used the generic named Holozoa to include the three groups). These investigators found that mesomycetozoeans were phylogenetically located after the divergence with the fungi and basal to the choanoflagelates and metazoans. This finding is very important to understand since later in this review we will address why *Rhinosporidium seeberi*'s new phylogenetic home, is so important to those interested in the origins of animals and fungi.

With the aid of molecular tools it has been also possible to determine that only ~20% of the microbial diversity is known and that approximately 80% of the microbial communities have yet to be described (Hugenholtz *et al.*, 1998; Torsvick *et al.*, 1996; Ward *et al.*, 1990; Whitman *et al.*, 1998). More importantly, these studies revealed that some of the DNA and RNA sequences they described came from numerous uncultured microorganisms of well-known prokaryotes and eukaryotes and also from novel microbial communities (Ludwing *et al.*, 1997; Patterson, 1999; Roger, 1999; Rondon *et al.*, 2000; Wainright *et al.*, 1993). These findings came as a surprise; however, the existences of microorganisms that have resisted cultivation from environmental and medical samples have been well known for many years. Classical examples of uncultivated prokaryotes in medical bacteriology are *Mycobacterium leprae* and *Treponema pallidum*, uncultivated eukaryotes of medical mycological species, such as *Lacazia loboi*, *R. seeberi* and *Pneumocystis carinii* and in parasitology, several uncultivated protists such as *Toxoplasma gondii*, *Plasmodium* spp. and others. What we know about these classical pathogens mostly came from morphological and other studies carried out with clinical samples, rather than from cultural analysis.

The early DNA studies done on *P. carinii* surprised many (Edman, 1988). These studies revealed that the rDNA and protein sequences placed this pathogen within the Fungi (Cushion, 1998; Hughes, 1989). These data caused a great controversy that continues even today. While *P. carinii* received a great deal of attention by the medical community and other investigators, *L. loboi* and *R. seeberi* remained unstudied until 1998 (Mendoza *et al.*, 2001). The phylogenetic analysis of the 18S SSU rDNA sequence of *L. loboi* showed that it is more likely a dimorphic fungus phylogenetically related more closely to *Paracoccidioides brasiliensis*, whereas the analyses done on *R. seeberi* concluded that this pathogen was part of the class Mesomycetozoea, a new group of microbes very closely related to choanoflagellates and animals, but not distant from the fungi (Herr *et al.*, 1999a; Herr *et al.*, 2001; Mendoza *et al.*, 2001; Mendoza *et al.*, 2002).

The data generated during the last 10 years using molecular tools have irreversibly shaped the way we approach the fungal molecular taxonomy (see Chapter 9). We have witnessed the creation of new phylogenetic species. *Coccidioides posadasii* is perhaps the best example of this group (Fisher *et al.*, 2002). Molecular studies have been a determining factor for the proposal to eliminate the artificial phylum Deuteromycota, a proposal accepted in the 9th edition of the Dictionary of the Fungi (Kirk *et al.*, 2001) as well as many other important fungal molecular characterizations that will not be discussed in this chapter. This review will focus on the role that molecular technology has played resolving the relationships of the two orphan uncultivated microbes studied in medical mycology: *L. loboi* and *R. seeberi.*

2. *Lacazia loboi*

Jorge Lobo (1930) reported the first case of an infection caused by an organism later called *L. loboi* (Taborda *et al.*, 1999). He reported that this organism was morphologically similar to *P. brasiliensis* but clinically different in the diseases it causes. *L. loboi*'s infections are characterized by the formation of chronic cutaneous and subcutaneous parakeloidal-nodular lesions in different areas of the human anatomy, particularly the ears, face, and arms (Lacaz *et al.*, 2002; Rippon, 1988a).The disease lobomycosis (Jorge Lobo's disease) is limited to some Latin American countries. However, the infection has been also reported in dolphins around the coasts of Florida and in the Gulf of Mexico. In addition, two anomalous cases were also reported. One case involved an American tourist visiting Venezuela (Burns *et al.*, 2000) and a case of a French aquarium attendant that was near a dolphin with the disease (Symmer, 1983).

In the infected tissues, *L. loboi* develops abundant thick-walled yeast-like cells usually 5 to 10 μm in diameter single, or as branched chains of cells connected by short tubules. All attempts to culture this pathogen have failed (Lacaz *et al.*, 2002; Rippon, 1988a). The strains of *L. loboi* isolated from cases of Jorge Lobo's disease by some investigators, proved later to be common laboratory contaminants, or strains of *P. brasiliensis* (Lacaz *et al*., 2002). Thus, the taxonomic relationship of *L. loboi* with the other fungal pathogens remained a mystery for almost 70 years. Due to this frustrating fact and that this organism does not respond to most antifungal drugs, it was speculated that it might be a protist or a lower fungus (Rippon, 1988a). This controversy contributed to the several different names under which this pathogen has been known. The last binomial, *L. loboi,* was just recently proposed and accepted by most mycologists (Taborda *et al*., 1999).

Despite several decades of taxonomic research using traditional methodologies, the relationship of *L. loboi* to other living things remained unanswered (Lacaz *et al*., 2002). The first attempts to use molecular biology to address the phylogenetic relationships of this pathogen occurred in the early 90's. However, all early attempts failed (Mendoza *et al*., 2001). The first organized study to investigate *L. loboi* using molecular tools was published by Haubold *et al*., (1998; 2002). These investigators, however, more likely amplified two fungal contaminants from the tissues of a dolphin with lobomycosis. This was demonstrated later when a search in GenBank, using Basic Local Alignment Search Tool (BLAST) of both sequences, identified them as closer to *Candida cleridarum* (AF035675) and *Cladosporium sphaerospermum* (AF035674) respectively. The latter sequence showed 100% identity with *C. sphaerospermun*, which might suggest that their sequence belongs to this species. Mendoza *et al*. (2001) noted that the amplification of contaminant DNA from insects and fungi was common when dealing with total genomic DNA from fresh or fixed host tissue containing *L. loboi* or *R. seeberi*, which in part explains Haubold *et al*. (1998) results.

In 1998 a team composed by researchers from Michigan State University, East Lansing, MI, the Center for Disease Control and Prevention, Atlanta, GA, the University of California, Berkeley, CA, and the Institute "Lauro de Souza Lima", Bauru, Brazil, presented a research proposal to investigate the relationship of *L. loboi* with the other microbes. Their proposal was simple. They isolated genomic DNA from fixed and fresh tissue, kept at -80°C, from patients with proven lobomycosis. The strategy was to use the universal primers NS1 and NS8 to amplify by PCR the 18S SSU rDNA sequence in the genomic DNA isolated from the infected tissues. This strategy gave no results. The amplicons obtained were always the 18S SSU rDNA of the host, the common contaminant fungi, and even from insects. Intrigued with the results, they evaluated by light and electron microscopy the samples from which the genomic DNA was extracted. They found that most yeast-like cells in the infected tissue had a degenerate cytoplasm. This finding was in agreement with Vilani-Moreno and Opromolla (1997). These investigators found that only 36 to 57% of the yeast-like cells were alive in the infected tissues. These data implied that more that ~50% of the cells possessed a degenerate cytoplasm. Thus, these cells might harbor great quantities of proteases and endonucleases that transform cells into a cocktail of lytic enzymes. This and the fact that other endonucleases are also present during storage at -80°C of host tissues for long periods of time, contributed to the degradation of the small amount of genomic DNA coming from viable cells of *L. loboi*.

To circumvent this drawback, we flew to Bauru, Brazil, to extract genomic DNA from a biopsy taken from a patient with lobomycosis within a few seconds of its collection. The tissue was frozen using liquid nitrogen, and immediately ground. The powder was treated with sodium dodecyl sulphate and proteinase K and the DNA extracted with phenol and chloroform. The universal primers amplified a 1769 bp amplicon. The fragment was sequenced and a BLAST analysis showed that its sequence had 97% identity and two gaps with those sequences from *P. brasiliensis*. Similar characteristics were found with the other dimorphic fungal pathogens in the Onygenales (Herr *et al.*, 2001).

Phylogenetic analysis using the sequences of two 18S SSU rDNA amplicons from two different patients with lobomycosis supported the BLAST results. The analysis indicated that the two sequences of *L. loboi* were the sister clade to *P. brasiliensis,* and that this clade was sister to the cluster formed by the sequences in GenBank from *Histoplasma capsulatum*, *Blastomyces dermatitidis* and *Emmonsia parva*; all dimorphic pathogens in the Onygenales (Herr *et al.*, 2001, Mendoza *et al.*, 2001) (Figure 1). The 18S SSU rDNA of *L. loboi* formed long branches between these sequences and that of *P. brasiliensis*. More recently, Tarcha *et al.*, (2001) sequenced the ITS1, 5.8S, and ITS2 from the DNA of two different patients with lobomycosis. Their data strongly agree with the 18S SSU rDNA analysis of Herr *et al.* (2001), and again placed *L. loboi* as the sister clade to two other sequences of *P. brasiliensis* used in this new study (Figure 2). The long branches encountered in previous 18S SSU rDNA analyses between *L. loboi* and *P. brasiliensis* were also present in this new study, proving that the two ITS sequences deposited in GenBank by Haubold *et al.* (1998),

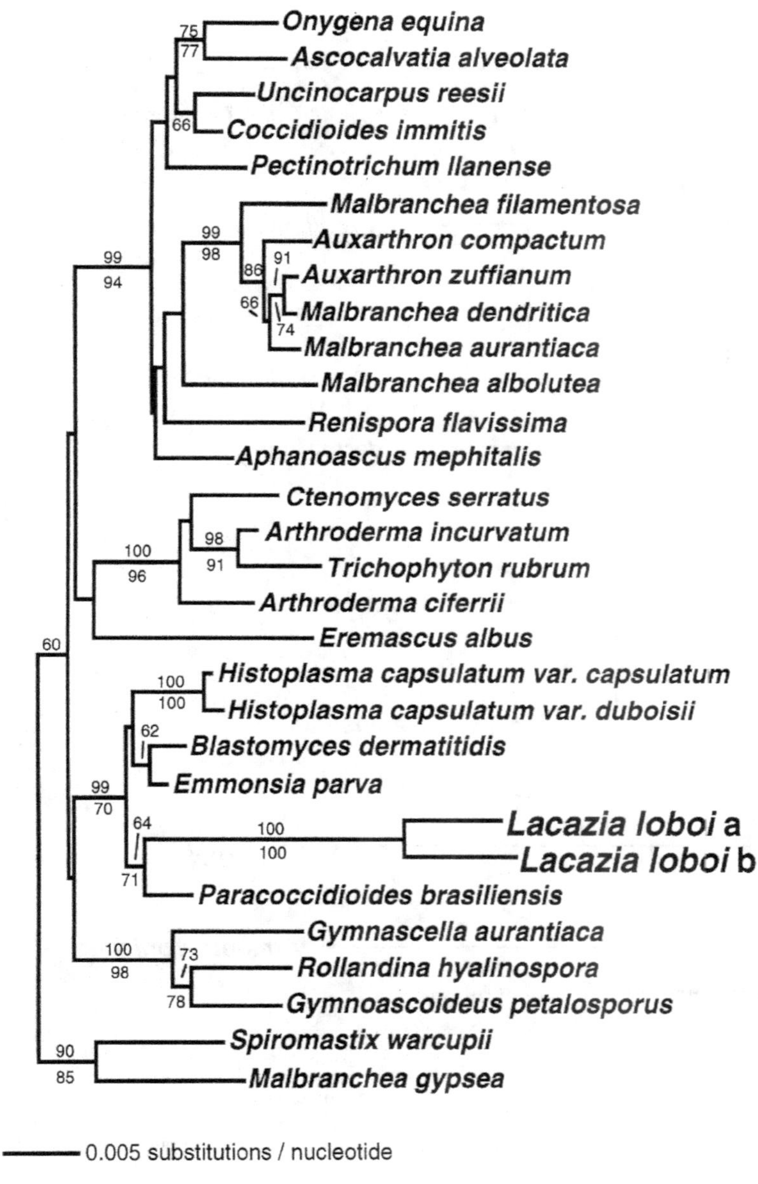

Figure 1. Phylogenetic analysis of two 18S SSU rDNA from *Lacazia loboi* and 28 other 18S SSU rDNA from different fungi in the Onygenales. This neighbor-joining tree showed *L. loboi* as the sister clade to *P. brasiliensis*. Note the long branches between *L. loboi* and *P. brasiliensis*. In turn, the clade formed by *L. loboi* and *P. brasiliensis* is the sister taxa to the other typical dimorphic fungal pathogens (modified from Herr *et al.*, 2001).

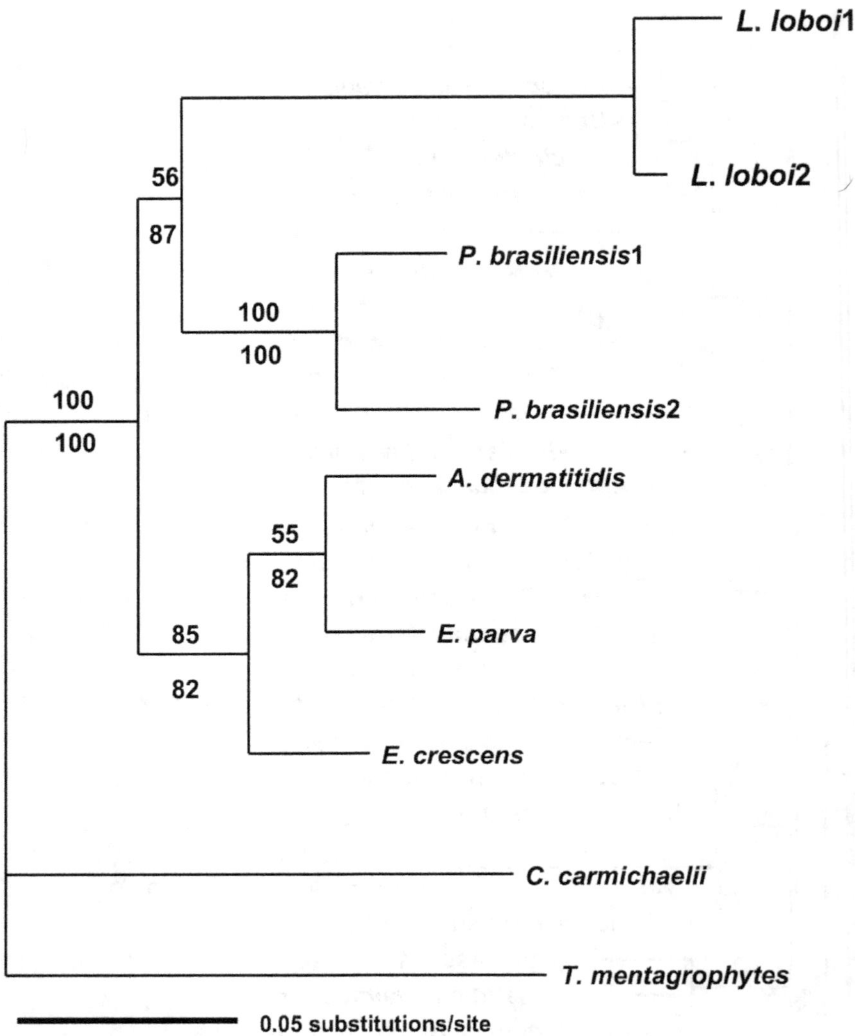

Figure 2. Phylogenetic analysis using two *Lacazia loboi* internal transcriber spacer (ITS) sequences and seven other Onygenales' ITS sequences, including two *Paracoccidioides brasiliensis* strains. *Lacazia loboi* is, in this neighbor-joining tree, the sister clade to *P. brasiliensis*, confirming analyses using the 18S SSU rDNA. The long branches between *L. loboi* and *P. brasileinsis* are also present in this analysis.

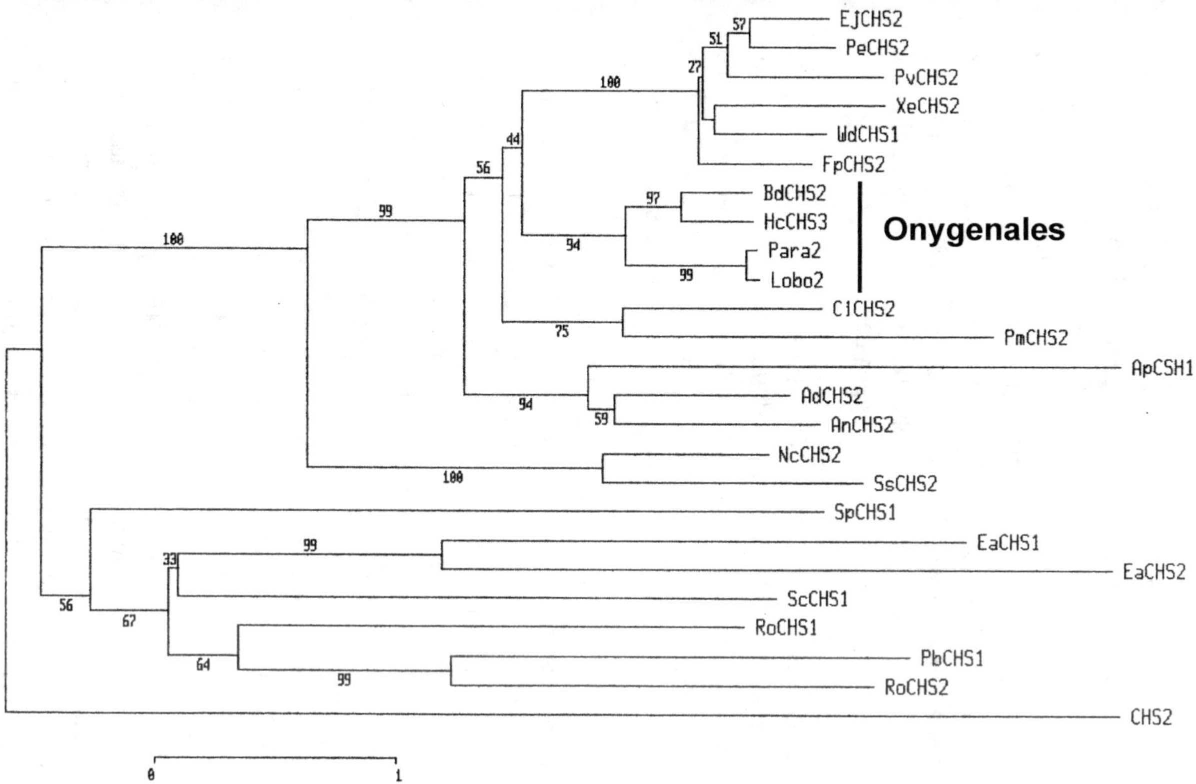

Figure 3. Neighbor-joining tree showing aligned deduced amino acid sequences from CHS2 sequences in 25 fungi. *Lacazia loboi*-CHS2 (Lobo2) sequence is the sister taxon to *Paracoccidioides brasiliensis*-CHS2 (Para2) and a member, in a well supported clade, of the two dimorphic pathogenic fungi used in this analysis: *Blastomyces dermatitidis* (BdCHS2) and *Histoplasma capsulatum* (HcCHS3). Numbers on the branches are percentages of bootstrap-resampled data.

were more likely contaminants. The nucleotide difference observed between the ITS sequences were similar to that found between the two 18S SSU rDNA sequences in previous studies, suggesting strong nucleotide substitution in these DNA molecules, or perhaps several populations of *L. loboi* (Figure 2).

With 97% identity between the genera *Lacazia* and *Paracoccidioides*, one can argue that these two microbes should be in the same genus. However, the presence of long branches between them, and the fact that the 18S SSU rDNA of *B. dermatitidis*, *E. parva*, and *H. capsulatum* all have two gaps and 96% identity with *L. loboi,* supports the placement of this uncultivated pathogen as a genus separate from the pathogen *Paracoccidioides*. These results are exciting since they suggest that *L. loboi* may be a dimorphic fungus with a mycelial phase in nature, much like the dimorphic Onygenales.

Although phylogenetic data from chitin synthase genes (*CHS*) strongly support DNA analyses, additional studies using other molecules, including the conserve proteins actin, β-tubulin, and elongation factor 1α, could further help in resolving this issue (Herr *et al.*, 2001). Chitin synthase genes have been successfully used for phylogenetic analysis (Bowen *et al.*, 1992). A 600 bp fragment of the *LlCHS2* gene from *L. loboi* was amplified and used with 24 other *CHS* sequences (Herr *et al.*, 2001). The sequences were aligned using CLUSTAL analysis and the samples investigated using Phylogenetic Analysis Using Parsimony (PAUP) with 24 other CHS deduced amino acid sequences. Again, the data showed strong support for the phylogenetic analysis performed using DNA sequences (Figure 3). However, the long branches between *L. loboi* and *P. brasiliensis* were not present using the LlCHS2 deduced amino acid sequences. Interestingly, in previous analyses the two *L. loboi*'s rDNA sequences showed 98% identity between them. Based on the minor variation found within the *LlCHS2* nucleotide sequences (Herr *et al.*, 2001), the differences between the two rDNA sequences may be due to rapid nucleotide substitution of the 18S SSU rDNA and ITS molecules rather than different populations of *L. loboi.* However, additional sequence analyses using other *L. loboi*'s proteins are necessary to investigate this anomaly. It is also possible that the difference in nucleotide substitutions between strains could reflect its adaptability to both, human or dolphin hosts or to the inability of the strain to grow in culture, although additional molecular data will be needed to answer this and other interesting questions.

Of interest to our research was the report that some strains of *P. brasiliensis* from patients with typical paracoccidioidomycosis were non-cultivable (Lacaz *et al.*, 2002). Although several hypotheses have been suggested to explain this phenomenon, we believe that the placement of *L. loboi* near *P. brasiliensis* may imply that some uncultivated strains of *P. brasiliensis* could be somehow closer to both *L. loboi* and *P. brasiliensis*. If this idea is correct, then *Lacazia* is not a separate genus but is more likely a species of *Paracoccidioides*. This hypothesis, suggested by many including Jorge Lobo, the first investigator who described the disease 70 years ago, could also indicate that the genus *Paracoccidioides* may include several species in addition to *P. brasiliensis*. We hope that future molecular studies of uncultured strains recovered from typical cases of paracoccidioidomycosis might solve this puzzle. Meanwhile, the data so far generated by molecular analysis suggests new breakthroughs concerning this elusive pathogen. We believe that the recent development of an animal model for lobomycosis (Madeira *et al.*, 2000) could be used as a reliable and constant source of DNA. This, in turn, could be decisive for new molecular studies to further understand the relationship of this unique pathogen with the other Onygenales including *P. brasiliensis*.

3. *Rhinosporidium seeberi*

Seeber first reported this pathogen, which produces spherical cells containing endospores, in Argentina in 1900 (Seeber, 1900). Three years later, O'Kinealy (cited by Ashworth, 1923) also reported the disease in India. As in the case of *L. loboi*, the first reports on its taxonomic placement stimulated controversy. For instance, Seeber believed that this pathogen was a sporozoa close to the coccidia Polysporea and also, perhaps related to the fungus *Coccidioides immitis,* while Ashworth (1923) argued that this unique pathogen was a lower fungus closely related to the members of the families Chytridineae and Olpidiaceae. Since then, the majority of investigators consider *R. seeberi* to be a protozoan or a fungus. Curiously, early investigators also noted the phenotypic similarities that some fish pathogens, such as *Ichthyophonus* and *Dermocystidium,* had with *R. seeberi* (Carini, 1940; Dunkerly, 1914).

The binomial *R. seeberi* was suggested by Ashworth (1923) to unify *Rhinosporidium kinealyi* (Minchin and Fantham, 1905) and *Coccidium seeberia* (cited by Ashworth, 1923), used by several authors until that time. Rhinosporidiosis has been reported in humans and animals of all continents except Australia. The disease, which is more prevalent in India and Sri Lanka (Kennedy *et al.*, 1995; Rippon, 1988b), is characterized by the formation of polyps more often found in the nostrils, mouth, and the ocular conjuctiva of its host. Although rhinosporidiosis has also been reported as multifocal skin lesions, the pathogen does not invade the internal organs. *R. seeberi* forms spherical sporangia that contain endospores in infected tissue at various stages of development. These phenotypic structures measure 60 to 450 μm in diameter and are pathognomonic of rhinosporidiosis. The mature sporangium contains thousands of 7 to 15 μm diameter endospores that are eventually released into the host's tissues through a pore (Mendoza *et al.*, 1999; Vanbreuseghen, 1973). Once in the tissues, the endospores increase in size eventually forming mature sporangia that then initiate another round of endospore formation during the *in vivo* life cycle.

All attempts to cultivate *R. seeberi* have failed (Rippon, 1988b), and reports claiming success in cultivating this organism have not been substantiated by others. Recently Mendoza *et al.* (1999) studying the mechanisms of endospores release from fresh sporangia, concluded that cultivation of *R. seeberi,* as suggested by others (Levy *et al.*, 1986), was more likely motivated by the observation of endospore release from mature sporangia on aqueous media rather than *in vitro* multiplication. Failure to cultivate this microbe has resulted in several bizarre suggestions that the ingestion of tapioca could lead to rhinosporidiosis or that *R. seeberi* was a cyanobacterium in the genus *Microcystis* (Ahluwalia, 1999). These and other new proposals proved to be incorrect (Ahluwalia *et al.*, 2001; Fredericks *et al.*, 2000; Mendoza *et al.*, 2002). Its intractability to culture and the fact that *R. seeberi* does not respond to any antifungal drugs, clearly suggested that this pathogen was a protista, but all attempts to prove this and other hypotheses using traditional approaches met with failure.

In 1998 four institutions, Michigan State University, East Lansing, MI, the Center for Disease Control and Prevention, Atlanta, GA, the University of California, Berkeley, CA, and the University of Paradeniya, Sri Lanka, proposed the use of molecular approaches to identify the phylogenetic connections of *R. seeberi* with the other microbes. One of the drawbacks in working with *R. seeberi* was the difficulty in finding fresh tissue. However, this problem has been solved since Dr. Sarath Arseculeratne has provided tissues from two patients. In order to isolate genomic DNA from tissues harbouring *R. seeberi*, we used a method similar to the one used for the analysis of *L. loboi*. However, we also purified sporangia from the infected tissues in order to eliminate as much host genomic DNA (Herr *et al.*, 1999). The primers NS1 and NS8 were used to amplify by PCR the 18S SSU rDNA from the genomic DNA samples isolated from the two patients with rhinosporidiosis. PCR amplification from the genomic DNA of both patients produced products of 1790 bp. After sequencing and BLAST analyses of the two strains, the data indicated that *R. seeberi* was phylogenetically related to a novel group of fish pathogens described as the DRIP (after *Dermocystidium*, rosette agent, *Ichthyophonus*, *Psorospermium*) clade by Ragan *et al.* (1996), later changed to Mesomycetozoa by Herr *et al.* (1999). Surprisingly, all of the organisms described by Ragan *et al.* (1996) were fungi-like fish pathogens found only in aquatic environments. Using their original data, we found that the closest relatives of *R. seeberi* were *Dermocystidium* species and the rosette agent, now known as *Sphaerothecum destruens* (Arkush *et al.*, 2003), both spherical and endosporulating fish pathogens, with a long history of inclusion within the protozoa and the fungi (Mendoza *et al.*, 2002).

Phylogenetic analysis of these two sequences showed that *R. seeberi* was the sister clade of two *Dermocystidium* species, and this clade was sister to *S. destruens* (Herr *et al.*, 1999a) (Figure 4). These two phylogenetic groups in turn, were sister to *I. hoferi* and *P. haeckeli*. This original research was soon confirmed by an almost identical study published a few months after the study by Herr *et al.* (1999a) (Fredericks *et al.*, 2000). Fredericks *et al.* (2000) came to the same conclusions using DNA from a dog with rhinosporidiosis. They reported that the phylogenetic affinities of *R. seeberi* with the other microbes were identical to those discussed recently by Herr *et al.* (1999a). Interestingly, both groups emphasized that the fish pathogens, described by Ragan *et al.* (1996), not only have phylogenetic similarities in their rDNA sequences with *R. seeberi*, but they also shared several phenotypic features in common. For instance, they possess spherical structures, some with endospores, a long history of inclusion within the fungi and protozoa, most are intractable to culture, and all are related to aquatic environments. *R. seeberi* phenotypic, biological, and molecular characteristics in common with this group strongly supports the belief that this human pathogen is found in wet environments.

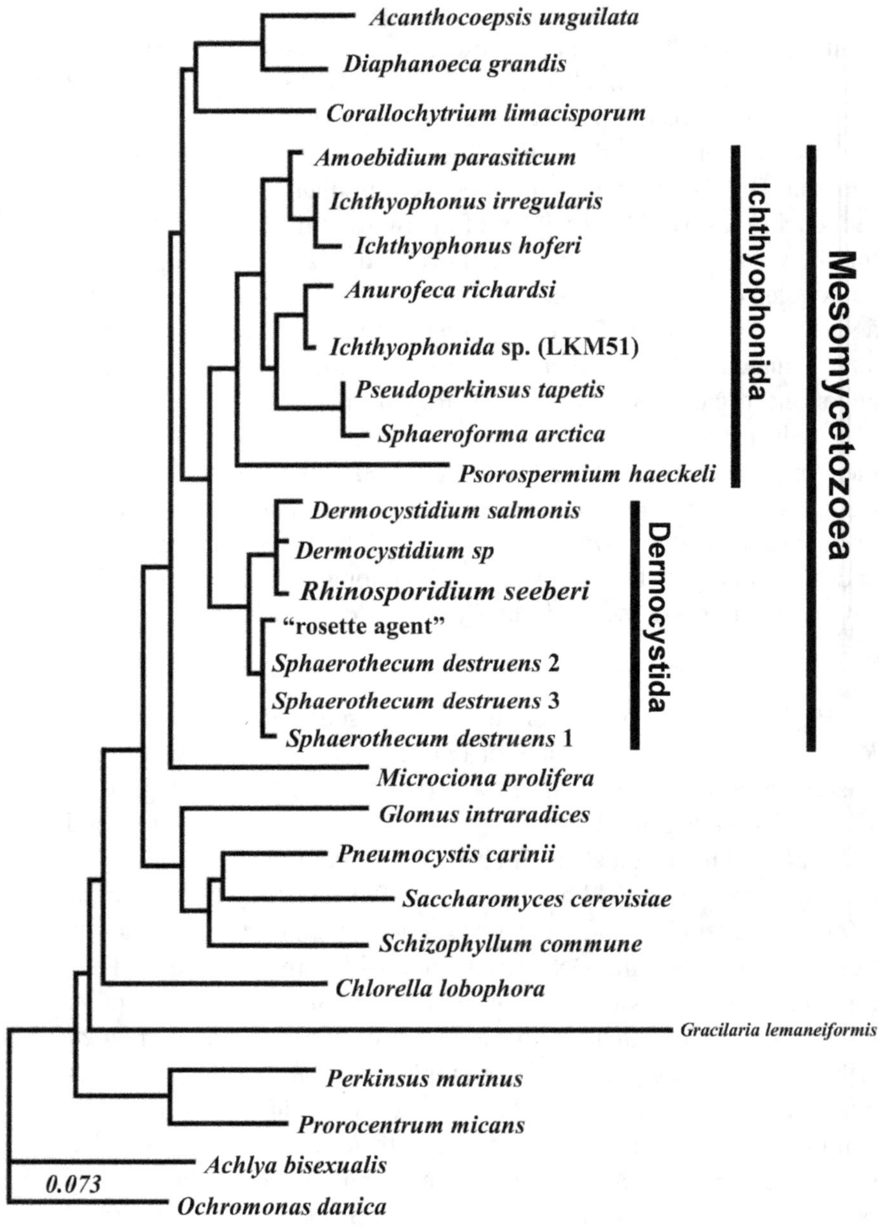

Figure 4. Phylogenetic analysis of the 18S SSU rDNA from several microbes and fifteen members of the Class Mesomycetozoea, made by neighbor joining using distances estimated by maximun likelihood. The topologies and branch support are similar to those obtained by parsimony analysis. In this tree *Rhinosporidium seeberi* is placed with the mesomycetozoeans in the Order Dermocystida. *Rhinosporidium seeberi* is the sister clade to *Dermocystidium* spp, their closest relatives (modified from Mendoza *et al.*, 2002).

Herr *et al.* (1999a) introduced the name Mesomycetozoa, which means between fungi and animals, without knowing that Cavalier-Smith (1998) had proposed the class Ichthyosporea to rename the DRIP's of Ragan *et al.* (1996). In his study, Cavalier-Smith also proposed two orders, Dermocystida and Ichthyophonida, to further accommodate this group of microbes. Recently, however, Mendoza *et al.* (2002) argued that the class Ichthyosporea was invalid. They based their position on the fact that the definitions used by Cavalier-Smith (1998) to erect the class does not reflect the main features of the group, and that not all members of the class are related to fish hosts. Thus, Mendoza *et al.* (2002) introduced a change to their original proposal on the Mesomycetozoa of Herr *et al.* (1999a), and erected the class Mesomycetozoea to replace the Ichthyosporea of Cavalier-Smith (1998). In recent reviews, Mendoza *et al.* (2001, 2002) retained the class Mesomycetozoea, the two original orders of Cavalier-Smith (1998), and created two new families, the family Rhinosporidiacae and Ichthyophonacae. Based upon these phylogenetic proposals, *R. seeberi* phylogenetic affinities are:

Kingdom= Protozoa (Cavalier Smith, 1998)
Phylum= Nomonada (Cavalier-Smith, 1998)
Class= Mesomycetozoea (Mendoza *et al.*, 2002)
Order= Dermocystida (Cavalier-Smith, 1998)
Family= Rhinosporidacae (Mendoza *et al.*, 2001)
Genus= *Rhinosporidium* (Minchin and Fantham, 1905)
Species= *seeberi* (Ashworth, 1923)

Two life cycles for the members of the order Dermocystida and Ichthyophonida have been proposed (Mendoza *et al.*, 2002). This proposal was based on the known parasitic stages of some members of the class. For instance, the Dermocystida, where *R. seeberi, S. destruens* and *Dermocystidium* spp. are placed, is characterized by the formation of spherical bodies with endospores. The endospores are retained within the host's tissues, and the cycle is repeated inside the host. However, in the genera *Dermocystidium* and *Sphaerothecum*, some of the endospores are released outside the host and develop round uniflagellated motile cells that later encyst to produce spherical endosporulating structures. Arkush *et al.* (1998) and others (Arkush *et al.*, 2003) succeeded in replicating the *in vitro* formation of flagellated cells from these two fish pathogens. Because *R. seeberi* is classified in the Dermocystida, it was thought that it might have a flagellated stage as well. However, *in vitro* experiments, using the protocols of Arkush *et al.* (1998) with the viable *R. seeberi* endospores from infected humans and dogs, did not lead to the observation of formation of flagellated cells (D. McMeekin and L. Mendoza, unpublished data). Because *R. seeberi* is phylogenetically more closely related to *Dermocystidium*, this suggests that perhaps *R. seeberi* lost its flagellated stage or that the conditions for zoospore induction in *Rhinosporidium* require special factors. In addition, the Order Ichthyophonida is characterized by the formation of oval or round endosporulating bodies. The endospores usually are released but then form amoeba-like cells that later encyst and repeat the cycle. The formation of amoeba-like cells in the mesomycetozoeans illustrates the taxonomic heterogenicity of the group.

Based on the early findings that the 18S SSU rDNA sequences of *R. seeberi* from humans and dogs were identical (Herr *et al.*, 1999a; Fredericks *et al.*, 2000), it was thought that *Rhinosporidium* more likely was a monotypic genus (Mendoza *et al.*, 2002). More recently, however, phylogenetic analysis of the ITS1, 5.8S, and ITS2 sequences from humans, dogs, and swans with rhinosporidiosis indicated that the genus *Rhinosporidium* may possess several strains with important changes in their ITS sequences (V. Silva and L. Mendoza, unpublished data). For instance, all ITS sequences from humans were almost identical. However, the ITS sequences from dogs and swans possessed larger ITS sequences that those in humans and their similarities were low. More importantly, the DNA similarity of strains from dogs and swans was also low with long 50 bp gaps between them. These unexpected results suggest that the genus *Rhinosporidium* may possess several host specific strains (new species?). We presently are investigating more strains of *R. seeberi* from different hosts to validate these preliminary results. However, if *R. seeberi* indeed harbours several host specific species, this finding could well support early reports that rhinosporidiosis can not be experimentally reproduced in animals infected with *R. seeberi* from humans.

From only six members in 1999, the class Mesomycetozoea has increased in number at an accelerated rate (Mendoza *et al.*, 2002). Presently, the class comprises 12 members, some of them previously studied within the fungi. It is predicted that more members will be added in the near future, especially fungal-like microbes with obscure taxonomic backgrounds. Of interest to the mesomycetozoeans and to

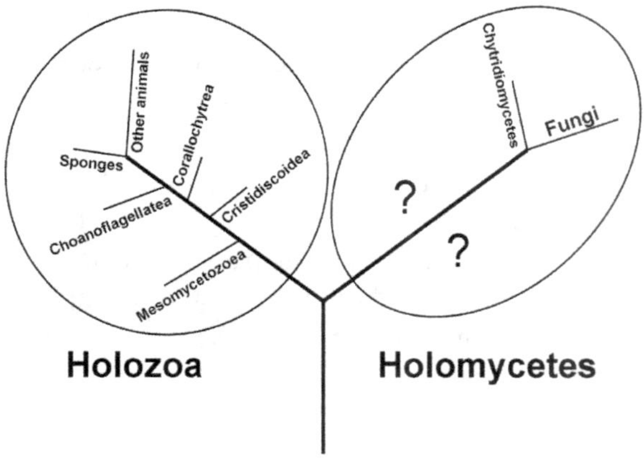

Figure 5. The tree was constructed using data from the proposal of Cavalier-Smith (1998), the finding of Lang *et al.* (2002), Mendoza *et al.* (2002), and Ragan *et al.* (1996). The figure depicts the putative tree relationships between the Mesomycetozoea, home to *Rhinosporidium seeberi*, and the other microbes basal to the multicellular animals (Holozoa). The classes Cristidiscoidea and Corallochytrea have been placed in the Holozoa based on the Cavalier-Smith's proposal (1998) and on recent DNA analyses (Mendoza, *et al.*, 2002). The term Holomycetes is proposed to name any future basal unicellular microbes linked to the multicellular fungi plus the Chytridiomycetes.

the genus *Rhinosporidium,* was the proposal of four classes in the Phylum Neomonada by Cavalier-Smith (1998): Choanoflagellatea, Corallochytrea, Mesomycetozoea (Ichthyosporea), and Cristidiscoidea. All of them are located just below the animal branch and after the divergence of fungi (Cavalier-Smith, 1998; Lang *et al.*, 2002, Mendoza *et al.*, 2002). A recent report of a new genus and species, *Capsaspora owczarzaki*, placed this unique organism as a sister clade with the Mesomycetozoea (Hertel *et al.*, 2002), but distant from the choanoflagellates and corallochytres. We believe that to maintain simplicity within these groups, new classes will be needed to phylogenetically accommodate new members that do not fit in any of the current classes. Lang *et al.* (2002) and more recently Burger *et al.* (2003), introduced the term Holozoa to accommodate the animals and the basal unicellular protist microbes located after the divergence of fungi. The fungi, on the other hand, seem to have only a basal group of microbes with uniflagellated cells, the Chytridiomycetes. However, are the chytridiomycetes really the only basal group of the Fungi? Based on the diversity of unicellular microbes found below the animals, one could speculate that a similar situation could be expected in the fungi. If this is the case, an epithet to name any basal unicellular protistal microbe plus the fungi and the chythridiomycetes is necessary. We are proposing the term Holomycetes to fulfill this task (Figure 5).

In summary, five years after Herr *et al.* (1999a) reported the phylogenetic affinities of *R. seeberi*, more new information has appeared on this enigmatic spherical pathogen and its relationship with the other microbes than during the past 100 years of taxonomic studies. This new information would not be possible without the aid of phylogenetic and molecular analyses.

4. Future Molecular Strategies to Further Study *L. loboi* and *R. seeberi*

The use of molecular approaches to address the status of *L. loboi* and *R. seeberi* finally revealed their phylogenetic relationships. However, these results are only preliminary to more broad applications to answer other questions related to these two uncultivated microbes. For instance, the molecular characterization of the homologous gp43 protein of *P. brasiliensis* in *L. loboi* is a very interesting study. Based on the protective features of this molecule (Taborda *et al.*, 1998), investigations of *L. loboi*'s gp43 homologues could lead to the development of vaccine strategies to protect communities in endemic areas. The characterization of more strains of *L. loboi*, using phylogenetic analysis, would be of importance to better characterize this pathogen. The development of specific molecular probes, to investigate the presence of *L. loboi* in environmental samples, could be of paramount importance to map its ecological niche in the endemic areas. Each of these new molecular studies would be of importance to decipher *L. loboi*'s ecological and epidemiological features, still elusive.

Regarding *R. seeberi*, the observations made over the next years should be exciting. The molecular characterization of the antigenic proteins, recently reported by Herr *et al.* (1999b), is an interesting research proposal that could lead to novel targets for drug and vaccine development. The investigation of new strains of *R. seeberi* from a variety of hosts would be significant to confirm the existence of host specific strains, or even new species in the genus *Rhinosporidium*. And finally, perhaps the sequencing of *R. seeberi*'s entire genome would contribute to our understanding of its intractability to culture and add more data about the origin of the multicellular animals and fungi.

5. Acknowledgments

We thank Dr. Libero Ajello for his comments and suggestions to the manuscript. The authors also thank Dr. John W. Taylor for some of the figures used in this study.

References

Ahluwalia, K.B. 1999. Culture of the organisms that causes rhinosporidiosis. J. Laryngol. Otol. 113: 523-528.

Ahluwalia, K.B. Mendoza, L., Herr, R.A., and Ajello, L. 2001. Causative agent of rhinosporidiosis. J. Clin. Microbiol. 39: 413-415.

Alexopoulus, C.J., Mims, C.W., Blackwell, M. 1996. Introductory Mycology. 4th ed. John Wiley & Sons Inc., USA.

Arkush, K.D., Mendoza, L., Adkison, M.A., and Ajello, L. 2003. Observations of the life stages of *Sphaerothecum destruens* n. g., n. sp., a mesomycetozoean fish pathogen formally referred to as the rosette agent. J. Eukaryot. Microbiol. 50: 421-429.

Arkush, K.D., Frasca, Jr. S., and Hedrick, R.P. 1998. Pathology associate with the rosette agent, a systemic protist infecting salmonid fishes. J. Aquat. Anim. Health 10: 1-11.

Ashworth, J.H. 1923. On *Rhinosporidium seeberi* (Wernicke, 1903) with special reference to its sporulation and affinities. Trans. R. Soc. Edinb. 53: 301-342.

Baldauf, S.L., and Palmer, J.D. 1993. Animals and fungi are each other's closest relatives: congruent evidence from multiple proteins. Proc. Natl. Acad. Sci. USA 90: 11558-11562.

Baldauf, S.L., Roger, A.J., Wenk-Siefert, I., and Doolittle, W.F. 2000. A kingdom-level phylogeny of eukaryotes based on combined protein data. Science 290: 972-977.

Berbee, M.L., and Taylor, J.W. 2001. Fungal molecular evolution: gene trees and genologic time. In: The Mycota: Systematics and Evolution. D. McLaughlin, E. McLaughlin, and P. Lemke, eds. Springer-Verlag, Berlin. p. 229-245.

Bowen, A.R., Chen-Wu, J.L., Momany, M., Young, R. Szaniszlo, P.J., and Robbins, P.W. 1992. Classifications of fungal chitin synthases. Proc. Natl. Acad. Sci. USA 89: 519-523.

Burger, G., Forget, L., Zhu, Y., Gray, M.W., and Lang, B.F. 2003. Unique mitochondrial genome architecture in unicellular relatives of animals. Proc. Natl. Acad. Sci. USA 100: 892-897.

Burns, R.A., Roy, J.S., Woods, C., Padhye, A.A., and Warnock, D.W. 2000. Report of the first human case of lobomycosis in the United States. J. Clin. Microbiol. 38: 1283-1285.

Carini, A. 1940. Sobre un parásito selmehante ao "*Rhinosporidium*", encontrado em quistos da pele de una "phyla" Arq. Inst. Biol. São Paulo 11: 93-96.

Cavalier-Smith, T. (1998). Neomonada and the origin of animal and fungi. In: Evolutionary Relationships among Protozoa, vol 65. Kluwer, Dordrecht. p. 375-407.

Copeland, H.F. 1956. The Classification of Lower Organisms. Pacific Books, Palo Alto, California.

Cushion, M.T. 1998. *Pneumocystis carinii*. In: Topley & Wilson's Microbiology and Microbial Infections. Vol 4. L. Ajello R.J. Hay Eds. Arnold, London-Sydney. p.645-683.

Dick, M.W. 2001. Straminipilous Fungi: Systematics of the Peronosporomycetes Including Accounts of the Marine Straminipilous Protist, the Plasmodiophorids and Similar Organisms Kluwer Academic Publishers, London.

Dunkerly, J.S. 1914. *Dermocystidium pusula* Perez, parasitic on trutta fario. Zool. Anz. 44: 179-182.

Edman, J.C., Kovacs, J.A., Masur, H., Santi, D.V., Elwood, H.J., and Sogin, M.L. 1988. Ribosomal RNA sequence shows *Pneumocystis carinii* to be a member of the fungi. Nature 334: 519-522.

Fisher, M.C., Koenig, G.L. White, T.J. and J. W. Taylor. 2002. Molecular and phenotypic description of *Coccidioides posadasii* sp. nov., previously recognized as the non-California population of *Coccidioides immitis*. Mycologia 94: 73-84.

Fredericks D.N., Jolley, J.A., Lepp, P.W., Kosek, J.C., and Relman, D.A. 2000. *Rhinosporidium seeberi*: a human pathogen from a novel group of aquatic protistan parasites. Emerg. Infect. Dis. 6: 273-282.

Haubold, E.M., Aronson, J.F., Cowan, D.F., McGinnis M.R., and Cooper Jr., C.R. 1998. Isolation of fungal rDNA from bottlenose dolphin skin infected with *Loboa loboi*. Med. Mycol. 35: 389-398.

Haubold, E.M., Cooper Jr. C.R., Wen, J.W., McGinnis M.R., and Cowan D.F. 2002. Comparative morphology of *Lacazia loboi* (syn. *Loboa loboi*) in dolphins and humans. Med. Mycol. 38: 9-14.

Herr, R.A., Tarcha, E.J., Taborda, P.R., Taylor, J.W., Ajello, L., and Mendoza, L. 2001. Phylogenetic analysis of *Lacazia loboi* places this previously uncharacterized pathogen within the dimorphic Onygenales. J. Clin. Microbiol. 39: 309-314.

Herr, R.A., Ajello, L., Taylor, J.W., Arseculeratne, S.T., and Mendoza, L. 1999a. Phylogenetic analysis of *Rhinosporidium seeberi*'s 18S small subunit ribosomal DNA groups this pathogen among members of the protoctistan Mesomycetozoa clade. J. Clin Microbiol. 37: 2750-2754.

Herr, R.A., Mendoza, L., Arseculeratne S.N., and Ajello, L. 1999b. Immunolocalization of an endogenous antigenic material of *Rhinosporidium seeberi* expressed only during mature sporangial development. FEMS Immunol. Med. Microbiol. 23: 205-212.

Hertel, L.A., Bayne, C.J., and Loker, E.S. 2002. The symbiont *Capsaspora owczarzaki*, nov. gen. nov. sp., isolated from three strains of the pulmonate snail *Biomphalaria glabrata*, is related to members of the Mesomycetozoea. Int. J. Parasitol. 32: 1183-1191.

Hughes, W.T. 1989. *Pneumocystis carinii*: Taxing taxonomy. Europ. J. Epidemiol. 5: 265-269.

Hugenholtz, P., Goebel B.M., and Pace, N.R. 1998. Impact of culture-independent studies on the emerging phylogenetic view of bacterial diversity. J. Bacteriol. 180: 4765-4774.

Katz, L.A. 1999. The tangled web: gene genealogies and the origin of eukaryotes. Amer. Naturalist 154: S136-S145.

Kendrick, B. 1985. The Fifth Kingdom. Stan Brown Printers Limited, Ontario, Canada.

Kennedy, F.A., Buggage, R.R., and Ajello, L. 1995. Rhinosporidiosis: a description of an unprecedented outbreak in captive swans (*Cygnus* spp) and a proposal for revision of the ontogenic nomenclature of *Rhinosporidium seeberi*. J. Med. Vet. Mycol. 33: 157-165.

Kirk, P.M., Cannon, P.F., David, J.C., and Stalpers, J.A. 2001. Ainsworth & Bisby's Dictionary of the Fungi, 9th ed. CAB International, Wallingford, Oxfordshire, UK.

Lacaz, C.S., Porto, E., Costa-Martins, J.E., Heins-Vaccari, E.M., and de Melo, N.T. 2002. Tratado de Micologia Médica Lacaz. Sarvier, São Paulo, Brazil.

Lang, B.F., O'Kelly, C., Nerad, T., Gray, M.W., and Burger, G. 2002. The closest relatives of animals. Curr. Biol. 12: 1773-1778.

Levy, M.G., Meuten, D.J., and Breitschwerdt, E.B. 1986. Cultivation of *Rhinosporidium seeberi in vitro*: interaction with epithelial cells. Science 234: 474-476.

Ludwing, W., Bauer, S.H., Bauer, M., Held, I., Kirchhof, G., Schulze, R., Huber, I.R., Spring, S.,Hartmenn, A., and Schleifer, K.H. 1997. Detection and *in situ* identification of representative of widely distributed new bacterial phylum. FEMS Microbial Ecol. 153: 181-190.

Lobo J.O. 1930. Nova specie de blastomicose. Brasil Med. 44: 1227.

Madeira, S., Opromolla, D.V.A., and Belone, A.F.F. 2000. Inoculation of Balb/C mice with *Lacazia loboi*. Rev. Inst. Med. Trop. S. Paulo. 42: 239-243.

Margulis, L., and Schewartz, K.V. 1998. Five Kingdoms. An Illustrate Guide of the Phyla of Life on Earth. W.H. Freeman and Company, New York.

Mendoza, L., Herr, R.A., Arseculeratne, S.N., and Ajello, L. 1999. *In vitro* studies on the mechanisms of endospore release by *Rhinosporidium seeberi*. Mycopathologia 148: 9-15.

Mendoza, L., Ajello, L., and Taylor, J.W.2001. The phylogenetic connections of *Lacazia loboi* and *Rhinosporidium seeberi* have been finally resolved by the used of molecular tools. Rev. Iberoam Micol.18: 95-98.

Mendoza, L., Taylor, J.W., and Ajello, L. 2002. The class Mesomycetozoea: a heterogeneous group of microorganisms at the animal-fungal boundary. Annu. Rev. Microbiol. 56: 315-344.

Minchin, E.A., and Fantham, H.B. 1905. *Rhinosporidium kinealyi*, n.g., n.sp., a new sporozoon from the mucous membrane of the septum nasi of man. Quart. J. Microbiol. Sci. 49: 521-532.

Patterson, D.J. 1989. Stramenopiles: chromophytes from a protistan perspective. In: The Chromophyte Algae, Problems and Perspectives. G.C. Green, B.S.C. Leadbeater, and W.L. Diver, eds. Clarendon Press, Oxford, UK. p. 357-379.

Patterson, D.J. 1999. The diversity of eukaryotes. Amer. Naturalist 65: S96-S124.

Ragan, M.A., Goggin, C.L., Cawthorn, R.J., Cerenius, L., Janieson, A.V.C., Plourdes, M., Tand, T.G., Soderhall K., and Gutell, R.R. (1996). A novel clade of protistan parasites near the animal-fungal divergence. Proc. Natl. Acad. Sci. USA. 93: 11907-11912.

Rippon, J.W. 1988a. *Lacazia loboi*. In: Medical Mycology, 3rd ed. W.B. Saunders Company, Philadelphia, PA. p. 353-361.

Rippon, J.W. 1988b. *Rhinosporidium seeberi*. In: Medical Mycology, 3rd ed. W.B. Saunders Company, Philadelphia, PA. p. 381-432.

Roger, A.G. 1999. Reconstructing early events in eukaryotic evolution. Amer. Naturalist 154: S146-S163.

Rondon, M.R., August, P.R., and Bettermann, A.D. 2000. Cloning the soil metagenome a strategy for accessing the genetic and functional diversity of uncultured microorganisms. Appl. Environ. Microbiol. 66: 2541-2547.

Seeber, G.R. 1900. Un nuevo esporozoario parásito del hombre. Dos casos encontrados en polipos nasales. Thesis. Universidad Nacional de Buenos Aires, Imprenta Librería Boullosa, Buenos Aires, Argentina.

Symmer, W.S. 1983. A possible case of Lobo's disease acquired in Europe from a bottle nosed dolphin. Bull. Soc. Path. Exot. 46: 777-764.

Taborda, C.P. Juliano, M.A., Puccia, R., Franco, M., and Travassos, L.R. 1998. Mapping of the T-cell epitope in the major 43-kilodalton glycoprotein of *Paracoccidioides brasiliensis* which induces a Th-1 response protective against fungal infection in BALB/c Mice. Infect. Immun. 66: 786-793.

Taborda, P.R., Taborda, V.A., and McGinnis, M.R. 1999. *Lacazia loboi* gen. Nov., comb. Nov., the etiologic agent of lobomycosis. J. Clin. Microbiol. 37: 2031-2033.

Tarcha E.J., Herr, R.A., Taylor, J.W., Ajello, L., and Mendoza, L. 2001. Phylogenetic analysis of the ITS1, 5.8S and ITS sequences of *Lacazia loboi* confirms the placement of this unique fungal pathogen within the Onygenales. 101st General Meeting of the American Society for Microbiology. p 381. Orlando, Florida.

Torsvik, V.R., Sorheim, B., and Goksoyr, J. 1996. Total bacterial diversity in soil and sediment communities a review. J. Ind. Microbiol. 17: 170-178.

Vanbreuseghen, R.1973. Ultrastructure of *Rhinosporidium seeberi*. Int. J. Dermatol. 12: 20-28.

Vilani-Moreno, F.R., and Opromolla, D.V.A. 1997. Determinacão da viabilidade do *Paraccidioides loboi* em biópsias de pacientes portadores de doença de Jorge Lobo. An. Bras. Dermatol. (Rio de Janeiro) 72: 433-437.

Wainright, P.O., Hinkle, G., Sogin, M.L., and Stickel, S.K. 1993. Monophyletic origins of the metazoa: an evolutionary link with fungi. Science 260: 340-342.

Ward, D.M., Weller, R., and Bateson, M.M. 1990. 16S rDNA sequences reveal numerous uncultured microorganisms in natural community. Nature 345: 63-65.

Whitman, W.E., Coleman, D.C., and Wiebe, W.J. 1998. Prokaryotes: the unseen majority. Proc. Natl. Acad. Sci. USA 95: 6578-6583.

Whittaker, R.H. 1969. New concepts of kingdoms or organisms. Evolutionary relations are better represented by new classifications than by the traditional two kingdoms. Science 163: 150-160.

From: Pathogenic Fungi: Structural Biology and Taxonomy
Edited by: G. San-Blas and R.A. Calderone

Chapter 9

Gene Genealogical Analyses of Human Fungal Pathogens

Jianping Xu

Abstract

The last several years have seen increasing use of DNA sequence data in population genetic analyses. In this chapter, I summarized the principles of gene genealogical analyses and how such analyses have been used to address evolutionary and population genetic issues of three human fungal pathogens *Cryptococcus neoformans, Histoplasma capsulatum* and *Coccidioides immitis*. Within each of the species, gene genealogical analyses provided unambiguous evidence for significant divergence, recent dispersion, clonality and recombination. In addition, hybridization was detected in both *C. neoformans* and *H. capsulatum*. While gene genealogical analyses so far have been focused on haploid fungal species, the approach described here should be applicable to diploid organisms such as *Candida albicans*.

1. Introduction

The last two decades saw significant increases in human fungal infections. Consequently, our understandings of fungal infectious disease epidemiology are improving. Many potential risk factors have been analyzed, including geographic location, age, sex, nutrition, life style, and other underlying conditions such as infections with the human immunodeficiency virus. Parallel to our understandings of host factors are studies of fungal factors that make these pathogens virulent. While most of our attention has been on specific genes and genetic pathways of these fungal pathogens on their virulence, population studies are beginning to make increasingly important contributions. In this chapter, I will review recent research advances in population studies of fungal pathogens, focusing on results obtained with DNA sequencing.

Compared to other types of markers (e.g. multilocus enzyme electrophoresis or MLEE, random amplified polymorphic DNA or RAPD, amplified fragment length polymorphisms or AFLP, restriction fragment length polymorphisms or RFLP, PCR-RFLP and PCR fingerprinting) that have been applied to analyzing fungal populations, DNA sequence markers have many advantages. First, nucleotides in a DNA sequence are unambiguous. Such certainty is essential for many analyses. Second, nucleotides in a given DNA fragment typically share extended evolutionary history. Such sharing cannot be assumed between genetic markers in different parts of genomes as those obtained with other methods. Third, DNA sequences can be easily stored in and retrieved from public databases such as GenBank. Existence of such public databases makes data sharing among investigators possible. Fourth, many analytical tools for DNA sequences are available. Indeed, methods have been developed to infer a variety of processes governing the changes in populations and species (Swofford, 2003). Lastly, DNA sequencing technology is now widespread and affordable in many laboratories throughout the world.

2. Gene Genealogy and its Applications

Recent advances in molecular biology, genomics and biotechnology have made DNA sequencing a relatively inexpensive data gathering exercise. Typically, using gene genealogies to infer a population process involves the following steps:

1. Select a set of strains appropriate to address the specific question(s).
2. Select a set of genes or DNA fragments containing polymorphic nucleotide sites.
3. Obtain DNA sequence for each gene from each of the strains.

4. Align and construct a genealogy for each gene and compare genealogies from different genes.

Within a species, different strains, genes and analytical methods may be needed to address different questions. The basic gene genealogical analysis is to determine the divergence and phylogeny among alleles of the same gene and to compare phylogenies of different genes. Using gene genealogies, a variety of fundamental questions have been addressed in several human fungal pathogens. For example, first, are fungal populations clonal or sexual or a combination of both? Second, what are the geographical patterns of genetic variations? Is there evidence of recent dispersals? Third, how much divergences are there among strains and populations? Significant divergence may indicate speciation. Fourth, are genealogies of different genes congruent? Significant incongruence indicates that different genes may have different evolutionary histories. The following basic tests and underlying principles are used to address the above questions.

2.1. Clonality and Recombination

Clonality is the result of asexual reproduction. During asexual reproduction, the whole genome is inherited intact from one generation to the next and, baring mutations, they should have identical genomic sequences. As a result, different genes should produce the same phylogenetic pattern for the same set of strains. In contrast, recombining genomes are reshuffled and reconstituted in every generation. Consequently, genealogies of different genes may be different.

2.2. Biogeography and Dispersal

Gene genealogies can be used to infer patterns of relationships among geographic strains and populations. If dispersals are limited or non-existent, strains from the same geographic areas will be clustered on individual gene genealogies. The closer geographically, the more similar strains' gene sequences should be. Conversely, lack of a positive correlation between geographic distance and sequence divergence would be consistent with frequent dispersals.

2.3. Divergence and Speciation

The patterns and degrees of genetic divergence among gene sequences can be used to define phylogenetic species. Phylogenetic species are widely used in organisms where sexual crossing is not observable or not possible in laboratory settings. To call a group of strains a phylogenetic species, significant divergences are needed from all analyzed loci between this group and other strains and populations.

Gene genealogies have been most heavily used in three fungal pathogens *Cryptococcus neoformans*, *Histoplasma capsulatum*, and *Coccidioides immitis*. In the following sections, I will review gene genealogical studies in each of the three species.

3. *Cryptococcus neoformans*

C. neoformans (=*Filobasidiella neoformans*) is an important fungal pathogen of humans and other mammals throughout the world. It is a basidiomycetous yeast typically with a haploid nucleus (Casadevall and Perfect, 1998). From the 1980's to early 1990's, due to the high number of immunocompromised hosts, the incidence of cryptococcal disease increased dramatically. However, the incidence has been decreasing since the mid-1990s in the developed countries perhaps because of improved therapies for HIV and other immunocompromising conditions. Elsewhere, particularly in Africa, the incidence of cryptococcosis remains high and is expected to continue to rise due to increasing incidences of AIDS.

Using commercial monoclonal antibodies to capsular epitopes, strains of *C. neoformans* are classified into five distinct serotypes - A, B, C, D, and AD. While reacting with species-specific antibodies, a small number of strains do not react with any of the four serotype-specific antibodies (A, B, C, or D). These serotypes are proposed to be grouped into three varieties: *C. neoformans* var. *neoformans* representing serotype D strains, *C. neoformans* var. *grubii* represents serotype A strains, and *C. neoformans* var. *gattii* represents serotypes B and C strains (Kwon-Chung *et al.*, 1982; Franzot *et al.*, 1999). However, this three-variety classification system ignores the classification of serotype AD strains.

Table 1. Geographic and serotype distributions of *Cryptococcus neoformans* strains used in the study by Xu *et al.* (2000)

Geographic origin	Serotype						**Total**
	A	D	AD	B	C	UT	
New World							
USA North Carolina	2	1	1		1	2	**7**
New York City	1	2					**3**
California	1				1		**2**
Cuba	1						**1**
Colombia					2		**2**
Brazil	3						**3**
Old World							
Asia							
China	1	1		1			**3**
India	1		1	1	1		**4**
Kuwait	1		1				**2**
Europe							
Denmark		1					**1**
Italy	2	2					**4**
Africa							
Tanzania	1						**1**
Zaire				1			**1**
Total	**14**	**7**	**3**	**3**	**5**	**2**	**34**

To understand the evolutionary relationships among strains, geographic populations, serotypes and varieties, a series of gene genealogy-based studies were conducted. The first analyzed 34 strains from various locations around the world (Xu *et al.*, 2000; Table 1). These strains included 14 serotype A strains, 7 serotype D strains, 3 serotype B strains, 5 serotype C strains, 3 serotype AD strains and 2 strains whose serotypes could not be determined (UT). Fragments of four genes were analyzed for each strain: (1) the mitochondrial large ribosomal subunit RNA or mtLrRNA; (2) the internal transcribed spacer (ITS) region of the nuclear rRNA, including ITS1, 5.8S rRNA subunit and ITS2; (3) the orotidine monophosphate pyrophosphorylase gene or URA5; and (4) the diphenol oxidase gene or LAC.

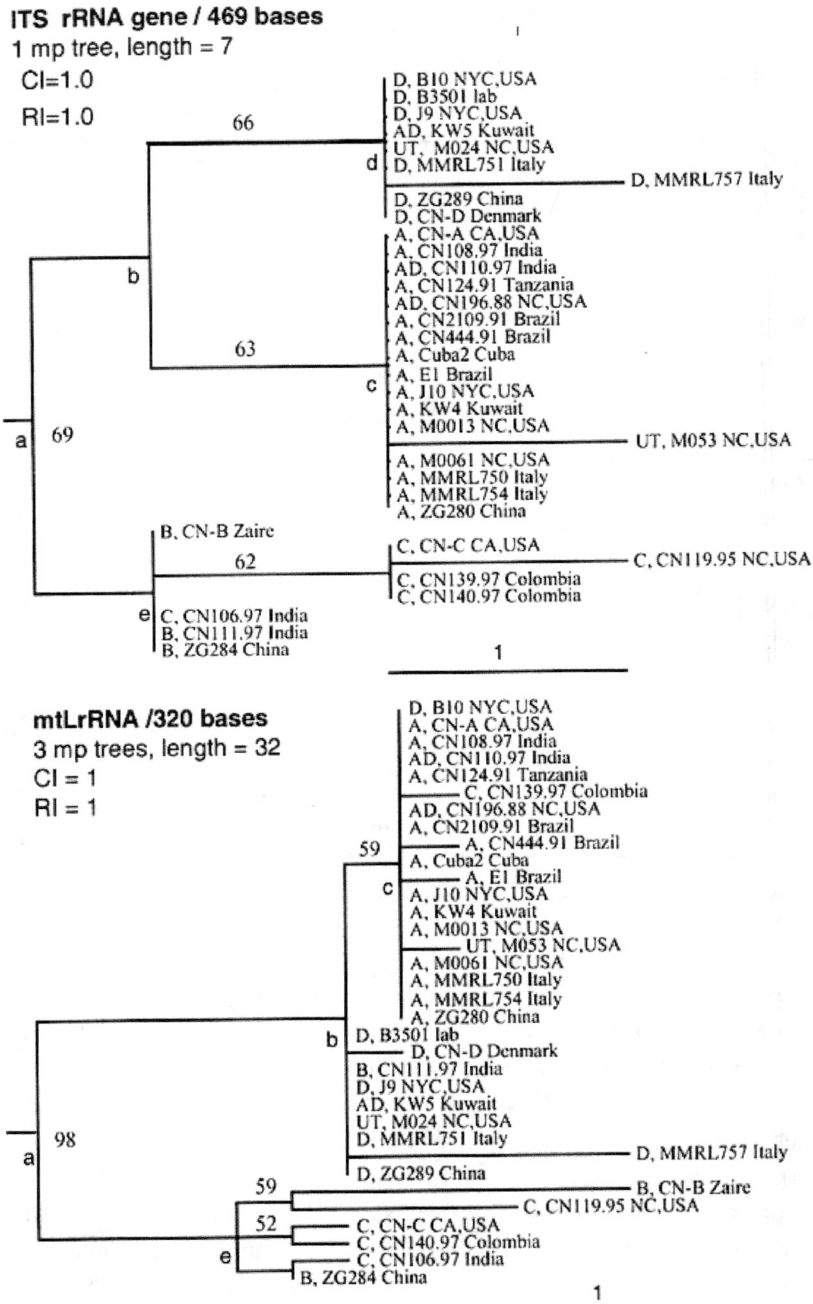

Figure 1. One most parsimonious tree for 34 isolates of *C. neoformans* from each of the four gene regions sequenced. CI, consistency index; RI, retention index. Numbers above each branch are bootstrap values >50% and based on 500 replicates. For URA5 and LAC trees, branches with >50% of bootstrap values were also strict consensus branches. Strain designation indicates serotype, isolate name, and geographic origin (CA, California; NYC, New York City; NC, North Carolina, all from USA). With the exception of five strains (see text), all major phylogenetic

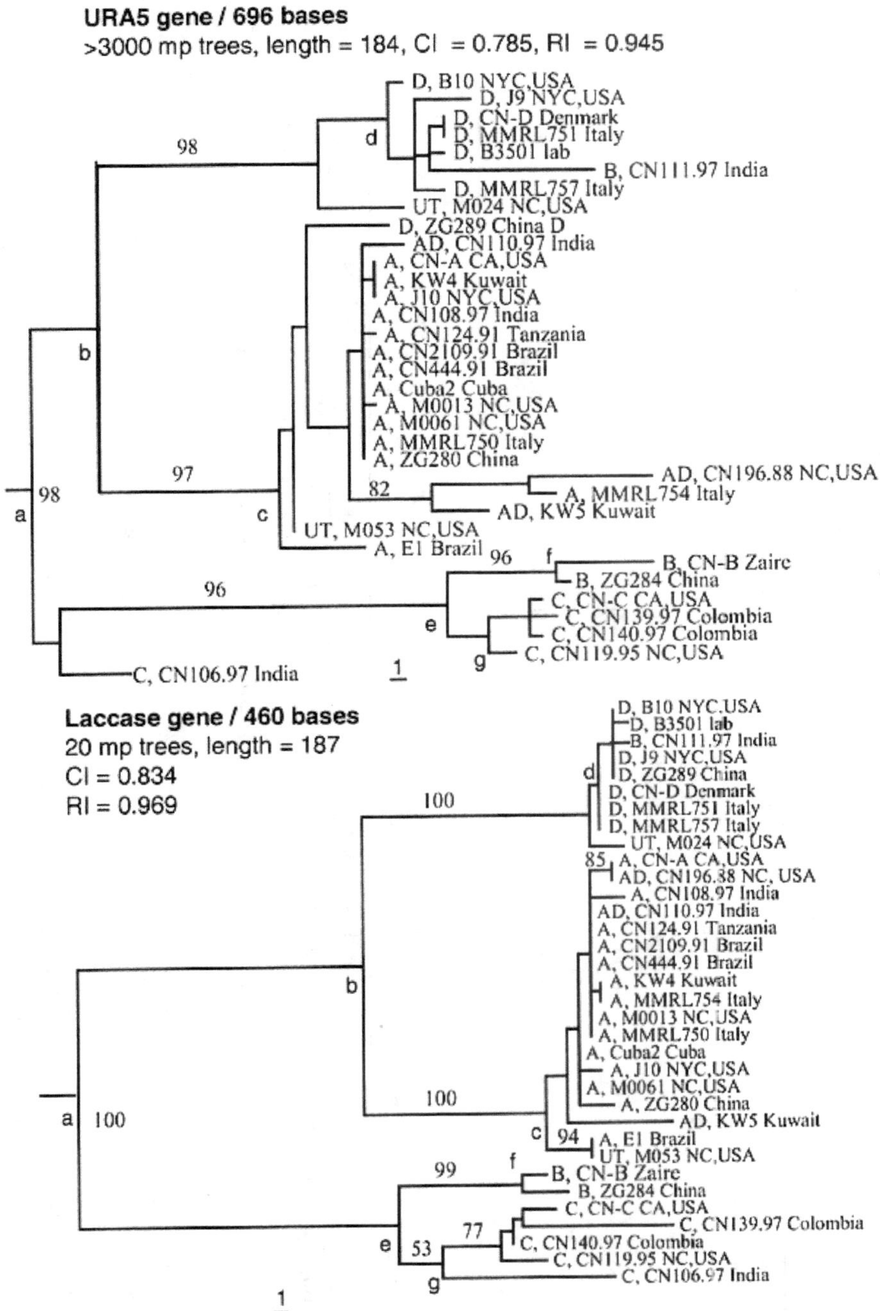

groups correspond to traditional classifications. Of the two serologically untypable strains, one (M0024) clustered consistently with the serotype D group and the other (M0053) clustered consistently with the serotype A group. Two of the three strains of serotype AD, CN110.97 and CN196.88, clustered consistently with the serotype A group; while the other (KW5) lacked a consistent affinity with any of the serotypes. Scale bar represents one nucleotide substitution. Reproduced with permission from Xu *et al.* (2000).

Table 2. Kimura-2 parameter distance among gene lineages in *Cryptococcus neoformans* (modified from Xu *et al*., 2000)

Serotype group	Genes				Time since divergence In million years [1]
	MtLrRNA	ITS	URA5	LAC	
Between serotypes A, D group and B, C group	0.02037	0.00310	0.04748	0.11296	37
Between serotypes A and D	0.00225	0.00205	0.03035	0.04685	18.5
Within serotype A	0.00032	0.00013	0.00580	0.01348	4.5
Within serotype D	0.00225	0.00026	0.01033	0.00121	3.3
Between serotypes B and C	0.01178	0.00078	0.01038	0.03158	9.5
Within serotype B	0.01178	0.00078	0.00425	0.00637	2.6
Within serotype C	0.01178	0.00028	0.00370	0.02610	6.4

Table 3. Genetic distance between pairs of strains of *Cryptococcus neoformans* from the same or different continents (Jukes-Cantor distance $\pm$ standard deviation) [1] (modified from Xu *et al*., 2000)

	North America	**South America**	**Europe**	**Asia**	**Africa**
North America (N=12)	0.0415 ± 0.0316	0.0434 ± 0.0335	0.0365 ± 0.0280	0.0407 ± 0.0318	0.0535 ± 0.0397
South America (N=5)		0.0492 ± 0.0354	0.0494 ± 0.0295	0.0450 ± 0.0334	0.0491 ± 0.0394
Europe (N=5)			0.0298 ± 0.0215	0.0399 ± 0.0282	0.0604 ± 0.0374
Asia (N=9)				0.0459 ± 0.0325	0.0529 ± 0.0406
Africa (N=2)					0.0913

[1]Students t-tests demonstrated no significant difference in genetic dissimilarity between pairs of strains from the same or different continents ($p>0.05$ in all comparisons).

The genealogical patterns among all strains for each of the four genes are shown in Figure 1. Three issues were addressed in this study: (1) divergence and speciation, (2) biogeography and dispersal, and (3) hybridization.

3.1. Divergence and Speciation

Phylogenetic analysis of each of the four genes indicated considerable divergence among lineages, which corresponded to the current classification of *C. neoformans* into three varieties. The amount of divergence differs among genes. The LAC gene showed the greatest divergence, followed by URA5, mtLrRNA, and ITS (Table 2). Within individual protein-coding genes LAC and URA5, introns were more polymorphic than exons. These results suggest that serotypes A, B, C, and D have diverged from each other for a long time. Each corresponds to a good phylogenetic species.

3.2. Biogeography and Dispersal

Our analyses indicated little geographic pattern of genetic variation. When all isolates were included and analyzed according to their geographic origins, we saw no positive correlation between genetic distance and geographic distance. On average, strains from the same geographic area (e.g. a continent) were no more similar to each other than to strains from other geographic areas (continents) (Table 3). The same is true if strains in each serotype were analyzed separately. The results are consistent with recent dispersals of *C. neoformans* throughout the world.

3.3. Hybridization

Interestingly, we observed several inconsistent gene genealogical placements (Figure 1). Five strains were identified to have inconsistent genealogies: B10, ZG289, CN111.97, CN139.97 and KW5. The simplest explanation was hybridization. Based on sequence divergence, we were able to estimate a relative time frame such hybridization events might have occurred. Our analysis indicated recent hybridizations for all five strains (Xu *et al.*, 2000).

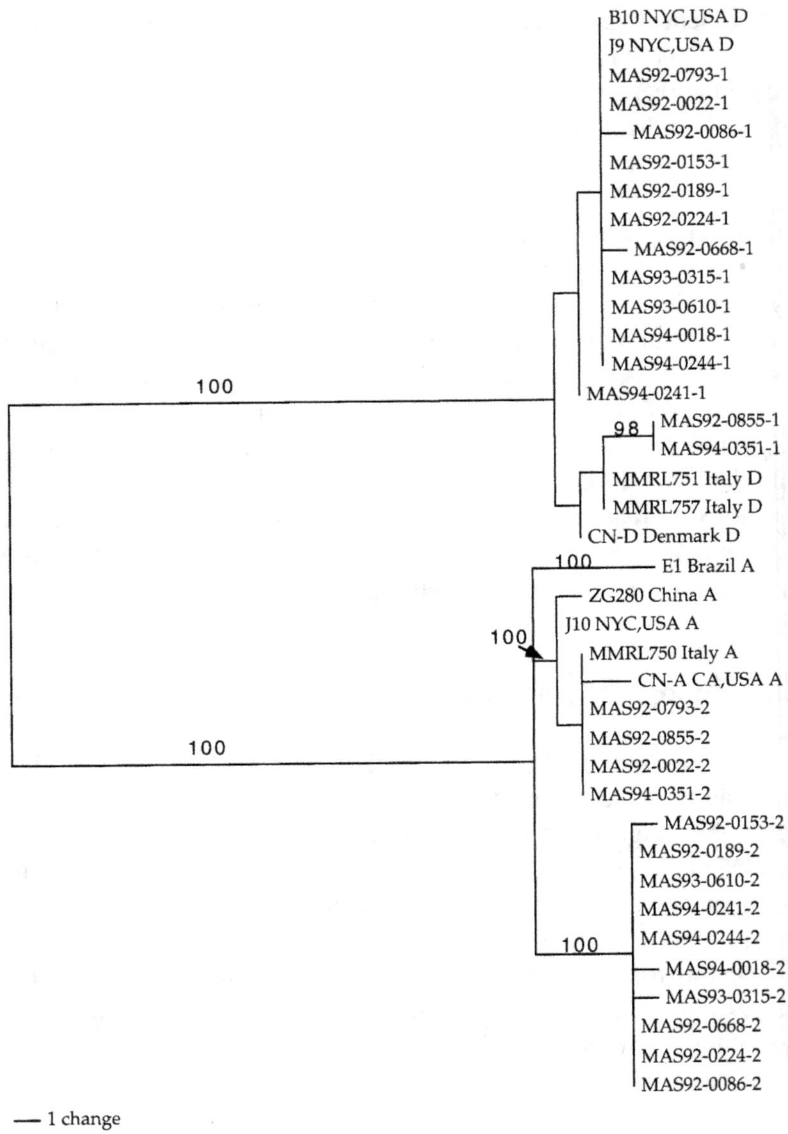

Figure 2. One of 10 most parsimonious trees for the 28 LAC sequences from 14 strains of serotype AD in *C. neoformans*. For comparison, five representative sequences from serotype A (E1, CN-A, MMRL750, J10 and ZG280) and five from serotype D (B10, CN-D, J9, MMRL751 and MMRL757) were included in this figure. These ten sequences were reported in an earlier study (Xu *et al.*, 2000b). Numbers above branches are bootstrap values >50% and based on 1000 replicates. Designations for strains of serotypes A and D included the isolate name, geographic origin (CA, California; NYC, New York City, both in the USA), and serotype. For the 28 serotype AD sequences, designations correspond to those in Table 1 and are followed by '-1' or '-2' to indicate alleles within each strain (Table 2). Mid-point rooting is used for this phylogeny but the tree topology is identical to that when serotype B or C sequences were used as outgroups. Scale bar represents one nucleotide substitution (Xu *et al.*, 2002). Reproduced by permission.

3.4. Recent Multiple Hybridizations are Responsible for the Origins of Serotype AD Strains

Strains of serotype AD often showed intermediate genotypic and phenotypic features between strains of serotypes A and D. During our preliminary screening, direct sequencing of PCR products from serotype AD strains often failed to obtain clear chromatograms and DNA sequences. Such results suggested sequence heterogeneity within individual strains. Because the current three-variety system does not accommodate serotype AD strains, we decided to investigate the origins of serotype AD strains. To obtain clear DNA sequences from these strains, we first cloned PCR products and then selected individual clones for sequencing. A portion (537 bp) of the LAC gene from each of the 14 strains of serotype AD was cloned and selected clones from each strain was sequenced (Xu *et al.*, 2002). We found that each strain contained two different alleles and sequences for both alleles were obtained. These sequences were compared to those from serotypes A, B, C, and D using gene genealogical analyses. The analysis indicated that each of the 14 serotype AD strains contained two phylogenetically distinct alleles: one allele highly similar to the serotype A group and the other to the serotype D group (Figure 2). In addition, we inferred that at least three recent and independent hybridization events had occurred to generate the 14 strains of serotype AD (Xu *et al.*, 2002).

3.5. Clonality and Recombination

Despite extensive population surveillance and strain typing studies of *C. neoformans*, one important question remained unresolved for a long time: are individual serotypes asexual, sexual or both? Because all microorganisms can reproduce asexually through simple mitotic divisions, it is therefore expected that microbial populations will have a clonal component. Indeed, in *C. neoformans*, most previous researches indicated that there were clones and clonal lineages within each of the serotypes, consistent with asexual reproduction and clonal dispersal. For example, epidemiological surveys in four areas of the US revealed only a few multilocus enzyme electrophoretic genotypes, and most genotypes were represented by multiple strains that were often distributed over wide geographic areas (Brandt *et al.*, 1996). The question then becomes: does recombination occur at all in natural populations? Here, we define recombination as genetic exchange among a group of closely related strains (i.e. serotype). Recombination is different from hybridization between groups of strains (i.e. serotypes) that have diverged from each other for millions of years.

The failures to detect recombination in *C. neoformans* in previous studies might reflect a sampling problem. Almost all previous studies used clinical serotypes A, B, C and D strains. Clinical strains may not be representative of environmental populations. One consistent bias is mating type alleles. More than 90% of all clinical and environmental isolates of *C. neoformans* of serotype D possess the α mating type (MATα), and nearly 100% of isolates of serotype A are MATα. For example, among 355 clinical and environmental strains of serotype A, D, or AD, which were obtained from four areas in the US, only one strain contained only the mating type **a** (MAT**a**) allele at the STE20 gene, which is located within the mating type region, and this strain had serotype AD (Yan *et al.*, 2002). Clinical isolates of serotype A or D strains are predominantly MATα, likely because pathogenicity is linked to this locus (Kwon-Chung *et al.*, 1992). The predominance of MATα among environmental isolates may similarly reflect a relationship of this locus to fitness, although there is no experimental evidence to support this hypothesis. Another possibility is related to haploid fruiting ability of MATα strains. Under starvation conditions in the laboratory, MATα strains produce hyphae, basidia, and haploid fruiting (Wickes *et al.*, 1996). Because of their small size, basidiospores are more effectively dispersed than the encapsulated vegetative yeast cells. Thus, prolific haploid fruiting of MATα strains may explain the dominance of this mating type in patients and in the environment.

To get a more balanced sample for analysis, we examined strains of serotype AD. Most strains of serotype AD possess both MATa and MATα loci (e.g. Yan *et al.*, 2002). In addition, in one study, a significant proportion (14 of 19) of serotype AD isolates contained the MAT**a** allele from serotype A parents (Aa) (Yan *et al.*, 2002).

Because strains of serotype AD are recent hybrids of serotypes A and D, they may contain genes from both serotypes. Individual alleles from each gene can be cloned and sequenced and easily identified into serotypes A and D groups (see above). Genealogies of different genes can be then compared within individual serotype clusters. Building on

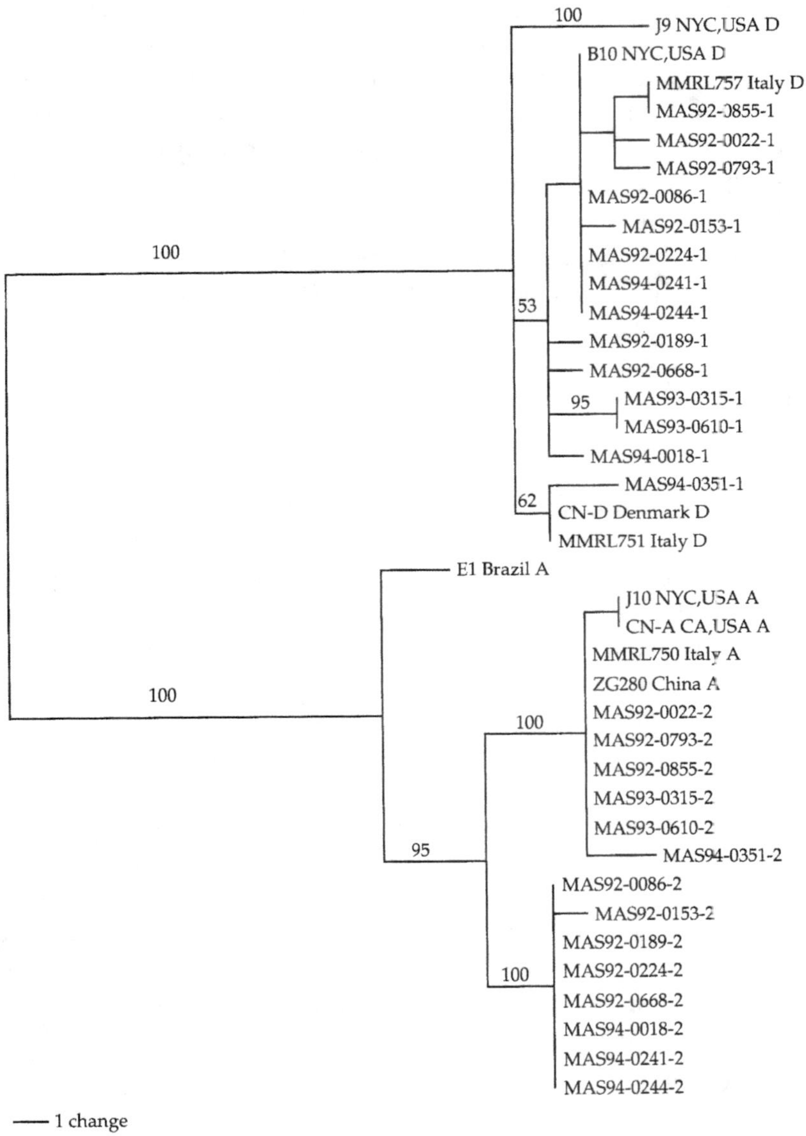

Figure 3. One of 12 most parsimonious trees for the 28 URA5 sequences from 14 strains of serotype AD in *C. neoformans*. Other descriptions are similar to those in Figure 2 (Xu *et al.*, 2003). Reproduced by permission.

results obtained from our previous studies (Xu *et al.*, 2000; 2002), we cloned and sequenced a portion (642 bp) of URA5 gene from the same 14 strains of serotype AD that we already have data on the LAC gene. Similar to results from the LAC gene, we identified that each of the 14 strains of serotype AD contained two phylogenetically distinct alleles: one allele was highly similar to those from serotype A strains and the other to alleles from serotype D strains. However, within both the serotype A allelic group and the serotype D allelic group, there was significant incongruence between genealogies derived from URA5 and LAC (Figures 2 and 3). These results suggest recombination in natural populations of both serotypes A and D (Xu and Mitchell, 2003).

4. *Histoplasma capsulatum*

H. capsulatum Darling is an ascomycetous species. It occurs throughout the world and causes infections in a variety of mammalian species, including humans. It is a dimorphic fungus: in filamentous forms when grown at ambient temperatures and in unicellular yeast forms when grown at 37°C. Its natural reservoir is soil. Hosts acquire it by accidentally inhaling its airborne spores or hyphal fragments.

To understand the evolutionary relationships among strains and populations of *H. capsulatum*, Kasuga *et al.*, (1999) analyzed genealogical relationships among 46 geographically diverse isolates. These 46 strains represent three previously characterized varieties: *H. capsulatum var. capsulatum*, *H. capsulatum var. duboisii* and *H. capsulatum var. farciminosum.* Portions of four protein-coding genes were sequenced for each of the 46 strains: arf1 (ADP-ribosylation factors); H-anti3 (H-antigen precursors); ole3 (delta-9 fatty acid desaturases) and tub1 (alpha-tubulins). These gene sequences were analyzed separately and together.

Data from the combined four genes identified 33 multilocus genotypes from the 46 strains. Phylogenetic analysis revealed several important findings.

4.1. Divergence and Speciation

Each of the four genes revealed six statistically robust clusters that were essentially the same for all four genes. The authors suggested that each of the six clusters could be called a phylogenetic species: two in North America, one in Central America, two in South America and one in Africa. Interestingly, strains from a previously named variety *H. capsulatum var. farciminosum* all clustered into one of the six clades that consists mainly of isolates of *H. capsulatum var. capsulatum*.

4.2. Clonality and Recombination

Within the North American clade Hcc2, several strains clearly showed genealogical incongruence among the four sequenced genes. This result was consistent with a recombining population structure of this clade. A similar pattern existed in one of the South American clades. The sample sizes for the remaining four clades were all small and data were insufficient for inferences of clonality and recombination.

4.3. Biogeography and Dispersal

While clades were mainly identified based on their geographic locations (a result consistent with geographic speciation), their data also provided evidence for recent dispersals. Within each clade, there was little evidence of genetic isolation by geographic distance among strains.

4.4. Hybridization

While most strains were unambiguously clustered into one of six clades, three strains showed inconsistent placements. If the clades were considered individual species, then these three strains were likely the results of hybridization among phylogenetic species. Conversely, the existence of such hybrid strains suggests that these clades are not reproductively isolated from each other in nature. Therefore, such hybridizations argue against classifying these clades as different species.

5. *Coccidioides immitis*

C. immitis is the etiological agent of coccidioidomycosis. It is endemic to the lower Sonoran areas of Southwestern US. Infection of this fungus is caused by the inhalation of asexual spores in the soil. Once in the host, spores germinate and form spherules and endospores and often result in a systemic mycosis. This fungus is highly virulent. It is the only eukaryote regulated under the United States Anti-Terrorism and Effective Death Penalty Act. Koufopanou *et al.*, (1997) carried out a genealogical survey of strains of *C. immitis*. A summary of their study is presented below.

They analyzed a total of 17 strains covering the entire known species range from California to Texas, Arizona, Mexico and Argentina. Fragments of five genes were sequenced from each strain: (i) the chitin synthase gene CHS1; (ii) the orotidine 5'-monophosphate decarboxylase gene pyrG; (iii) the T-cell reactive protein gene tcrP; (iv) a serine proteinase gene; and (v) a complement fixation antigen gene CTS2. Their genealogical analysis identified the results summarized below.

5.1. Divergence and Speciation

Genealogies of all five genes showed that strains could be partitioned into two consistent groups: one centered in California and the other outside California. The degree of divergence varied among the genes. The authors suggested that these two groups were reproductively isolated from each other and as such, they should be considered distinct taxa.

5.2. Biogeography and Dispersal

The Californian isolates were clustered together in one branch of the genealogical trees, separate from strains in other geographic areas. This result suggested that geographic separation played an important role in the species history. However, little geographic pattern was observed for strains from Arizona, Texas, Mexico, and Argentina. In fact, all the five Argentina strains shared the same multilocus genotype and identical to one of the two Mexican strains. Further research using microsatellite markers suggested recent dispersal of *C. immitis* from Texas to South America, likely associated with human migration (Fisher *et al.*, 2001).

5.3. Clonality and Recombination

While there was no evidence of gene flow between the two groups, genealogical placements of strains within each group were highly inconsistent. Inconsistent genealogies suggested recombination within both groups. This finding was surprising giving that mating and recombination have never been observed in laboratory crosses.

6. Conclusions and Perspectives

In this chapter, I reviewed how gene genealogies could be used to address fundamental issues regarding the natural histories of human pathogenic fungi. As was shown for each of the three species, results from gene genealogical analyses were unambiguous and definitive. DNA sequence variations were commonly found in each species. Within each of the three species, divergent lineages were found, indicating cryptic speciation. However, there were abundance evidences of recent dispersals. With increasing human migrations and other types of transportations, the speed and range of pathogen dispersals will likely increase. Gene genealogical analysis offers a powerful tool to track down sources and routes of pathogen dispersals.

Among the 70,000 or so fungal species identified so far, over 20% are classified into the Division Fungi Imperfecti - fungi without identifiable sexual reproduction. *C. neoformans* and *C. immitis* were originally placed in Fungi Imperfecti. Many other human fungal pathogens are still classified into this Division, including *Candida albicans*, the common agent of yeast infection. Genealogical analyses reviewed above showed unambiguous evidence of recombination in both *C. neoformans* and *C. immitis* (and *H. capsulatum*). Analyses of other human fungal pathogens using a similar approach will likely reveal similar evidence of recombination.

All three species described above are haploids, with the exception of serotype AD strains in *C. neoformans*. Gene genealogical analyses for haploid organisms are relatively straightforward. In contrast, genealogical analyses of diploid species such as *C. albicans* are more difficult, both technically and analytically. Technically, genealogical analyses of diploids require that both alleles of each locus sequenced individually. These would require PCR, cloning, allele screening, and sequencing of representative alleles. Because these steps can introduce mistakes, appropriate care is needed throughout the experimental processes. In addition, because diploids may undergo mitotic recombination, the analytical inferences of gene genealogies will be a lot more difficult. Nonetheless, these difficulties will be overcome in the foreseeable future.

A thorough understanding of the natural histories of human pathogens will help preventions and managements of human infectious diseases. For example, a population genotype database such as those stored in GenBank will allow tracing pathogen sources much easier and more effective. In addition, the identification of divergent lineages within individual species will help reveal genetic bases for differences in virulence and drug susceptibilities. With increasing availability and affordability of DNA sequencing, the gene genealogical approach will be moved into the center stage of fungal population genetics research.

7. Acknowledgements

Research in laboratory is supported by grants from McMaster University, the Natural Sciences and Engineering Research Council (NSERC) of Canada, the Premier's Research Excellence Award, Genome Canada, the Canadian Foundation for Innovation (CFI), and the Ontario Innovation Trust (OIT).

References

Brandt, M.E., Hutwagner, L.C., Klug, L.A., Baughman, W.S., Rimland, D., Graviss, E.A., Hamill, R.J., Thomas, C., Pappas, P.G., Reingold, A.L., Pinner, R.W., and the CDC Cryptococcal Disease Active Surveillance Group. 1996. Molecular subtype distribution of *Cryptococcus neoformans* in four areas of the United States. J. Clin. Microbiol. 34: 912-917.

Casadevall, A., and Perfect, J.R. 1998. *Cryptococcus neoformans*. American Society for Microbiology Press, Washington, DC.

Fisher, M.C., Koenig, G.L., White, T.J., San-Blas, G., Negroni, R., Alvarez, I.G., Wanke, B., and Taylor, J.W. 2001. Biogeographic range expansion into South America by *Coccidioides immitis* mirrors New World patterns of human migration. Proc. Natl. Acad. Sci. USA. 98: 4558-4562.

Franzot, S.P., Salkin, I.F., and Casadevall, A. 1999. *Cryptococcus neoformans* var. *grubii*: separate varietal status for *Cryptococcus neoformans* serotype A isolates. J. Clin. Microbiol. 37: 838-840.

Kasuga, K., Taylor, J.W., and White, T.J. 1999. Phylogenetic relationships of varieties and geographical groups of the human pathogenic fungus *Histoplasma capsulatum* Darling. J. Clin. Microbiol. 37: 653-663.

Koufopanou, V., Burt, A., and Taylor, J.W. 1997. Concordance of gene genealogies reveals reproductive isolation in the pathogenic fungus *Coccidioides immitis*. Proc. Natl. Acad. Sci. USA. 94: 5478-5482.

Kwon-Chung, K.J., Bennett, J.E., and Rhodes, J.C. 1982. Taxonomic studies on *Filobasidiella* species and their anamorphs. Antonie van Leeuwenhoek 48: 25-38.

Kwon-Chung, K.J., Edman, J.C., and Wickes, B.L. 1992. Genetic association of mating types and virulence in *Cryptococcus neoformans*. Infection and Immunity. 60: 1869-1874.

Swofford, D.L. 2001. PAUP4.0: Phylogenetic Analysis Using Parsimony. Mass: Sinauer Associates.

Wickes, B.L., Mayorga, M.E., Edman, U., and Edman, J.C. 1996. Dimorphism and haploid fruiting in *Cryptococcus neoformans*: Association with the α-mating type. Proc. Natl. Acad. Sci. USA. 93: 7327-7331.

Xu, J., Vilgalys, R.J., and Mitchell, T.G. 2000. Multiple gene genealogies reveal recent dispersion and hybridization in the human pathogenic fungus *Cryptococcus neoformans*. Molecular Ecology. 9: 1471-1481.

Xu, J., and Mitchell, T.G. 2003. Comparative gene genealogical analyses of strains of serotype AD identify recombination in populations of serotypes A and D in the human pathogenic yeast *Cryptococcus neoformans*. Microbiology 149: 2147-2154.

Xu, J., Luo, G., Vilgalys, R., Brandt, M.E., and Mitchell, T.G. 2002. Multiple origins of hybrid strains of *Cryptococcus neoformans* with serotype AD. Microbiology 148: 203-212.

Yan, Z., Li, X., and Xu, J. 2002. Geographic distribution of mating type alleles of *Cryptococcus neoformans* in four areas of the United States. J. Clin. Microbiol. 40: 965-972.

From: Pathogenic Fungi: Structural Biology and Taxonomy
Edited by: G. San-Blas and R.A. Calderone

Chapter 10

Polysaccharides F1SS: Taxonomic and Evolutionary Characters for Ascomycetes

A. Prieto, O. Ahrazem,
M. Bernabé, and J.A. Leal

Abstract

The alkali-extractable and water-soluble F1SS polysaccharides from the fungal cell wall seem to be a reliable taxonomic and evolutionary character. The number of polysaccharidic structures that may be formed by a combination of different sugars and linkages can be used for the delimitation of genera, grouping of taxa in higher ranks, establishment of connections among meiotic and strictly mitotic genera, and the elaboration of fungal evolutionary schemes. In addition, the immunogenic potential of fungal cell wall and cell surface polysaccharides may result in the establishment of sensitive immunological techniques for the rapid identification of pathogenic fungi.

1. Introduction

The clinician needs rapid and reliable methods for the identification of the human pathogenic fungi in order to establish an early medical treatment. Although the Onygenales taxonomic group of fungi contain the largest number of pathogenic fungi, human pathogenic fungi may be represented from any taxonomic group due to the increase in patients with a deficient or suppressed immune system. Therefore, the characters used in the identification and systematics of pathogens should cover all fungi. The systematics of pathogenic and opportunistic fungi has been recently reviewed by Guarro *et al.* (1999). Fungal taxonomy is mainly based on morphological characters. Some of these characters may be useful only for the definition of low level taxa (spore size or morphology) or high level taxa (the organs that produce the meiospores -basidium/ascus- or the ascoma). The rank at which a character varies is defined as "the resolution level" (Kohn, 1992). Since the variation or evolution of a trait in a lineage is a continuum, transition states of that trait will be usually found in transition taxa, which may be included in any of the related taxa or in their own taxa, depending on the taxonomist criterion and on the selected character or characters. As a result, a great number of taxa are ill defined or misplaced. The common feature of a large group of Ascomycetes, the Plectomycetes, is a closed ascoma (cleistothecium) and globose, evanescent asci. Because the Plectomycetes include transition genera, families, even orders, many taxa were transferred to the Pyrenomycetes, Discomycetes and Loculoascomycetes (Cain, 1972; Malloch, 1981). In addition, morphological characters of unrelated taxa may reach similar morphology in the process of adaptation to the same environment-convergent evolution. Therefore, the groups based on such characters are heterogeneous.

Taxonomists now include features such as cell wall structure, the sculpturing of the surface of the spores, and chemical and biochemical features, which are very useful when morphology is insufficient for a correct classification of a taxon. These characters have been used in current taxonomic revisions (Hawksworth, 1994; Frisvad *et al.*, 1998). The ideal characters should not only be tools for grouping homogeneous taxa, but they should also show the polarity of the evolutionary pathway of related taxa. At present, it seems that these characters do not exist, or perhaps they exist but their importance is minimized.

The cell wall, according to Bartnicki-García (1984) "…constitutes a quintessential ingredient in the life and the evolution of most living organisms". Since the fungal cell wall is mainly composed by polysaccharides (Bartnicki-García, 1968; Sietsma and Wessels, 1977), these polymers might be of great relevance in fungal morphology, as Bartnicki-García (1987) postulated: "The nature of these polymers in any particular fungus is not capricious but it is related to its taxonomical position and thus reflects its evolutionary history". He reported that the groups of fungi arranged by dual combination of cell wall polysaccharides followed the boundaries of traditional taxa (Bartnicki-García, 1968). Recently, a minor component of the fungal cell wall, the F1SS polysaccharide, has been proposed as a chemotaxonomic marker at the genus level (Leal, 1994; Leal *et al.*, 2001; Bernabé *et al.*, 2002). The comparison of the structures of the F1SS polysaccharides from different taxa reveals an evolutive pattern that is in agreement with certain fungal evolutionary theories (Savile, 1968; Barr, 1976; Cavalier-Smith, 1987).

In this chapter, we review the literature on the utilisation of F1SS polysaccharides as chemotaxonomic characters in fungal taxonomy, include unpublished results, and propose a fungal evolutionary scheme based on the evolution of the structure of these polysaccharides.

2. F1SS Polysaccharides

The cell wall alkali-extractable material (F1) consists of water insoluble glucans (F1I) and water-soluble polysaccharides (F1S), usually a mannan or heteropolysaccharide (F1SS polysaccharide), that is frequently accompanied by an α-(1→4)-glucan (Leal *et al.*, 1992a). The F1SS polysaccharides are purified from the α-glucan, taking advantage of their high solubility in water (Leal *et al.*, 2001) and 50% ethanol (Ahrazem *et al.*, 2002b), and by filtration on columns of Sepharose CL-6B. They have an average molecular mass of around 60kDa, as determined by gel permeation chromatography. The characterisation of these polymers involves the determination of linkage types by methylation analysis, NMR spectroscopy and other methods used in carbohydrate chemistry. Details are given in the original publications.

Highly specific antibodies have been raised in rabbits against these polysaccharides, and used to stain, by immunofluorescence techniques, mycelia containing the antigens (Domenech *et al.,* 1996; 1999; Ahrazem *et al.,* 1999). The staining of the mycelia reveals the superficial localisation of the antigenic polysaccharide, which probably constitutes the glycidic moiety of a cell wall glycoprotein or glycopeptide that is hydrolysed in the extraction process. Glycoproteins with polysaccharidic moieties similar to the F1SS polysaccharides have been reported (Gander, 1974; Jikibara *et al.,* 1992). The complex carbohydrates of these molecules serve different biological functions, one of the most important being their participation in recognition phenomena (Albersheim *et al.,* 1984).

F1SS polysaccharides from all fungi previously investigated always contain mannose at concentrations which range from traces to 100%. Attached to the mannan backbone are single residues or chains of unique or different monosaccharides. The galactofuranose chains linked to the mannan backbone can be removed by mild acid hydrolysis, and the mannan purified (Domenech *et al.,* 1996; Prieto *et al.,* 1997; Ahrazem *et al.,* 1999).

The arrangement of F1SS polysaccharides according to their structural complexity reveals an evolutionary scheme in which the more simple structures belong to fungi classified in the Onygenales and some Sordariales, followed by those from the Eurotiales, and then the complex polysaccharides of the Hypocreales (Bernabé *et al.*, 2002).

3. F1SS Polysaccharide: Chemotaxonomic Marker

In the subsequent sections, the F1SS polysaccharides found in different taxa are reviewed. The strains investigated are reported in the original publications. The information has been ordered following the systematics arrangement of the last edition of the Dictionary of the Fungi (Kirk *et al.,* 2001). The structure of the repeating unit of these polysaccharides and the genera from which they have been obtained are reported in Table 1. When several genera have the same structure, they usually are congeneric, a teleomorph and its anamorphic form genus, or closely related. When several polysaccharides are obtained from species of a genus, the genus is considered to be heterogeneous.

4. Saccharomycetes

This group of fungi comprises the Ascomycetes which do not produce a dikaryon. Vegetative cells reproduce by budding or fission and mycelium is generally absent or poorly developed. Sexual reproduction takes place by fusion of free cells or gametangial tips. Usually, asci are morphologically identical to vegetative cells.

4.1. Saccharomycetales

4.1.1. Dipodascaceae

F1SS polysaccharides from fungi of the genus *Geotrichum* and its teleomorphic states, *Galactomyces* and *Dipodascus* (structure **11**), consist of a chain of →6)-α-D-Man*p*-(1→ partially substituted at *O*-2 by a disaccharide of α-D-Gal*p*-(1→2)-α-D-Man*p*-(1→ (Ahrazem *et al.*, 2002d). A similar galactomannan was obtained from *Trichosporon fermentans* (=*Geotrichum fermentans*) and *Candida lipolytica* (Barreto-Bergter and Gorin, 1983).

4.1.2. Saccharomycetaceae

Several genera of this family have been exhaustively studied because of their importance as human pathogens or in industry. Although the mannans described in the literature have not been obtained and purified in the same way as the F1SS polysaccharides, since they are water-soluble components of the cell wall, they may be considered similar to F1SS polysaccharides. The structural differences among these polysaccharides, revealed by NMR spectroscopy, have been used for identification of yeast species (Gorin and Spencer, 1970). The structure of the *Saccharomyces cerevisiae* mannan consists of a backbone of →6)-α-D-Man*p*-(1→ residues to which side chains of →2)-α-D-Man*p*-(1→ and →3)-α-D-Man*p*-(1→ are linked (Peat *et al.*, 1961). The detailed structure of the side chains was further determined (Lee and Ballou, 1965; Vinogradov *et al.*, 1998). In *Candida* species, a variety of related mannan structures occur. In *Candida albicans* the mannan has the same basic structure as that from *S. cerevisiae* but, in addition, it contains β-(1→2)-linked oligomannosyl side chains (Kobayashi *et al.*, 1992). The studies on yeast mannans have been reviewed on numerous occasions (Phaff, 1971; Barreto-Bergter and Gorin, 1983; Fleet, 1991).

5. Plectomycetes

The plectomycetes are a group of fungi which produce globose, evanescent asci irregularly distributed within a closed ascomatal cavity, the cleistothecium, or are naked. The systematics of the group is unsettled and has been the object of several rearrangements. Benny and Kimbrough (1980) accepted six orders, Elaphomycetales, Onygenales, Eurotiales, Ascosphaerales, Microascales and Ophiostomatales. Malloch (1981) redistributed the group in four orders belonging to three traditional classes: Pleosporales (Loculoascomycetes); Diaporthales and Hypocreales (Pyrenomycetes) and Pezizales (Discomycetes). Von Arx (1987) reduced the Eurotiales, Gymnoascales and Onygenales to the single order Eurotiales, which included four families. Currah (1994) realigned the orders Eurotiales and Onygenales. Phylogenetic analyses based on DNA sequences show that the plectomycetes form a well-supported monophyletic group, which comprises most of the cleistothecial ascomycetes, with three orders: Ascosphaerales, Onygenales and Eurotiales (Berbee and Taylor, 1994; Geiser and Lobuglio, 2001).

For specialists in medical mycology, this group is of special interest since it includes most of the human and animal pathogenic fungi that produce cutaneous or systemic infections.

5.1. Onygenales

5.1.1. Arthrodermataceae

The F1SS polysaccharide consists of a chain of →6)-α-D-Man*p*-(1→ with some residues substituted at *O*-2 by α-D-Man*p* (structure **2**) in species of *Ctenomyces, Trichophyton* and *Microsporum* (Guarro *et al.*, 1993; Jiménez-Barbero *et al.*, 1995; Ikuta *et al.*, 1997; Ahrazem *et al.*, 2002b) and in *Epidermophyton floccosum* (Takeda *et al.*, 1981). In *M. fulvum, M. nanum, M. persicolor, T. equinum, T. simi* and other strains of *E. floccosum* (Ahrazem *et al.*, 2002b), the main chain is partially substituted at *O*-2 by the disaccharide β-D-Gal*f*-(1→6)-α-D-Man*p*-(1→ or by terminal residues of mannopyranose (structure **10**).

5.1.2. Onygenaceae

The F1SS polysaccharide from most species of *Aphanoascus* consists of a chain of →6)-α-D-Man*p*-(1→, fully substituted at *O*-2 by terminal mannopyranose (structure **3**) (Leal *et al.*, 1992a; Jiménez-Barbero *et al.*, 1993). A similar structure, but with some of the terminal mannopyranose residues substituted at *O*-6 by α-D-galactofuranose (structure **12**), has been found in *A. terreus* and *A. saturnoideus* (Ahrazem *et al.*, 2002b). The existence of two structures in the polysaccharide from species of *Aphanoascus* reveals that there are two lineages in the genus. This dichotomy was also observed in the characters of the anamorphs and ascospore morphology (Cano and Guarro, 1990). The last structure has also been found in the mycelial form of *Paracoccidiodes brasiliensis* (Ahrazem *et al.*, 2003), in *Ascocalvatia alveolata* and two isolates of *Onygena equina* (Ahrazem *et al.*, 2002b). The two last genera are related morphologically (Currah, 1985) and by rDNA analysis (Sugiyama *et al.*, 1999). In *Amauroascus* and *Auxarthron,* whose morphology is very similar (Currah, 1994), the mannose of the main chain is partially substituted at *O*-2 by methyl-3-*O*-mannose (structure **4**) (Ahrazem *et al.*, 2002b). The polysaccharide from *Onygena piligena* contains rhamnose (Ahrazem *et al.*, 2002b). This isolate was later identified (Sigler and Carmichael, 1976) as *Geomyces panorus* (anamorphic Myxotrichaceae). The polysaccharide from *Monascella botryosa* (Ahrazem *et al.*, 2002b) has the mannose chain substituted at *O*-2 by α-galactopyranose (structure **7**). Guarro and von Arx (1986) considered *M. botryosa* as an intermediate between the Pezizales and the Eurotiales, and that its classification in Monascaceae or Onygenaceae was a matter of choice. Currah (1994) recommended the use of molecular techniques to ascertain the taxonomic relatedness of *Monascella.* The fact that the polysaccharide isolated from some species belonging to the Pezizales also has structure **7** (Table 1) points to a relationship of *Monascella* with this order.

5.1.3. Myxotrichaceae

F1SS polysaccharides from members of the Myxotrichaceae contain rhamnose in addition to mannose and galactofuranose, except *Myxotrichum deflexum, M. setosum* and the related anamorphic species belonging to the form genera *Oidiodendron* and *Malbranchea* (Ahrazem *et al.*, 2002b). Rhamnose has not been found in F1SS

Table 1. Repeating unit of F1SS polysaccharide from Ascomycetes

	Repeating unit	Genera
1[a]	[→6)-α-D-Man*p*-(1→]$_m$	*Ancestor*
2	{→6)-α-D-Man*p*-(1→6)-α-D-Man*p*-(1→6)-α-D-Man*p*-(1→}$_m$ 2 ↑ 1 α-D-Man*p*	*Ctenomyces* *Epidermophyton* *Microsporum* *Trichophyton* *Curvularia*
3	{→6)-α-D-Man*p*-(1→}$_n$ 2 ↑ 1 α-D-Man*p*	*Aphanoascus* *Arachnotheca* *Chrysosporium*
4	[→6)-α-D-Man*p*-(1→6)-α-D-Man*p*-(1→]$_n$ 2 2 ↑ ↑ 1 1 CH_3→3)-α-D-Man*p* α-D-Man*p*	*Amauroascus* *Malbranchea* *Auxarthron* *Gymnoascus*
5	{[→6)-α-D-Man*p*-(1→]$_m$[→6)-α-D-Man*p*-(1→]$_p$}$_n$ 2 ↑ 1 β-D-Gal*f*	*Neurospora* *Cephalotheca* *Apodus* *Copromyces* *Gelasinospora* *Sordaria*

	Structure	Genera
6	{[→6)-α-D-Man*p*-(1→]$_n$→6)-α-D-Man*p*-(1→}$_m$ 4 ↑ 1 β-D-Gal*p*	*Paecilomyces*
7	[→6)-α-D-Man*p*-(1→6)-α-D-Man*p*-(1→]$_n$ 2 ↑ 1 α-D-Gal*p*	*Monascella* *Ascobolus* *Peziza*
8	[→6)-α-D-Man*p*-(1→]$_n$ 2 ↑ 1 α-D-Glc*p*	*Ceratocystis* *Phoma*
9	[→6)-α-D-Man*p*-(1→]$_n$ 3 ↑ 1 α-L-Rha*p*	*Ophiostoma* *Cephalotheca*
10	{→6)-α-D-Man*p*-(1→[→6)-α-D-Man*p*-(1→]$_n$→6)-α-D-Man*p*-(1→}$_m$ 2 (first residue) ← 1 α-D-Man*p* 2 (last residue) ← 1 β-D-Gal*f*-(1→6)-α-D-Man*p*	*Trichophyton* *Microsporum* *Epidermophyton*

Table 1, Continued

11

```
{→6)-α-D-Manp-(1→[→6)-α-D-Manp-(1→]n}m
                     2
                     ↑
                     1
  α-D-Galp-(1→2)-α-D-Manp
```

Dipodascus
Geotrichum
Galactomyces

12

```
{[→6)-α-D-Manp-(1→]m[→6)-α-D-Manp-(1→]p}n
         2                     2
         ↑                     ↑
         1                     1
α-D-Galf-(1→6)-α-D-Manp    α-D-Manp
```

Ascocalvatia
Aphanoascus
Onygena
Chrysosporium
Paracoccidioides

13

```
[→6)-α-D-Manp-(1→]n
         3
         ↑
         1
α-D-Galf-(1→2)-α-D-Galf
```

Apodus
Podospora

14

```
{→6)-α-D-Manp-(1→6)-α-D-Manp-(1→}m
      2                 2
      ↑               ( ↑         )
      1               ( 1         )
  β-D-[Galf]n         ( α-D-Glcp  )
      5               ( 4         )
      ↑               ( ↑         )
      1               ( 1         )
  β-D-Galf            ( α-D-Glcp  )
                      ( 6         )
                      ( ↑         )p
                        1
```

Chaetosphaeria

15

```
{[→6)-α-D-Manp-(1→]m→6)-α-D-Manp-(1→6)-α-D-Manp-(1→}n
                         2                 2
                         ↑                 ↑
                         1                 1
                      β-D-Galf        [α-D-Manp]w
                         5                 2
                         ↑                 ↑
                         1                 1
β-Galf-(1→5)-β-Galf-(1→[6)-β-Galf-(1]p→[5)-β-Galf]q     α-D-Manp
```

Byssochlamys
Paecilomyces
Talaromyces
Thermoascus

	Structure	Genera
16	$[\rightarrow 5)$-β-D-Gal*f*-$(1\rightarrow]_n\rightarrow$(α-mannan)	*Eupenicillium* *Penicillium* *Aspergillus* *Geosmithia*
17	$\{\rightarrow 5)$-β-D-Gal*f*-$(1\rightarrow 6)$-β-D-Gal*f*-$(1\rightarrow\}_m\rightarrow$(α-mannan)	*Arachniotus* *Gymnascella* *Geosmithia*
18	$\{\rightarrow 6)$-β-D-Gal*f*-$(1\rightarrow[\rightarrow 5)$-β-D-Gal*f*-$(1\rightarrow]_2\rightarrow 5)$-β-D-Gal*f*-$(1\rightarrow\}_n\rightarrow$(α-mannan)	*Penicillium*
19	$\{\rightarrow 6)$-β-D-Gal*f*-$(1\rightarrow 5)$-β-D-Gal*f*-$(1\rightarrow 5)$-β-D-Gal*f*-$(1\rightarrow\}_n\rightarrow$(α-mannan)	*Neosartorya* *Aspergillus*
20[a]	$[\rightarrow 6)$-β-D-Gal*f*-$(1\rightarrow]_n\rightarrow$(α-mannan)	

Table 1, Continued

21

```
[→3)-β-D-Galf-(1→2)-β-D-Galf-(1→]m→(α-mannan)
```

Talaromyces
Penicillium

22

```
[→3)-β-D-Galf-(1→5)-β-D-Galf-(1→]m→(α-mannan)
```

Chaetosartorya
Aspergillus

23

```
[→5)-β-D-Galf-(1→6)-β-D-Galf-(1→]m→(α-mannan)
                            2
                            ↑
                            1
     α-D-Glcp-(1→2)-α-D-Galf
```

Talaromyces
Penicillium
Geosmithia

24

```
{[→5)-β-D-Galf-(1→]n[→5)-β-D-Galf-(1→]p}m→(α-mannan)
                                6
                                ↑
                                1
                              α-D-Glcp
```

Penicillium

25

```
                                                        ↓
                                                        6
β-D-Galf-(1→[6)-β-D-Galf-(1→5)-β-D-Galf-(1→]n→2)-α-D-Manp
                                                        1
                                                        ↓
                                                        6
                                                    α-D-Manp
                                                        1
                                                        ↓
                                                        6
                                  [β-D-Galf-(1→2)-α-D-Manp]m
                                                        1
```

Cryphonectria
Coniella
Cryptodiaporthe
Cryptosporella
Endothia
Greeneria
Harknessia
Melanconis
Melanconiella
Ophiovalsa
Plagiostoma
Pilidiella

26

```
                                                                  ↓
                                                                  6
β-D-Galf-(1→[6)-β-D-Galf-(1→5)-β-D-Galf-(1→5)-β-D-Galf-(1→]n→2)-α-D-Manp
                                                                  1
                                                                  ↓
```

Diaporthe
Hercospora
Leucostoma
Mazzantia
Melanconis
Valsa
Valsella

27

```
                                                        ↓
                                                        6
β-D-Galf-(1→[6)-β-D-Galf-(1→5)-β-D-Galf-(1→]n→2)-α-D-Manp
                                                        1
                                                        ↓
                                                        6
                                                    α-D-Manp
                                                        1
                                                        ↓
                                                        6
             β-D-Galf-(1→[→5)-β-D-Galf-(1→]m→2)-α-D-Manp
                                   6                    1
                                   ↓                    ↓
                                   1
                                α-D-Galp
```

Apiognomonia
Amphiporthe
Ditopella
Discula
Gnomoniella
Phragmoporthe

Table 1, Continued

```
28   [→5)-β-D-Galf-(1→6)-β-D-Galf-(1→]n[→5)-β-D-Galf-(1→6)-β-D-Galf-(1→]m→(α-mannan)
                         2
                         ↑
                         1                                                      Discula
                     α-L-Rhap

29b  {→6)-β-D-Galf-(1→[→5)-β-D-Galf-(1→]m→5)-β-D-Galf-(1→}n→(α-mannan)      Sesquicillium
           2
           ↑
           1
       α-D-Glcp

30b  {[→6)-β-D-Galf-(1→5)-β-D-Galf-(1→]2→6)-β-D-Galf-(1→[→5)-β-D-Galf-(1→]2 }n→(α-mannan)
           2                                  2
           ↑                                  ↑
           1                                  1                           Nectria (Bionectria)
       α-D-Glcp                           α-D-Glcp
           6
           ↑
           1
       α-D-Glcp

31b  {→6)-β-D-Galf-(1→5)-β-D-Galf-(1→5)-β-D-Galf-(1→5)-β-D-Galf-(1[→6)-β-D-Galf-(1→5)-β-D-Galf-(1→]2}n
           2                                6                                2
           ↑                                ↑                                ↑
           1                                1                                1     Nectria (Bionectria)
       α-D-Glcp                       α-D-GlcpNAc                        α-D-Glcp   Sesquicillium
```

32 [→6)-β-D-Gal*f*-(1→]$_n${→6)-β-D-Gal*f*-(1→}$_m$→(α-mannan)

- first branch: α-D-Glc*p*-(1→2)
- second branch: α-D-Glc*p*A-(1→2)

Calonectria
Cylindrocladium
Myrothecium

33 [→6)-β-D-Gal*f*-(1→]$_n${→6)-β-D-Gal*f*-(1→}$_m$→(α-mannan)

- first branch: α-D-Glc*p*-(1→2)
- second branch: β-D-Man*p*-(1→4)-α-D-Glc*p*A-(1→2)

Gibberella
Nectria
Fusarium

34 [→6)-β-D-Gal*f*-(1→]$_n${→6)-β-D-Gal*f*-(1→}$_m$→(α-mannan)

- first branch: α-D-Glc*p*-(1→2)
- second branch: α-D-Glc*p*-(1→4)-α-D-Glc*p*A-(1→2)

Myrothecium

Table 1, Continued

35

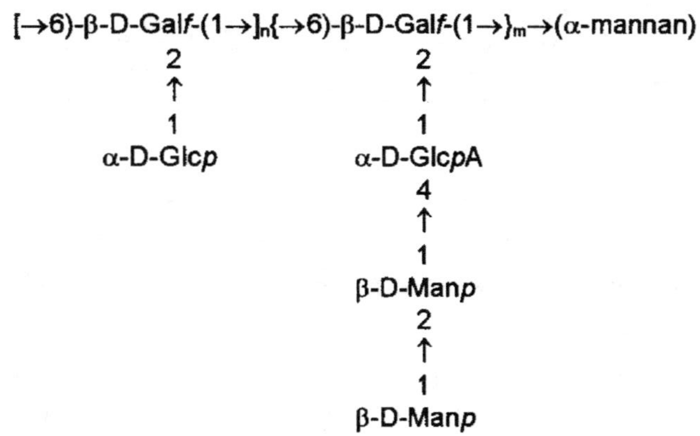

Gibberella
Fusarium
Penicillium

36

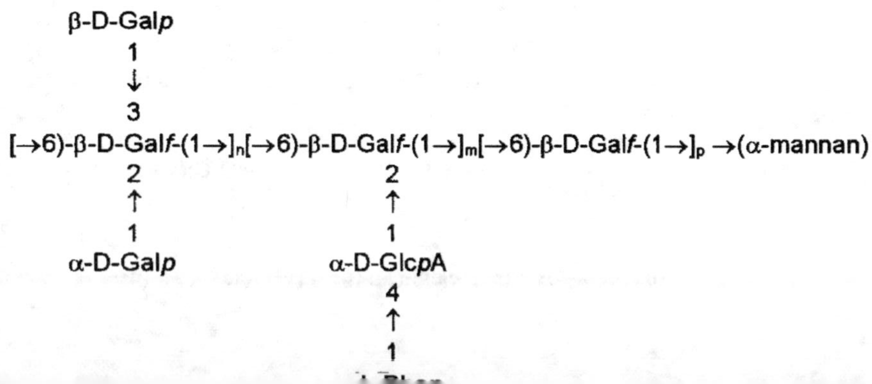

Melanospora
Microthecium
Pteridiosperma
Sphaerodes

37	↓ 6 [α-D-Glc*p*-(1→4)-α-D-Glc*p*A-(1→2)-β-D-Gal*f*-(1→6)-β-D-Gal*f*-(1→2)-α-D-Man*p*]$_m$ 1 ↓ 6 α-D-Man*p* 1 ↓ 6 [α-D-Man*p*-(1→3)-α-D-Man*p*]$_n$ 1 ↓ 6 α-D-Man*p* 1 ↓ 6 [β-D-Gal*f*-(1→6)-β-D-Gal*f*-(1→5)-β-D-Gal*f*-(1→6)-β-D-Gal*f*-(1→2)-α-D-Man*p*]$_w$ 1 ↓	*Hypocrea* *Trichoderma*

[a]These structures constitute the core or backbone of several polysaccharides,but they have not been found as such in any fungi until now.
[b]These polysaccharides contain small amounts of glucuronic acid.

polysaccharides from members of the Eurotiales or other families of Onygenales, except *Onygena piligena*. The results of the methylation analysis and the ^{1}H-NMR spectra show that the F1SS polysaccharides from members of this family may have different structures, indicating the existence of different lineages (Ahrazem *et al.*, 2002b). Currah (1985; 1994) considered the Myxotrichaceae a distinct evolutionary line derived from the inoperculate Discomycetes. From the analysis of SSU rDNA sequences, Sugiyama *et al.* (1999) reported two lineages in the Myxotrichaceae in a clade separated from other plectomycetous taxa. Rhamnomannans have been found in the cell wall of Microascales as *Pseudoallescheria boydii* (Pinto *et al.*, 2001), Ophiostomatales (Spencer and Gorin, 1971; Travassos *et al.*, 1974), *Cephalotheca (=Fragosphaeria) purpurea* and *C. reniformis* (Ahrazem *et al.*, 2002a), and in members of Pezizales (A. Prieto, unpublished).

Recent molecular studies (Leclerc *et al.*, 1994; Sugiyama *et al.*, 1999; Bialek *et al.*, 2000) showed that the families currently established within the Onygenales are not longer supported. On the basis of these and other data, Guarro and Cano (2002) indicated that the use of suprageneric ranks is unreliable in the taxonomy of Onygenales, since even several widely accepted onygenalean genera are polyphyletic. They suggest that a more natural scheme of classification should be based, at this moment, on genera that are monophyletic.

5.2. Eurotiales

5.2.1. Trichocomaceae

F1SS polysaccharides from members of this family, the anamorphic form genera *Aspergillus* and *Penicillium*, and some species of *Paecilomyces* consist of a short chain of →6)-α-D-Man*p*-(1→ more or less branched at *O*-2 (the core) to which chains of galactofuranose with different linkage types are attached:

1. Chains of →5)-β-D-Gal*f*-(1→ (structure **16**) in species of: *Eupenicillium, Emericella, Eurotium, Fennellia, Hemicarpenteles, Hemisartorya, Neosartorya, Petromyces, Warcupiella* (Leal and Bernabé, 1998), in species of the form-genera *Aspergillus, Penicillium* (Gómez-Miranda *et al.*, 1986; Leal *et al.*, 1992b; Leal and Bernabé, 1998) and in *Geosmithia namylowskii* (Prieto *et al.*, 2002).

2. Chains containing →5)-β-D-Gal*f*-(1→ and →6)-β-D-Gal*f*-(1→ in a 3 :1 ratio (structure **18**) in *Penicillium expansum, P. clavigerum* and *P. digitatum* (Parra *et al.*, 1994a); in a 2:1 ratio (structure **19**) in species of *Neosartorya* and *A. fumigatus* (Leal *et al.*, 1995); in a 1:1 ratio (structure **17**) in *Arachniotus* (Ahrazem *et al.*, 2001a), *Geosmithia putterillii* (Prieto *et al.*, 2002), *Gymnascella citrina* and *G. devroeyii* (Ahrazem *et al.*, 2002b); and short chains of → 5)-β-D-Gal*f*-(1→ with few or any →6)-β-D-Gal*f*-(1→ residues (structure **15**) in *Byssochlamys nivea, Thermoascus crustaceus, Talaromyces byssochlamydoides*, *T. leycettanus* and *Paecilomyces variotii* (Domenech *et al.*, 1999).
3. Chains of →3)-β-D-Gal*f*-(1→5)-β-D-Gal*f*-(1→ (structure **22**) in *Aspergillus wentii* and *Chaetosartorya chrysella* (Gómez-Miranda *et al.*, 2004).
4. Chains of →2)-β-D-Gal*f*-(1→3)-β-D-Gal*f*-(1→ (structure **21**) in *Talaromyces wortmannii, T. rotundus, Penicillium allahabadense* CBS 304.63 and *P. zacinthae* (Prieto *et al.*, 1995; Leal *et al.*, 1997; Leal and Bernabé, 1998).
5. A complex glucogalactomannan (structure **23**) with a structure similar to that described for varianose (Jansson and Lindberg, 1980) in most species of *Talaromyces* (Parra *et al.*, 1994b; Prieto *et al.*, 1995), in *Geosmithia viridis* (Prieto *et al.*, 2002) and in several species of *Penicillium* (Leal *et al.*, 1997).
6. Long chains of β-(1→6)-galactofuranose almost fully substituted at *O*-2 by single residues of glucose and disaccharides of glucuronic acid and β-mannopyranose (structure **35**) in *Penicillium vermoesenii* (Ahrazem *et al.*, 1999).

Several species have F1SS polysaccharides without galactofuranose. In *Gymnoascus demonbreunii* and *G. desertorum*, the F1SS polysaccharides are partially methylated comb-like mannans (structure **4**) (Ahrazem *et al.*, 2002b) and in *Paecilomyces fumosoroseus* and *P. farinosus* they consist of a →6)-α-D-Man*p*-(1→ chain substituted at *O*-4 mainly by β-D-Gal*p*, and with a minor proportion of D-Glc*p* (structure **6**) (Domenech *et al.*, 1996).

The order Eurotiales appears as a fairly homogeneous assemblage characterised by F1SS polysaccharides that contain long chains of galactofuranose with different linkage types. Several genera, with anamorphs in the form genera *Penicillium* and *Aspergillus,* have a similar polysaccharide (structure **16**), which confirms their close relatedness. Several structures have been obtained from the genus

Talaromyces (structures **15, 21** and **23**), which indicate that it is heterogeneous as reported by Pitt (1995) using other characters. The fact that several polysaccharides have been obtained from members of the form genera *Aspergillus, Paecilomyces* and *Penicillium* confirms their heterogeneity and shows the relationship with teleomorphic genera containing similar polysaccharides.

The species with polysaccharides that are different from this order (*Paecilomyces farinosus*, *P. fumosoroseus*, *Penicillium vermoesenii, Gymnoascus demonbreunii* and *G. desertorum*) should be removed from the Eurotiales.

Molecular analyses indicate that the polyphyletic anamorphic genus *Geosmithia* has its origins in *Eupenicillium, Talaromyces,* and the order Hypocreales (Ogawa *et al.,* 1997). Analysis of the F1SS polysaccharides from several species yielded structures **16**, **17** and **23**, which have been only found in members of the Eurotiales (Prieto *et al.,* 2002).

Arachniotus and *Gymnascella,* which were transferred to the Eurotiales (Currah, 1994), have a polysaccharide with structure **17**, which has the characters of the polysaccharides from the Eurotialean fungi. Therefore, the F1SS polysaccharides support the transfer of these genera to this order. However, the genus *Gymnoascus*, which was also transferred by this author, has structure **4**, which is comparable to those found in the polysaccharides from Onygenales (Ahrazem *et al.,* 2002b). Thus, according to our results, the genus should remain in this order.

6. Pyrenomycetes

This assemblage is recognized by the ascohymenial origin of the ascomata, their unitunicate asci with an operculum for the forcible discharge of the ascospores, and a flask-like ostiolate ascoma (the perithecium).

6.1. Diaporthales

Morphologically, the order is characterised by their brown to black perithecia usually aggregated into a pseudostroma. The ascogenous hyphae form a basal hymenium. Asci become detached from the hymeniun and fill the centrum cavity (Luttrell, 1951). The anamorphs are Coelomycetes. Uecker (1976) speculated on the possibility that the *Xylaria* and *Diaporthe* type of centrum might be the end of a continuum. A diagram of the major stages of the centrum development of *Diaporthe*, *Melanospora* and *Xylaria* was described by Jenssen (1983). The order Diaporthales is well defined morphologically; nevertheless, the familial classification of the order has varied from two to eight families. The number and configuration of these families differs depending on the author (Castlebury *et al.*, 2002).

In our laboratory, the FISS polysaccharides belonging to this order are being analysed, and their structure is almost determined. These isolates were kindly provided by Dr. A. Y. Rossman, from the Systematic Botany and Mycology Laboratory (USDA, Agricultural Research Service, Beltsville, MD, USA).

6.1.1. Melanconidaceae

F1SS polysaccharides from members of the family (A. Prieto, unpublished) are composed of a mannan core to which, at the *O*-2 position, the following chains are linked:

1. Single residues of β-D-Gal*f* or chains of →5)-β-D-Gal*f*-(1→6)-β-D-Gal*f*-(1→ (structure **25**), in *Coniella musaiensis* AR 3534, *Harknessia lythri* AR 3383, *Melanconiella chrysostroma* AR 3457, and *Melanconis alni* AR 3529.
2. A chain of →5)-β-D-Gal*f*-(1→5)-β-D-Gal*f*-(1→6)-β-D-Gal*f*-(1→ with terminal residues of β-D-Gal*f* (structure **26**), in *Melanconis desmazierii* AR 3525 and *M. marginalis* AR 3442.
3. Chains of →5)-β-D-Gal*f*-(1→6)-β-D-Gal*f*-(1→ and chains of →5)-β-D-Gal*f*-(1→ branched at some *O*-6 positions with terminal residues of α-D-Gal*p* (structure **27**) have been found in *Phragmoporthe conformis* AR 3632.

6.1.2. Valsaceae

Analogous structures to those described in Melanconidaceae have been found in species of this family:

1. Polysaccharides with structure **25** have been isolated from species of *Cryptodiaporthe salicella* AR 3455, *Endothia gyrosa* AR 3396, *Greeneria uvicola* OH 35, *Plagiostoma conradii* AR 3488, *Cryptosporella hypodermia* CBS 171.69, *Ophiovalsa betulae* AR 3524, *Cryphonectria macrospora* AR 3444, and *Pilidiella castaneicola* CBS 143.97.
2. In *Valsa abietes* CBS 185.42, *Diaporthe arctii* AR 3450, *Leucostoma nivea* AR 3512, *Hercospora tiliae* AR 3526, and *Mazzantia napelli* AR 3498, polysaccharide F1SS has structure **26**.
3. Structure **27** has been found in *Ditopella ditopa* AR 3423. The polysaccharides of *Apiognomonia veneta* CBS 897.79, *Amphiporthe castanea* CBS 392.93, and *Discula fraxinii* AR 2789 have a similar structure, but the proportion of the chain containing α-Gal*p* varies among the species. They might be considered as transitional states in a continuum between those taxa characterised by structures **25** and **27**.
4. The polysaccharides F1SS of *Discula destructiva* AR 2596 (structure **28**) and *Gnomonia gnomon* CBS 199.35 contain rhamnose, but their structures are different.

Most members of the order Diaporthales investigated may be included in three main groups according to the structure of their F1SS polysaccharides (Table 1, structures **25**, **26** and **27**). Each of these groups contains genera from the families Melanconidaceae and Valsaceae. This may indicate that these families are not homogeneous (Rogers, 1994). In some cases, species of the same genus (*Melanconis* and *Discula*) have different polysaccharides, suggesting that those genera are heterogeneous, which may explain the unstable concepts of certain genera (Castlebury *et al.,* 2002). The species included in a fourth small group contain F1SS polysaccharides with rhamnose, although with different structures. The polysaccharides described in this order are similar to those found in the Eurotiales, but the galactofuranose chains are shorter.

6.2. Hypocreales

The order *Hypocreales* mainly comprises species with perithecial ascocarps although some genera have cleisthotecia. Perithecia having light to bright colors may or may not be formed on a stroma. The asci are unitunicate. The ascospores are typically colorless or with light colors (green or yellow brown), can be one- to several-celled and have various shapes. Filiform ascospores are character of Clavicipitaceae.

The hypocrealean fungi also include a number of anamorphic genera and related conidial species, some of which do not reproduce sexually. They have colorless or light colored conidia that are produced from phialides. They can be saprobic, endophytes and parasites of animals, plants and other fungi.

The taxonomy of the order was first reviewed by Rogerson (1970), who accepted 115 genera and 26 synonyms. The 8th edition of the Dictionary of the Fungi (Hawksworth *et al.,* 1995) included 199 genera arranged in three families: Hypocreaceae, Clavicipitaceae and Niessliaceae. Rossman *et al.* (1999) examined the type species of all these genera, recognizing a total of 76 genera with 46 synonyms in five families, while excluding all others. They accepted the families Bionectriaceae, Hypocreaceae and Nectriaceae, which correspond to the three major clades described using molecular techniques (Rehner and Samuels, 1994; 1995; Ogawa *et al.,* 1997), comprising 56 genera (with 43 generic synonyms). The Niessliaceae account for 14 genera with two synonyms and the Clavicipitaceae have 6 genera with one synonym. Additional studies are required to elucidate the taxonomic position of these two families (Rossman *et al.,* 1999). They are now considered to be either part of or closely related to the *Hypocreales* (Rossman, 1996; Rossman *et al.,* 1999).

6.2.1. Bionectriaceae

The F1SS polysaccharides isolated from species of the group *Sesquicillium/Bionectria* consist of an α-(1→6)-mannan core to which long chains of β-(1→5) and (1→6)-galactofuranose are attached, partially substituted by α-glucopyranose and small proportions of glucuronic acid (structures **29** and **30**). In addition, other species contain N-acetyl glucosamine linked to the galactofuranose chains (structure **31**) (Ahrazem *et al.,* 2001b).

6.2.2. *Nectriaceae*

The F1SS polysaccharides from species of this group have a small α-(1→6)-mannan core and long chains of β-(1→6)-galactofuranose, almost fully substituted at *O*-2 by:

- single residues of glucose or glucuronic acid (structure **32**) in *Calonectria/Cylindrocladium* and some *Myrothecium* species (Ahrazem *et al.*, 1997; Ahrazem *et al.*, 2000b);
- single residues of glucose and di- or trisaccharides of glucuronic acid and β-mannopyranose (structure **33**) in *Fusarium/Gibberella* (Jikibara *et al.*, 1992; Ahrazem *et al.*, 2000a) and *Nectria cinnabarina* (Ahrazem *et al.*, 2001b);
- single residues of glucose and disaccharides of glucuronic acid and β-glucopyranose (structure **34**) in species of *Myrothecium* (Ahrazem *et al.*, 2000b).

6.2.3. *Hypocreaceae*

Two related structures, with different complexity, have been described in the *Hypocrea/Trichoderma* group. In *H. psychrophila* and two species of *Trichoderma* (structure **37**), the *O*-3 positions of the α-(1→6)-mannan backbone are partially substituted by single residues of α-mannopyranose, and some of the *O*-2 positions are substituted by two types of side chains (up to 4 residues) which contain only β-galactofuranose, or β-galactofuranose, α-glucuronic acid and α-glucopyranose (Messner *et al.*, 1990; Rath *et al.*, 1995; Prieto *et al.*, 1997). In *H. gelatinosa*, the chain containing only galactofuranose is reduced to a single residue, and there is not substitution at *O*-3 (Prieto *et al.*, 2001). Intermediate forms between these two structures have been found in other species of the group (A. Prieto, unpublished).

6.2.4. *Ceratostomataceae*

Polysaccharides from *Melanospora tiffanii* CECT 2800, *M. zamiae* CBS 166.27, *M. zobellii* CBS 268.62, *Microthecium levitum* CBS 966.97, *Pteridiosperma ciliatum* CBS 437.74 and *Sphaerodes fimicola* CBS 422.87 have been characterised (A. Leal, unpublished). All the F1SS polysaccharides contain galactose (40-80%), mannose (tr.-8%) rhamnose (6-21%) and uronic acid (2-15%). Methylation analysis and

^{1}H-NMR spectra indicate that in these species F1SS polysaccharides have chains of β-(1→6)-galactofuranose, with some residues disubstituted at *O*-2 and *O*-3 by single α- and β-galactopyranose residues, respectively (structure **36**). Other residues of the main chain are substituted at *O*-2 by dimers or trimers of glucuronic acid and rhamnose. This structure is related to that described for other hypocrealean fungi (Ahrazem *et al.*, 1997; Ahrazem *et al.*, 2000a; Ahrazem *et al.*, 2000b), although the presence of rhamnose and disubstituted residues of galactofuranose are distinctive characters.

6.2.5. Clavicipitaceae

Chemical and methylation analyses of F1SS polysaccharides from species of *Atkinsoniella, Balansia, Beauveria, Claviceps, Cordyceps, Epicoccum* and *Harposporium* indicate that they do not contain glucuronic acid, and in most cases, are more or less branched galactomannans composed of a (1→6)-mannan backbone partially substituted at *O*-2 by chains containing β-(1→5), β-(1→6) and terminal galactofuranose (O. Ahrazem, unpublished).

Except in the family Clavicipitaceae, most polysaccharides isolated from species belonging to the Hypocreales have variable amounts of glucuronic acid residues attached to a galactofuranose chain.

The last edition of the Dictionary of the Fungi (Kirk *et al.*, 2001) includes family Ceratostomataceae, which has been traditionally a member of the order Sordariales, among the Hypocreales. Molecular studies on species of *Melanospora* and *Sphaerodes* indicate that these genera are closely related to or within the Hypocreales (Zhang and Blackwell, 2002). Species of this family have perithecial or cleistothecial ascocarps and deliquescent asci. The structure of the F1SS polysaccharide from *Sphaerodes fimicola* is similar to that described for *Neurospora* (Leal *et al.*, 1996), which indicates the relationship of the two genera. The structure of the polysaccharides of the other isolates from Ceratostomataceae has the characters of hypocrealean F1SS polysaccharides.

6.3. Sordariales

Members of Sordariales are perithecial or cleistothecial. The perithecium has black pigment and is generally superficial, non stromatic. The hamathecium contains paraphyses, which can be either sparse, broader at the base than at the tip and disperse among asci, or very abundant and cylindrical. The interascal tissue is inconspicuous or lacking at maturity. Unitunicate, cylindrical or ovoid asci may be formed in a basal cluster or in a basal hymenium. Ascospores are generally dark and unicellular. They may have a gelatinous sheath or appendages, and in many species a germinative pore is observed. Conidiogenesis is enteroblastic.

Several arrangements of the Sordariales have been published (Barr, 1990; Spatafora and Blackwell, 1994; Suh and Blackwell, 1999). On the basis of recent molecular data, the Dictionary of the Fungi (Kirk *et al.*, 2001) includes seven families in the order, and the Outline of Ascomycota 2003 (Eriksson *et al.*, 2003), eleven. Polysaccharides F1SS from species belonging to several families are under study.

6.3.1. Sordariaceae

In *Neurospora crassa* and *N. sitophila* the F1SS polysaccharide (structure **5**) is a galactomannan with a comb-like structure, which consists of a chain of →6)-α-D-Man*p*-(1→ which has 7 of 8 mannose residues substituted at *O*-2 by single residues of β-D-Gal*f* (Leal *et al.*, 1996). Similar structures have been found in the following isolates: *Apodus oryzae* CBS 376.74, *Copromyces octosporus* CBS 386.78, *Gelasinospora cratophora* CBS 245.89, *G. inversa* CBS 554.72, *Sordaria lappae* CBS 214.75, *S. fimicola* CBS 911.73 and 940.87 and *Diplogelasinospora inaequalis* CBS 436.74. The polysaccharide from *Diplogelasinospora groovesii* CBS 340.73 and *Gelasinospora saitoi* CBS 435.74 contain, in addition, terminal galactopyranose residues (A. Prieto, unpublished). The polysaccharide from *Apodus deciduus*, described by Ahrazem *et al.* (2002c), consists of a chain of →6)-α-D-Man*p*-(1→ substituted at *O*-3 by the disaccharide α-D-Gal*f*-(1→2)-α-D-Gal*f*-(1→ (structure **13**).

6.3.2. Chaetosphaeriaceae

The F1SS polysaccharide from *Chaetosphaeria endophytica* CBS 385.81B has structure **14** (M. Bernabé, unpublished), with the mannan core substituted at the *O*-2 positions by chains of →5)-β-D-Gal*f*-(1→ or by chains containing the repeating disaccharide →6)-α-D-Glc*p*-(1→4)-α-D-Glc*p*-(1→.

6.3.3. Cephalothecaceae

The polysaccharide from *Cephalotheca sulfurea* (structure **5**) is similar to that isolated from *N. crassa* (Ahrazem *et al.*, 2002a). In *Cryptendoxyla hypophloia* CBS 508.70 the polysaccharide F1SS is similar, but more complex (A. Prieto, unpublished). Those from *Fragosphaeria (=Cephalotheca) purpurea* and *F. reniformis* (structure **9**) are branched rhamnomannans (Ahrazem *et al.*, 2002a) comparable to those described for some Ophiostomatales. These data agree with the findings of Suh and Blackwell (1999) based on both SSU and LSU rDNA combined data, which indicated that *F. purpurea* was in a clade with *Ophiostoma piliferum*, while *C. sulfurea* falls within the Sordariales, although with low bootstrap support.

6.3.4. Lasiosphaeriaceae

The only isolate studied until now is *Podospora curvicolla* CBS 259.69. The ^{1}H-NMR spectrum of its F1SS polysaccharide resembles that of the polysaccharide from *Apodus deciduus* (structure **13**). The presence of a similar polysaccharidic structure in two families reflects their heterogeneity.

6.4. Microascales

Members of Microascales are perithecial or cleistothecial. Asci are irregularly distributed in the ascoma, thin-walled and evanescent. The perithecia form filiform beaks. The order includes plant (*Ceratocystis*) and animal parasites (*Pseudallescheria, Scedosporium*).

6.4.1. Microascaceae

A glucomannan with structure **8** has been described in several isolates of *Ceratocystis* (Spencer and Gorin, 1971). Other species of the genus, such as *C. ulmi* or *C. stenoceras*, contain a rhamnomannan (Spencer and Gorin, 1971; Travassos *et al.*, 1974), but these isolates were later renamed as *Ophiostoma*. A rhamnomannan with a different structure has been described in *Pseudallescheria boydii* (Pinto *et al.*, 2001).

7. Other Groups Under Study

The taxa included in the following sections belong to:

- Class Archiascomycetes. Proposed by Nishida and Sugiyama (1994), who consider that this group was originated before the Euascomycetes and Hemiascomycetes, and that perhaps it is not monophyletic. This assemblage is characterized by the production of asci without ascogenous hyphae. The Dictionary of the Fungi (Kirk *et al.*, 2001) places these fungi in the classes Schizosaccharomycetes and Taphrinomycetes.
- Class Loculoascomycetes, which are Ascomycota with fissitunicate asci that are born in unwalled locules.

The structure of the F1SS polysaccharides of these groups has not been completely ascertained, but the preliminary results show that F1SS polysaccharides might be useful chemotaxonomic characters for these fungi.

8. Schizosaccharomycetes

8.1. Schizosaccharomycetales: The Schizosaccharomycetaceae

This family includes the fission yeasts. The history, taxonomy and phylogenetic relationships of the family have been reviewed (Kurtzman and Sugiyama, 2001). The results of the methylation analysis and the ^{1}H-NMR spectra of F1SS polysaccharides from *Schizosaccharomyces octosporus* CECT 10684 and *S. pombe* CECT 10685, CECT 10812 and CECT 1378 suggest that they are branched glucogalactomannans having a chain of →6)-α-D-Man*p*-(1→ with most of the *O*-2 positions substituted by single residues of α-galactopyranose and small amounts of α-D-Glc*p*-(1→, →3)-α-D-Glc*p*-(1→ and →3)-α-D-Gal*p*-(1→. Similar results have been previously described (Bush *et al.,* 1974). Barreto-Bergter and Gorin (1983) proposed that the galactomannan from these species has a repeating unit similar to structure **7**. Nevertheless, in this structure, minor components that are detected in the methylation analysis are not included.

9. Taphrinomycetes

9.1. Taphrinales

9.1.1. Taphrinaceae

Two isolates of *Taphrina* have been investigated, *T. deformans* CBS 355.55 and *T. populina* CBS 337.55. The composition of their F1SS polysaccharide is different. Both contain a backbone of →6)-α-D-Man*p*-(1→ partially branched at *O*-3, but the side chains contain →5)-β-D-Gal*f*-(1→ in *T. deformans*, and →2)-α-D-Rha*p*-(1→ in *T. populina,* as detected by methylation analysis. These results suggest that a larger number of species must be studied in order to determine the F1SS polysaccharide character of the genus.

9.1.2. Protomycetaceae

Protomyces inouyei CBS 222.97 and *P. pachydermus* CBS 224.57 contain a rhamnomannan different from that found in *T. populina.* The results of methylation analysis are compatible with the existence of a backbone of →2)-α-D-Man*p*-(1→ partially substituted at *O*-4 by chains of →2)-α-D-Rha*p*-(1→.

10. Loculoascomycetes

Include fungi with asci that are bitunicate, fissitunicate or semifissitunicate, produced in locules inside a stroma. A historical revision has been recently published (Barr and Huhndorf, 2001).

10.1. Chaetothyriales: The Herpotrichiellaceae

F1SS polysaccharide from species of *Capronia* (*C. acutiseta* CBS 618.96, *C. dactylotrichia* CBS 604.96 and *C. pilosella* CBS 125.88) is a slightly branched glucogalactomannan rich in →2)-α-D-Man*p*-(1→ whose structural analysis is in progress.

10.2. Pleosporales

The order includes many important plant parasites. The ascoma is generally perithecium-like, with a thick peridium immersed in the stroma.

10.2.1. Pleosporaceae

F1SS polysaccharides from species of the genus *Pleospora betae* CECT 20035, and the anamorphic form-genera *Phoma cava* CECT 2876, *Alternaria tenuissima* CBS 106.52 and *Curvularia lunata* CECT 2130 are being studied. Most of them contain →6)-α-D-Man*p*-(1→ as their main component, with great variations in the branching degree of this residue. For example, in *A. tenuissima* the polysaccharide has only about 10% branching points, while in species from *Pleospora* and *Phoma*, the structure may be almost comb-like.

11. Correlation of F1SS Polysaccharides with Traditional Evolutionary Characters

Cleistothecial fungi with evanescent asci which lack forcible ascospore discharge were considered the ancestors of the more advanced Pyrenomycetes (Luttrell, 1951; Savile, 1968; Barr, 1976; Cavalier-Smith, 1987). Other authors considered cleistothecial fungi reduced forms, which represent the termination of a large number of unrelated and highly evolved taxa (Cain, 1956). In accordance with this theory, the Plectomycetes were distributed in four orders (Malloch, 1981). His arguments against fungal evolution from lack of forcible discharge (cleistothecial) to forcible discharge (perithecial) of ascospores were: "..The gradual evolution of the centrum to the forcible discharge of ascospores, through an ostiole, before the ostiole has evolved, is extremely unlikely. Evolutionary mechanisms just do not plan for the future!." In a later article (Malloch, 1987), he recognized that in certain families (Onygenaceae, Trichocomaceae, etc.), connections to perithecial or apothecial taxa is extremely tenuous and that they may have descended from primitive and now extinct ancestors. He summarized the uncertainty on the evolution of the Ascomycetes as follows: "Although most authors now accept the premise that the cleistothecial ascomycetes are evolved from more complex forms, it is possible premature to completely reject the opposite view".

Pyrenomycetous fungi form a continuum, ranging from those that discharge ascospores from the ascus (perithecial) to fungi that lack mechanisms for ascospore release (cleistothecial). The polarity of this morphological series has not been resolved. 18S ribosomal RNA gene analysis shows that species lacking a discharge mechanism are intercalated among species with forcible ascospores release (Berbee and Taylor, 1992). From these results, the authors concluded: "...that the capacity to shoot ascospores into the air has been lost or, less probably, gained repeatedly and independently".

The arrangement of F1SS polysaccharides according to their increasing structural complexity generated an evolutionary scheme (Bernabé *et al.,* 2002). We have updated this scheme and included the first genus of each group of fungi containing a similar structure (Table 1). The new scheme has at least three main radiations in the cleistothecial and perithecial Ascomycetes:

1. Fungi containing F1SS polysaccharides which are simple mannans or mannans substituted by single residues or short chains of different sugars (Table 1, structures **2-13**). These polysaccharides form a first radiation which consists of at least eleven fungal lineages (Figure 1). Most of the fungi of this group have rhexolytically dehiscing conidiation, the ascocarp is cleistothecial and the asci are evanescent. It is interesting to note that the genera *Cephalotheca, Ceratocystis, Neurospora* and *Ophiostoma* are included in the group. These fungi, previously classified in the Plectomycetes (except *Neurospora*), are now included in the Pyrenomycetes.
2. Fungi containing a small mannan core to which chains of β-galactofuranose are attached (Table 1, structures **15-28**). It was considered that these structures form a second radiation represented by the Eurotiales (Figure 1), which are characterized

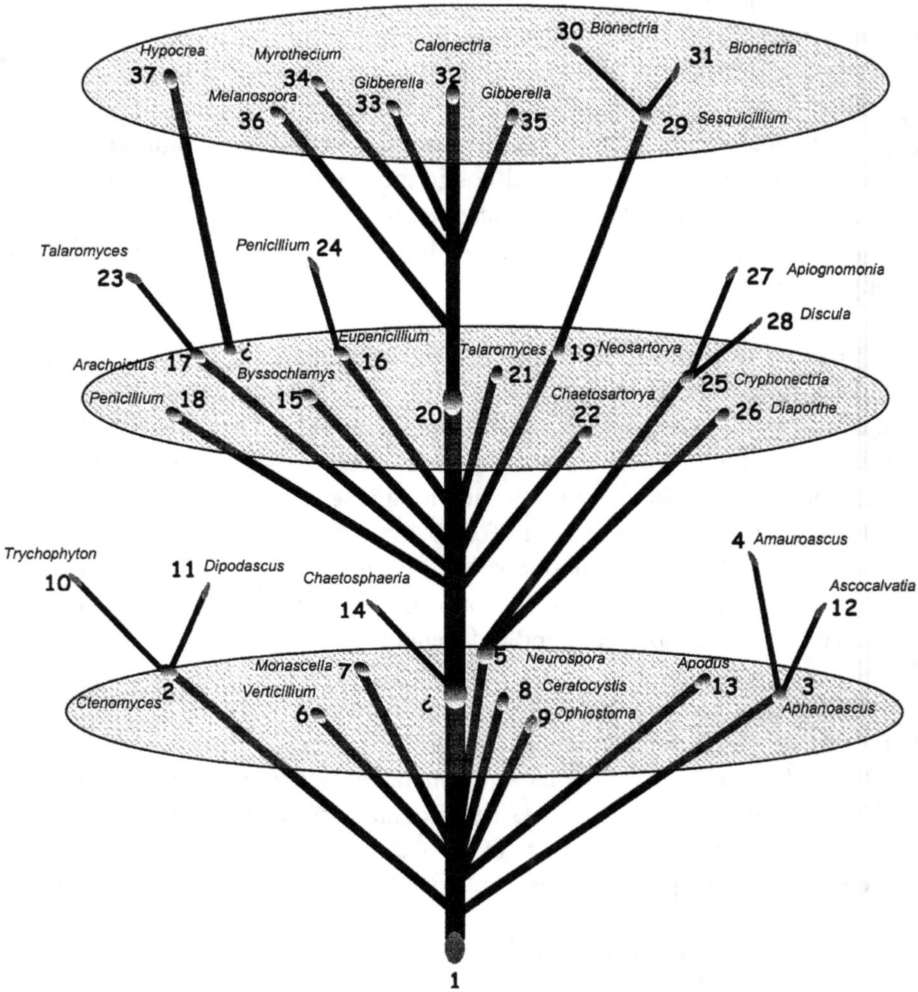

Figure 1. Evolutionary tree based on F1SS polysaccharides, which shows the first genus of each group and the number that represents their common structure (Table 1).

by their cleistothecial ascocarp, deliquescent asci and phialydic conidia. The major complexity of these polysaccharides may indicate that the Eurotiales are derived from a member of the first radiation, whose polysaccharide contains β-galactofuranose and may have also been the ancestor of the Sordariales. In this group are also included members of the Diaporthales, which are perithecial and may have evolved from certain members of the Sordariales, whose F1SS polysaccharide has shorter galactofuranose chains.

3. The complex polysaccharides of this group are composed of the mannan core and a chain of galactofuranose partially substituted by single residues or short chains of different sugars (structures **28** to **37**). Among these residues the presence of α-glucopyranuronic acid is character, although in some polysaccharides their low proportion may not allow their inclusion in the repeating unit. According to the galactofuranose chains, there are at least three main lineages in this radiation that include members of the Hypocreaceae, Bionectriaceae and Nectriaceae, and Ceratostomataceae. The two last families might have been derived from ancestors having polysaccharides with structures **19** and **20**, respectively (Figure 1). Most of the fungi included in this group are perithecial, with cylindrical asci, capable of forcible discharge of ascospores. The scheme of fungal evolution based on F1SS polysaccharides parallels those based on ascoma complexity, which consider the Plectomycetes the ancestors of the Pyrenomycetes (Savile, 1968; Barr, 1976; Cavalier-Smith, 1987).

Since polysaccharidic structures have been determined from a limited number of fungal taxa, the scheme is tentative and will be modified as new structures from other taxa are found. F1SS polysaccharides, in combination with other characters, might both contribute to the formation of stable natural taxa and help in the construction of reliable evolutionary data. However, these polysaccharides may fail to delineate certain taxa as occurs with most characters used in fungal classification. If this occurs, we should remember the following statement (Luttrell, 1981): "Yet, we have a remarkably profligate attitude towards characters. As soon as the slightest difficulty is encountered, there is a throwing up of the hands: the character has failed, the classification has collapsed. This is the "sky is falling" syndrome."

12. Acknowledgements

This work was supported by Grant BQU-2000-1501-C02-01 from DGICYT. We thank Dr. A. Y. Rossman for providing the diaporthalean strains.

References

Ahrazem, O., Blázquez, G., Prieto, A., Bernabé, M., and Leal, J.A. 2002a. Alkali-extractable and water-soluble polysaccharide (F1SS): chemotaxonomic and phylogenetic character for *Cephalotheca*. Mycol. Res. 106: 1187-1192.

Ahrazem, O., Gómez-Miranda, B., Prieto, A., Barasoaín, I., Bernabé, M., and Leal, J.A. 1999. Structural characterisation of a cell wall polysaccharide from *Penicillium vermoesenii*: chemotaxonomic application. Can. J. Bot. 77: 961-968.

Ahrazem, O., Gómez-Miranda, B., Prieto, A., Barasoaín, I., Bernabé, M., and Leal, J.A. 2000a. An acidic water-soluble cell wall polysaccharide: a chemotaxonomic marker for *Fusarium* and *Gibberella*. Mycol. Res. 104: 603-610.

Ahrazem, O., Gómez-Miranda, B., Prieto, A., Bernabé, M., and Leal, J.A. 2000b. Heterogeneity of the genus *Myrothecium* as revealed by cell wall polysaccharides. Arch. Microbiol. 173: 296-302.

Ahrazem, O., Leal, J.A., Prieto, A., Jiménez-Barbero, J., and Bernabé, M. 2001a. Chemical structure of a polysaccharide isolated from the cell wall of *Arachniotus verruculosus* and *A. ruber*. Carbohydr. Res. 336: 325-328.

Ahrazem, O., Prieto, A., Bernabé, M., and Leal, J.A. 2002b. F1SS polysaccharides: chemotaxonomic characters and evolutionary indicators for Plectomycetes. Stud. Mycol. 47: 37-48.

Ahrazem, O., Prieto, A., Gómez-Miranda, B., Bernabé, M., and Leal, J.A. 2001b. Comparison of cell-wall polysaccharides from *Nectria cinnabarina* with those from the group of *Nectria* with *Sesquicillium* anamorphs. Microbiology 147: 1839-1849.

Ahrazem, O., Prieto, A., Leal, J.A., Gómez-Miranda, B., Domenech, J., Jiménez-Barbero, J., and Bernabé, M. 1997. Structural elucidation of acidic fungal polysaccharides isolated from the cell-wall of genera *Cylindrocladium* and *Calonectria*. Carbohydr. Res. 303: 67-72.

Ahrazem, O., Prieto, A., Leal, J.A., Jiménez-Barbero, J., and Bernabé, M. 2002c. Fungal cell wall galactomannan isolated from *Apodus deciduus*. Carbohydr. Res. 337: 1503-1506.

Ahrazem, O., Prieto, A., Leal, J.A., Jiménez-Barbero, J., and Bernabé, M. 2002d. Fungal cell-wall galactomannans isolated from *Geotrichum* spp. and their teleomorphs, *Dipodascus* and *Galactomyces*. Carbohydr. Res. 337: 2347-2351.

Ahrazem, O., Prieto, A., San-Blas, G., Leal, J.A., Jiménez-Barbero, J., and Bernabé, M. 2003. Structural differences between the alkali-extracted water-soluble cell wall polysaccharides from mycelial and yeast phases of the pathogenic dimorfic fungus *Paracoccidioides brasiliensis*. Glycobiology 13: 743-747.

Albersheim, P., Darvill, A.G., Davis, K.R., Lau, J.M., McNeil, M., Sharp, J.K., and York, W.S. 1984. Why study the structures of biological molecules? The importance of studying the structures of complex carbohydrates. In: Structure, Function, and Biosynthesis of Plant Cell Walls. W.M. Dugger and S. Bartnicki-García, eds. American Society of Plant Physiologists, Rockville, Maryland, USA. p. 19-51.

Barr, M.E. 1976. Perspectives on the Ascomycotina. Mem. N. Y. Bot. Gd. 28: 1-8.

Barr, M.E. 1990. Prodromus to nonlichenized, pyrenomycetous members of class Hymenoascomycetes. Mycotaxon 39: 43-184.

Barr, M.E. and Huhndorf, S.M. 2001. Loculoascomycetes. In The Mycota, D.J. McLaughlin, E.G. McLaughlin, and P.A. Lemke, eds. (Berlin, Heidelberg: Springer-Verlag), p. 283-305.

Barreto-Bergter, E. and Gorin, P.A.J. 1983. Structural chemistry of polysaccharides from fungi and lichens. Adv. Carbohydr. Chem. Biochem. 41: 67-103.

Bartnicki-García, S. 1968. Cell wall chemistry, morphogenesis, and taxonomy of fungi. Annu. Rev. Microbiol. 22: 87-108.

Bartnicki-García, S. 1984. Kingdoms with walls. In: Structure, Function, and Biosynthesis of Plant Cell Walls. W.M. Dugger and S. Bartnicki-García, eds. American Society of Plant Physiologists, Rockville, Maryland, USA. p. 1-18.

Bartnicki-García, S. 1987. The cell wall: a crucial structure in fungal evolution. In: Evolutionary Biology of the Fungi. A.D.M. Rayner, C.M. Brasier, and D. Moore, eds. Cambridge University Press, New York, p. 389-403.

Benny, G.L. and Kimbrough, J.W. 1980. A synopsis of the orders and families of Plectomycetes with keys to genera. Mycotaxon 12: 1-91.

Berbee, M. and Taylor, J.W. 1992. Convergence in ascospore discharge mechanisms among pyrenomycete fungi based on 18S ribosomal RNA gene sequence. Mol. Phyl. Evol. 1: 59-71.

Berbee, M. and Taylor, J.W. 1994. From 18S ribosomal sequence data to evolution of morphology among the fungi. Can. J. Bot. 73 (Suppl. 1): S677-S683.

Bernabé, M., Ahrazem, O., Prieto, A., and Leal, J.A. 2002. Evolution of polysaccharides F1SS and proposal of their utilisation as antigens for rapid detection of fungal contaminants. EJEAFChe 1(1).(http://www.ejeafche.uvigo.es)

Bialek, R., Ibricevic, A., Fothergill, A., and Begerow, D. 2000. Small subunit ribosomal DNA sequence shows *Paracoccidioides brasiliensis* closely related to *Blastomyces dermatitidis*. J. Clin. Microbiol. 38: 3190-3193.

Bush, D.A., Horisberger, M., Horman, I., and Wursch, P. 1974. The wall structure of *Schizosaccharomyces pombe*. J. Gen. Microbiol. 81: 199-206.

Cain, R.F. 1956. Studies of coprophilous ascomycetes. II. *Phaeotrichum*, a new cleistocarpous genus in a new family, and its relationships. Can. J. Bot. 34: 675-687.

Cain, R.F. 1972. Evolution of the fungi. Mycologia 64: 1-14.

Cano, J. and Guarro, J. 1990. The genus *Aphanoascus*. Mycol. Res. 94: 355-377.

Castlebury, L.A., Rossman, A.Y., Jacklitsch, W.J., and Vasilyeva, L.N. 2002. A preliminary overview of the Diaporthales based on large subunit nuclear ribosomal DNA sequences. Mycologia 94: 1017-1031.

Cavalier-Smith, T. 1987. The origin of fungi and pseudofungi. In Evolutionary Biology of the Fungi. A.D.M. Rayner, C.M. Brasier, and D. Moore, eds. Cambridge University Press, Cambridge. p. 339-353.

Currah, R.S. 1985. Taxonomy of the Onygenales: *Arthrodermataceae*, *Gymnoascaceae*, *Myxotrichaceae* and *Onygenaceae*. *Mycotaxon* 24: 1-216.

Currah, R.S. 1994. Peridial morphology and evolution in the prototunicate Ascomycetes. In: Ascomycete Systematics: Problems and Perspectives in the Nineties. D.L. Hawksworth, ed. Plenum Press, New York. p. 281-293.

Domenech, J., Barasoaín, I., Prieto, A., Gómez-Miranda, B., Bernabé, M., and Leal, J.A. 1996. An antigenic water-soluble glucogalactomannan extracted from cell walls of *Paecilomyces fumosoroseus* and *Paecilomyces farinosus*. Microbiology 142: 3497-3503.

Domenech, J., Prieto, A., Barasoaín, I., Gómez-Miranda, B., Bernabé, M., and Leal, J.A. 1999. Galactomannans from the cell walls of species of *Paecilomyces* sect. *Paecilomyces* and their teleomorphs as immunotaxonomic markers. Microbiology 145: 2789-2796.

Eriksson, O.E., Baral, H.-O., Currah, R.S., Hansen, K., Kurtzman, C.P., Rambold, G., and Laessoe, T. 2003. Outline of Ascomycota-2003. Myconet 9: 1-89.

Fleet, G.H. 1991. Cell walls. In: The Yeasts. Yeast Organelles, A.H. Rose and J.S. Harrison, eds. Academic Press, London. p. 199-277.

Frisvad, J.C., Bridge, P.D., and Arora, D.K. 1998. Chemical Fungal Taxonomy. Marcel Dekker, Inc., New York.

Gander, J.E. 1974. Fungal cell wall glycoproteins and peptido-polysaccharides. Annu. Rev. Microbiol. 28: 103-119.

Geiser, D.M. and Lobuglio, K.F. 2001. The monophyletic Plectomycetes: Ascopsphaerales, Onygenales, Eurotiales. In: The Mycota, D.J. McLaughlin, E.G. McLaughlin, and P.A. Lemke, eds. Springer-Verlag, Berlin-Heidelberg. p. 201-219.

Gómez-Miranda, B., Moya, A., and Leal, J.A. 1986. Hyphal polysaccharides as potential phylogenetic markers for *Eupenicillium* species. Exp. Mycol. 10: 184-189.

Gómez-Miranda, B., Prieto, A., Leal, J.A., Ahrazem, O., and Bernabé, M. 2004. Differences among the cell wall galactomannans from *Aspergillus wentii* and *Chaetosartorya chrysella* and that of *Aspergillus fumigatus*. Glycoconj. J. In press.

Gorin, P.A.J. and Spencer, J.F.T. 1970. Proton magnetic resonance spectroscopy-An aid in identification and chemotaxonomy of yeast. Adv. Appl. Microbiol. 13: 25-89.

Guarro, J. and Cano, J. 2002. Phylogeny of Onygenalean fungi of medical interest. Stud. Mycol. 47: 1-4.

Guarro, J., Cano, J., Leal, J.A., Gómez-Miranda, B., and Bernabé, M. 1993. Composition of the cell wall polysaccharides in some geophilic dermatophytes. Mycopathologia 122: 69-77.

Guarro, J., Gene.J., and Stchigel, A.M. 1999. Developments in fungal taxonomy. Clin. Microbiol. Rev. 12: 454-500.

Guarro, J. and von Arx, J.A. 1986. *Monascella,* a new genus of Ascomycotina. Mycologia 78: 869-871.

Hawksworth, D.L. 1994. Ascomycete Systematics. Problems and Perspectives in the Nineties. Plenum Press, New York.

Hawksworth, D.L., Kirk, P.M., Sutton, B.C., and Pegler, B.N. 1995. Ainsworth and Bisby's Dictionary of the Fungi. International Mycological Institute. CAB International, Wallingford, UK.

Ikuta, K., Shibata, N., Blake, J.S., Dahl, M.V., Nelson, R.D., Hisamichi, K., Kobayashi, H., Suzuki, S., and Okawa, Y. 1997. NMR study of the galactomannans of *Trichophyton mentagrophytes* and *Trichophyton rubrum*. Biochem. J. 323: 297-305.

Jansson, P.E. and Lindberg, B. 1980. Structural studies of varianose. Carbohydr. Res. 82: 97-102.

Jensen, J.D. 1983. The development of *Diaporthe phaseolorum* var. *sojae* in culture. Mycologia 75: 1074-1091.

Jikibara, T., Takegawa, K., and Iwahara, S. 1992. Studies on the uronic acid-containing glycoproteins of *Fusarium* sp. M7-1:

III. The primary structures of the acidic polysaccharides of the glycoproteins. J. Biochem. (Tokyo) 111: 236-243.

Jiménez-Barbero, J., Bernabé, M., Leal, J.A., Prieto, A., and Gómez-Miranda, B. 1993. Chemical structure and conformational features of cell-wall polysaccharides isolated from *Aphanoascus mephitalus* and related species. Carbohydr. Res. 250: 289-299.

Jiménez-Barbero, J., Prieto, A., Gómez-Miranda, B., Leal, J.A., and Bernabé, M. 1995. Chemical structure of fungal cell-wall polysaccharides isolated from *Microsporum gypseum* and related species of *Microsporum* and *Trychophyton*. Carbohydr. Res. 272: 121-128.

Kirk, P.M., Cannon, P.F., David, J.C., and Stalpers, J.A. 2001. Ainsworth and Bisby's Dictionary of the Fungi. CAB International, Wallingford, UK.

Kobayashi, H., Shibata, N., and Suzuki, S. 1992. Evidences for oligomannosyl residues containing both b-1,2 and a-1,2 linkages as serotype A-specific epitope(s) in mannans of *Candida albicans*. Infect. Immun. 60: 2106-2109.

Kohn, L.M. 1992. Developing new characters for fungal systematics: an experimental approach for determining the rank of resolution. Mycologia 84: 139-153.

Kurtzman, C.P. and Sugiyama, J. 2001. Ascomycetous yeasts and yeastlike taxa. In: The Mycota. D.J. McLaughlin, E.G. McLaughlin, and P.A. Lemke, eds. Springer-Verlag, Berlin, Heidelberg. p. 179-200.

Leal, J.A. 1994. Water-soluble polysaccharides of fungal cell walls. In: Microorganisms in Ruminant Nutrition. R.A. Prins and C.S. Stewart, eds. Nottingham University Press

Leal, J.A. and Bernabé, M. 1998. Taxonomic applications of polysaccharides. In: Chemical Fungal Taxonomy, J.C. Frisvad, P.D. Bridge, and D.K. Arora, eds. Marcel Dekker, Inc., NY Basel Hong Kong. p. 153-181.

Leal, J.A., Gómez-Miranda, B., Bernabé, M., Cano, J., and Guarro, J. 1992a. The chemical composition of six species of *Aphanoascus*: The taxonomic significance of the presence of α-(1-2)-(1-6) mannan and α-(1-4) glucan. Mycol. Res. 96: 363-368.

Leal, J.A., Gómez-Miranda, B., Prieto, A., Domenech, J., Ahrazem, O., and Bernabé, M. 1997. Possible chemotypes from cell wall polysaccharides, as an aid in the systematics of the genus *Penicillium* and its teleomorphic states *Eupenicillium* and *Talaromyces*. Mycol. Res. 101: 1259-1264.

Leal, J.A., Guerrero, C., Gómez-Miranda, B., Prieto, A., and Bernabé, M. 1992b. Chemical and structural similarities in wall polysaccharides of some *Penicillium, Eupenicillium* and *Aspergillus* species. FEMS Microbiol. Lett. 69: 165-168.

Leal, J.A., Jiménez-Barbero, J., Gómez-Miranda, B., Parra, E., Prieto, A., and Bernabé, M. 1995. Structural investigation of cell-wall polysaccharides from *Neosartorya*: relationships with their putative anamorphs of *Aspergillus*. Carbohydr. Res. 273: 255-262.

Leal, J.A., Jiménez-Barbero, J., Gómez-Miranda, B., Prieto, A., Domenech, J., and Bernabé, M. 1996. Structural investigation of a cell-wall galactomannan from *Neurospora crassa* and *N. sitophila*. Carbohydr. Res. 283: 215-222.

Leal, J.A., Prieto, A., Ahrazem, O., Pereyra, T., and Bernabé, M. 2001. Cell wall polysaccharides: characters for fungal taxonomy and evolution. Rec. Res. Devel. Microbiol. 5: 235-248.

Leclerc, M.C., Phillipe, H., and Guého, E. 1994. Phylogeny of dermatophytes and dimorphic fungi based on large subunit ribosomal RNA sequence comparison. J. Med. Vet. Mycol. 32: 331-341.

Lee, Y.C. and Ballou, C.E. 1965. Preparation of mannobiose, mannotriose, and a new mannotetraose from *Saccharomyces cerevisae* mannan. Biochemistry 4: 257-264.

Luttrell, E.S. 1951. Taxonomy of the pyrenomycetes. University Mo. Stud. 24: 1-120.

Luttrell, E.S. 1981. The pyrenomycete centrum-Loculoascomycetes. In: Ascomycete Systematics. The Luttrelian Concept. D.R. Reynolds, ed. Springer-Verlag, New York. p. 124-137.

Malloch, D. 1981. The Plectomycete centrum. In: Ascomycete Systematics. The Luttrelian Concept. D.R. Reynolds, ed. Springer-Verlag, New York. p. 73-91.

Malloch, D. 1987. Ordinal relationships among reduced ascomycetes. Stud. Mycol. 30: 177-186.

Messner, R., Hagspiel, K., and Kubicek, C.P. 1990. Isolation of a β-glucosidase binding and activating polysaccharide from cell walls of *Trichoderma reesei*. Arch. Microbiol. 154: 150-155.

Nishida, H. and Sugiyama, J. 1994. Archiascomycetes: detection of a major new lineage within the Ascomycota. Mycoscience 35: 361-366.

Ogawa, H., Yoshimura, A., and Sugiyama, J. 1997. Polyphyletic origins of species of the anamorphic genus *Geosmithia* and the relationship of the cleistothecial genera: evidence from 18S, 5S and 28S rDNA sequence analysis. Mycologia 89: 756-771.

Parra, E., Jiménez-Barbero, J., Bernabé, M., Leal, J.A., Prieto, A., and Gómez-Miranda, B. 1994a. Structural investigation of two cell-wall polysaccharides of *Penicillium expansum* strains. Carbohydr. Res. 257: 239-248.

Parra, E., Jiménez-Barbero, J., Bernabé, M., Leal, J.A., Prieto, A., and Gómez-Miranda, B. 1994b. Structural studies of fungal cell-wall polysaccharides from two strains of *Talaromyces flavus*. Carbohydr. Res. 251: 315-325.

Peat, S., Whelan, W.J., and Edwards, T.E. 1961. Polysaccharides of baker's yeast. IV. Mannan. J. Chem. Soc. (London) 1: 29-34.

Phaff, H.J. 1971. Structure and biosynthesis of the yeast cell envelope. In: The Yeasts. Physiology and Biochemistry of Yeast. A.H. Rose and J.S. Harrison, eds. Academic Press, London. New York. p. 135-210.

Pinto, M.R., Mulloy, B., Haido, R.M., Travassos, L.R., and Barreto-Bergter, E. 2001. A peptidorhamnomannan from the mycelium of *Pseudallescheria boydii* is a potential diagnostic antigen of this emerging human pathogen. Microbiology 147: 1499-1506.

Pitt, J.I. 1995. Phylogeny in the genus *Penicillium*: a morphologist's perspective. Can. J. Bot. 73: S768-S777.

Prieto, A., Ahrazem, O., Gómez-Miranda, B., Bernabé, M., and Leal, J.A. 2002. Cell wall polysaccharides F1SS disclose the relatedness of the genus *Geosmithia* with *Eupenicillium* and *Talaromyces*. Can. J. Bot. 80: 410-415.

Prieto, A., Bernabé, M., and Leal, J.A. 1995. Isolation, purification and chemical characterization of alkali-extractable polysaccharides from the cell walls of *Talaromyces* species. Mycol. Res. 99: 69-75.

Prieto, A., Leal, J.A., Gómez-Miranda, B., Ahrazem, O., Jiménez-Barbero, J., and Bernabé, M. 2001. Structure of a cell wall polysaccharide isolated from *Hypocrea gelatinosa*. Carbohydr. Res. 333: 173-178.

Prieto, A., Leal, J.A., Poveda, A., Jiménez-Barbero, J., Gómez-Miranda, B., Domenech, J., Ahrazem, O., and Bernabé, M. 1997. Structure of complex cell wall polysaccharides isolated from *Trichoderma* and *Hypocrea* species. Carbohydr. Res. 304: 281-291.

Rath, J., Messner, R., Kosma, P., Altmann, F., März, L., and Kubicek, C.P. 1995. The a-D-mannan core of a complex cell wall heteroglycan of *Trichoderma reesei* is responsible for β-glucosidase activation. Arch. Microbiol. 164: 414-419.

Rehner, S.A. and Samuels, G.J. 1994. Taxonomy and phylogeny of *Gliocladium* analyzed by large subunit rDNA sequences. Mycol. Res. 98: 625-634.

Rehner, S.A. and Samuels, G.J. 1995. Molecular systematics of the Hypocreales: a teleomorph gene phylogeny and the status of their anamorphs. Can. J. Bot. 73: 816-823.

Rogers, J.D. 1994. Problem genera and family interfaces in the *Eupyrenomycetes*. In: Ascomycete Systematics: Problems and Perspectives in the Nineties. D.L. Hawksworth, ed. Plenum Press, New York. p. 321-331.

Rogerson, C.T. 1970. The hypocrealean fungi. Mycologia 62: 865-910.

Rossman, A.Y. 1996. Morphological and molecular perspectives on systematics of the Hypocreales. Mycologia 88: 1-19.

Rossman, A.Y., Samuels, G.J., Rogerson, C.T., and Lowen, R. 1999. Genera of *Bionectriaceae*, *Hypocreaceae* and *Nectriaceae* (Hypocreales, Ascomycetes). Stud. Mycol. 42: 1-248.

Savile, D.B.O. 1968. Possible interrelationships between fungal groups. In: The Fungi: An Advanced Treatise. G.C. Ainsworth and A.S. Sussman, eds. Plenum Press, New York. p. 649-675.

Sietsma, J.H. and Wessels, J.G.H. 1977. Chemical analysis of the hyphal wall of *Schizophyllum commune*. Biochim. Biophys. Acta 496: 225-239.

Sigler, L. and Carmichael, J.W. 1976. Taxonomy of *Malbranchea* and some other Hyphomycetes with arthroconidia. Mycotaxon 4: 349-388.

Spatafora, J.W. and Blackwell, M. 1994. Cladistic analysis of partial ssrDNA sequences among unitunicate perithecial ascomycetes and its implications on the evolution of centrum development. In: Ascomycete Systematics: Problems and Perspectives in the Nineties, D.L. Hawksworth, ed. Plenum Press, New York. p. 233-242.

Spencer, J.F.T. and Gorin, P.A.J. 1971. Systematics of the genera *Ceratocystis* and *Graphium*. Proton magnetic resonance spectra of the mannose-containing polysaccharides as an aid in classification. Mycologia 63: 387-402.

Sugiyama, J., Ohara, A., and Mikawa, T. 1999. Molecular phylogeny of onygenalean fungi based on small subunit ribosomal DNA (SSU rDNA) sequences. Mycoscience 40: 251-258.

Suh, S.-O. and Blackwell, M. 1999. Molecular phylogeny of the cleistothecial fungi placed in *Cephalothecaceae* and *Pseudoeurotiaceae*. Mycologia 91: 836-848.

Takeda, T., Kawarasaki, I., and Ogihara, Y. 1981. Studies on the structure of a polysaccharide from *Epidermophyton floccosum* and approach to a synthesis of the basic trisaccharide repeating unit. Carbohydr. Res. 89: 301-308.

Travassos, L.R., Gorin, P.A., and Lloyd, K.O. 1974. Discrimination between *Sporothrix schenckii* and *Ceratocystis stenoceras* rhamnomannans by proton and carbon-13 magnetic resonance spectroscopy. Infect. Immun. 9: 674-80.

Uecker, F.A. 1976. Development and cytology of *Sordaria humana*. Mycologia 68: 30-46.

Vinogradov, E., Petersen, B., and Bock, K. 1998. Structural analysis of the intact polysaccharide mannan from *Saccharomyces cerevisiae* yeast using ^{1}H and ^{13}C NMR spectroscopy at 750 MHz. Carbohydr. Res. 307: 177-83.

von Arx, J.A. 1987. A re-evaluation of the *Eurotiales*. Persoonia 13: 273-300.

Zhang, N. and Blackwell, M. 2002. Molecular phylogeny of *Melanospora* and similar pyrenomycetous fungi. Mycol. Res. 106: 148-155.

Index

D

H

www.ingramcontent.com/pod-product-compliance
Lightning Source LLC
LaVergne TN
LVHW011907200926
848484LV00002B/552